Intelligent Transportation Systems 2018

AA006002

Transportation Research Record: Journal of the Transportation Research Board

Volume 2672, Issue 19

ISBN: 978-1-5108-7824-2

Printed from e-media with permission by:

Curran Associates, Inc.
57 Morehouse Lane
Red Hook, NY 12571

Some format issues inherent in the e-media version may also appear in this print version.

Published by Sage Publications USA

ISBN (Print) 978-1-5108-7824-2
ISBN 2018 Printed Set (All Issues) 978-1-5108-7735-1

Printed by Curran Associates, Inc. (2019)

Additional copies of this publication are available from:

Curran Associates, Inc.
57 Morehouse Lane
Red Hook, NY 12571 USA
Phone: 845-758-0400
Fax: 845-758-2633
Email: curran@proceedings.com
Web: www.proceedings.com

Volume 2672 Number 19 December 2018

TRR
JOURNAL OF THE TRANSPORTATION RESEARCH BOARD

Contents

Intelligent Transportation Systems 2018

Articles

TRB

TRANSPORTATION RESEARCH BOARD

CORE SPONSORS

Transportation Departments of the 50 States and the District of Columbia

Federal Government

U.S. Department of Transportation
- Federal Aviation Administration (FAA)
- Federal Highway Administration
- Federal Motor Carrier Safety Administration (FMCSA)
- Federal Railroad Administration (FRA)
- Federal Transit Administration
- National Highway Traffic Safety Administration (NHTSA)
- Office of the Assistant Secretary of Transportation for Research and Technology (OST-R)

U.S. Air Force Civil Engineer Center (AFCEC)
U.S. Coast Guard (USCG)
U.S. Department of Energy
U.S. Department of the Interior (DOI)
U.S. Environmental Protection Agency (EPA)

Nongovernmental Organizations

American Public Transportation Association (APTA)
Association of American Railroads (AAR)
California Air Resources Board

The National Academies of

SCIENCES · ENGINEERING · MEDICINE

The **National Academy of Sciences** was established in 1863 by an Act of Congress, signed by President Lincoln, as a private, nongovernmental institution to advise the nation on issues related to science and technology. Members are elected by their peers for outstanding contributions to research. Dr. Marcia McNutt is president.

The **National Academy of Engineering** was established in 1964 under the charter of the National Academy of Sciences to bring the practices of engineering to advising the nation. Members are elected by their peers for extraordinary contributions to engineering. Dr. C. D. Mote, Jr., is president.

The **National Academy of Medicine** (formerly the Institute of Medicine) was established in 1970 under the charter of the National Academy of Sciences to advise the nation on medical and health issues. Members are elected by their peers for distinguished contributions to medicine and health. Dr. Victor J. Dzau is president.

The three Academies work together as the National Academies of Sciences, Engineering, and Medicine to provide independent, objective analysis and advice to the nation and conduct other activities to solve complex problems and inform public policy decisions. The Academies also encourage education and research, recognize outstanding contributions to knowledge, and increase public understanding in matters of science, engineering, and medicine.

Learn more about the National Academies of Sciences, Engineering, and Medicine at **www.national-academies.org.**

The **Transportation Research Board** is one of seven major programs of the National Academies of Sciences, Engineering, and Medicine. The mission of the Transportation Research Board is to increase the benefits that transportation contributes to society by providing leadership in transportation innovation and progress through research and information exchange, conducted within a setting that is objective, interdisciplinary, and multimodal. The Board's varied activities annually engage about 7,000 engineers, scientists, and other transportation researchers and practitioners from the public and private sectors and academia, all of whom contribute their expertise in the public interest. The program is supported by state transportation departments, federal agencies including the component administrations of the U.S. Department of Transportation, and other organizations and individuals interested in the development of transportation.

Learn more about the Transportation Research Board at **www.TRB.org**.

Article

TRR
JOURNAL OF THE TRANSPORTATION RESEARCH BOARD

Transportation Research Record
2018, Vol. 2672(19) 1–12

DOI: 10.1177/0361198118787374
journals.sagepub.com/home/trr
SAGE

Monte Carlo Simulation-Based Benefit-Cost Analysis Combined with Analytical Hierarchy Process to Support ITS Investment with Consideration of Connected Vehicle Technology

Samaneh Khazraeian[1] and Mohammed Hadi[2]

Abstract

Decisions to invest in alternative intelligent transportation system (ITS) technologies are expected to increase in complexity, particularly with the introduction of connected vehicles (CV) and automated vehicles (AV) in the coming years. Traditional alternative analyses based on deterministic return on investment analysis are unable to capture the risks and uncertainties associated with the investment problem. In addition, these methods cannot account for agency preferences and constraints that cannot be converted to dollar values. This study utilizes a combination of a stochastic return on investment and a multi-criteria decision analysis method referred to as the Analytical Hierarchy Process (AHP) to select between ITS deployment alternatives considering emerging technologies. The approach is applied in a case study of the selection between using CV data and point detector data to support the freeway traffic data collection and monitoring service. The four objectives specified in the AHP analysis are providing the required functions, providing the required performance, minimizing the risks and constraints, and maximizing the return on investment. A stochastic return-on-investment analysis using a Monte Carlo simulation was used to calculate the return on investment values for input to the AHP method.

Decisions to invest in alternative intelligent transportation system (ITS) technologies to support transportation system management and operations (TSMO) are expected to increase in complexity, particularly with the introduction of connected vehicles (CV) and automated vehicles (AV) in the coming years. A surveys done as a part of the United States Department of Transportation (USDOT) ITS Deployment Tracking Project (*1*) identified safety and mobility benefits, integration with existing technologies, availability of funding, and equipment price as the major factors in the decision-making process for investing in ITS technologies (*2*). The National CV Field Infrastructure Footprint Analysis document produced by the American Association of State Highway and Transportation Officials (AASHTO) [1] pointed out that public agencies will need to assess the use of connected vehicle probe data versus existing methods of data collection.

Although the rate of deployment of new technologies is highly uncertain, CV and AV technologies promise significant potential savings and opportunities for TSMO applications. Traditionally, ITS technology and associated strategy alternatives have been assessed using return-on-investment analyses that utilize methods like deterministic Net Present Value (NPV) and Discounted Cash Flow (DCF). For example, the traditional NPV method involves deterministic estimates of the present values of the current and future benefits and costs over the project's life. Discount rates are used to calculate the present values of the cash flows. These methods are unable to capture the risks and uncertainties associated with the investment problem (*3*). In addition, these methods cannot account for agency preferences and constraints that cannot be converted to dollar values.

There are three main techniques for incorporating uncertainty into the return-on-investment analysis: Black

[1]Department of Civil and Environmental Engineering, Florida International University, Miami, FL
[2]Department of Civil and Environmental Engineering, College of Engineering and Computing, Florida International University, Miami, FL

Corresponding Author:
Address correspondence to Samaneh Khazraeian: skhaz001@fiu.edu

Scholes method, binomial lattice, and Monte Carlo methods. The Black Scholes method (*4*) has several assumptions that may not be met in transportation system alternative analysis. For example, the method assumes that the volatility (standard deviation of price over time) is constant over the project life, returns to the project are normally distributed, and the project's underlying value is lognormally distributed. The binomial lattice method (*5*) can only consider one source of uncertainty. In addition, a reported weakness of this method is that it needs cumbersome computational effort to reach a certain level of accuracy (*6*). The Monte Carlo method (*7*) is capable of integrating multiple sources of uncertainty with no requirements on the distributions and has been used in various fields to account for uncertainty by expressing cost and benefit parameters as probability distributions rather than as fixed values (*8–11*). The Monte Carlo simulation method is selected for use to account for the uncertainty in traffic flow.

In addition, measures that are difficult to be assessed in dollar values cannot be accounted for using the economic return-on-investment analysis methods. Thus, these methods are not fully adequate for use in assessing technology alternatives (*12*). Multi-Criteria Decision Analysis (MCDA) methods have been proposed to account for both qualitative and quantitative factors in the decision-making process.

The simplest MCDA approach is the Simple Multi-Attribute Rating Technique (SMART) approach, also known as the Simple Additive Weighting (SAW) method, described by Fishburn (*13*). This approach involves identifying decision criteria, criterion weights, assessment of the values associated with each criterion, normalizing the values to a common scale, and obtaining "scores" based on the values. The result is a weighted score for each of the compared alternatives. When there are many evaluation criteria and the priorities of multiple stakeholders are to be considered, the SAW method cannot sufficiently capture the weights of different criteria (*14*). More advanced MCDA methods, compared with the SAW method, have been proposed and used successfully. These methods include the Analytic Hierarchy Process (AHP), Technique for Order Preference Similarity to Ideal Solution (TOPSIS) (*15*), Preference Ranking Organization Method for Enrichment of Evaluations (PROMETHEE) (*16*, *17*), Elimination et Choix Traduisant la Realité (ELECTRE) (*18–20*) and others (*21*). All of these methods require stakeholders to input their preferences and priorities with respect to various decision criteria and an assessment of each alternative to meet each criterion. Among these MCDA methods, the AHP method developed by Saaty (*22*) has been the most widely used in various disciplines including transportation engineering. According to the study by Macharis et al. (*23*), 33% of the reviewed publications on utilizing MCDA in the transportation field used AHP or a variant of AHP in their analysis. So, the AHP method was selected for use in this study.

This study utilizes a combination of a stochastic NPV and the AHP method to select between ITS deployment alternatives considering emerging technologies. The approach is applied to the selection between CV deployment and legacy detection technology to support the freeway traffic data collection and monitoring service which includes incident detection and travel time estimation. It should be noted that the application to data collection and monitoring service is utilized as a case study and the methodology can be applied to any other ITS services. The stochastic NPV analysis utilizes a Monte Carlo approach to account for uncertainty in the estimation of the benefits and costs of utilizing CV versus the legacy systems. The AHP method involves the consideration of various criteria and measures, weighted by agency priorities and preferences. This approach considers these priorities and preferences, in addition to accounting for measures that cannot be converted to dollar values. Neither of these considerations is possible with the NPV analysis. The result from the NPV analysis is included as one of the four objectives defined for the AHP evaluation. The other three objectives are the provision of the required functions, achieving the required performance, and minimizing the risks and constraints. The results of the selection between the alternatives are expected to be dependent on the input parameters, which are different for different agencies and locations.

Methodolgy

A four level decision-making hierarchy according to the AHP method is defined for the purpose of alternative selection of this study as shown in Table 1. The top level of the decision hierarchy or simply the goal for the case study service is "To select between CV and existing detection technologies for providing traffic data collection and monitoring service." The two middle levels of the hierarchy include the overall objectives and the sub-criteria associated with each of the objectives. And the lower level is the alternative level. The monetizable measures are assessed in the stochastic NPV analysis using Monte Carlo simulation and the NPV results are included as a criterion in the AHP analysis. The non-monetizable measures are included as additional measures in the AHP analysis. The NPV and AHP analysis, as conducted in this study, are discussed in more detail in the following subsections.

NPV Analysis

Maximizing the return on investment is one of the criteria included in this study's MCDA analysis, as indicated in

Table 1. Criteria for Selection between Emerging and Existing Traffic Monitoring and Data Collection Technologies

Goal	Objectives	Sub-criteria	Alternatives
Selecting between CV and existing detection technologies for providing traffic data collection and monitoring service	1-Provision of the required functions	1.1-Ability to measure travel time 1.2-Ability to measure volume and/or density 1.3-Ability to measure new measures (acceleration noise, stops, breaks, wiper on/off, lights on/off, emission, potential for crashes, etc.)	CV data Point detectors
	2-Achieving the required performance	2.1-Mean absolute error in measurement 2.2-Standard deviation of error of measurement 2.3-Detection timeliness	
	3-Minimizing the risks and constraints	3.1-Technology uncertainty 3.2-Standard stability 3.3-Uncertainty in business model 3.4-Data archiving and processing and technical skill concerns 3.5-Funding (initial and recurring) 3.6-Data availability and sharing concern 3.7-Security and privacy concerns	
	4-Maximizing the return on investment based on NPV	4.1-Median present worth (based on Monte Carlo analysis) 4.2-15% present worth (based on Monte Carlo analysis)	

Table 1. As stated earlier, this study utilizes a stochastic approach based on Monte Carlo analysis to estimate the NPV. This section describes the assessment of the return on investment of utilizing connected vehicle data versus traditional point-detector data to detect freeway incidents and collect travel time information using a case study of a 15-mile long freeway facility in Florida. Unlike the traditional return-on-investment analysis which calculates deterministic point estimates of the NPV of the project alternatives, the NPV analysis of this study employs a Monte Carlo simulation to account for the uncertainties involved in the assessment and to generate a distribution for each alternative's NPV. To calculate the NPV, the alternative benefits and costs need to be calculated first, as discussed in the following subsections.

Traffic Data Collection and Monitoring Benefits. The incremental benefits of traffic monitoring technology alternatives are assumed to result from: 1) faster incident detection that results in lower delays, and 2) more accurate traveler time estimation that results in better diversion decisions.

Benefits due to faster congestion/incident detection: To evaluate the two alternatives of this study (CV versus point detectors), a base case was defined and the benefit, in terms of delay reduction, of each alternative was calculated relative to the base case, as shown in Equation 1. The defined base case is "No detection technology" that involves detection utilizing service patrol, police calls, CCTV monitoring, and other "manual" methods but no infrastructure-based incident detection. More information about detection by manual methods is provided is Reference 29:

$$\begin{aligned} &\textit{Incident Detection Benefit}_i \\ &\quad = (TD_{base} - TD_{alternative})^{*}VOT^{*}IF_i \end{aligned} \tag{1}$$

where, TD_{base} is the total delay of the base alternative in veh-hr, $TD_{alternative}$ is total delay of the alternative in veh-hr, VOT is the value of time in dollars, and IF_i is the total number of incidents for the ith year. The benefits of using CV vehicle data versus point detector data were calculated using the difference between the benefits in delay when utilizing CV data for detection (TD_{CV}) and the benefits in delay when using detector data for detection ($TD_{detector}$) in veh-hr. Delay saving of incident detection is assessed using queuing theory equations (*24*) that calculate total delay TD based on demand, capacity, and incident duration as follows:

$$TD = t_R t_Q(\lambda - \mu_R)/2 \tag{2}$$

$$t_Q = t_R(\mu - \mu_R)/(\mu - \lambda) \tag{3}$$

where

λ = Arrival rate (demand) (vph)

μ = Arrival rate (demand) (vph)

μ_R = Reduced capacity rate under metering for ramps or under non-recurrent conditions for the mainline (vph)

t_R = Event duration (hr)

t_Q = Time in queue duration (hr)

An important input to the calculation of delay in the above equations is the reduction in the incident duration due to the reduction in the detection time resulting from applying the alternative detection technology. The incident detection time based on CV data and traffic detector were derived based on simulation results by the authors of this study, as shown in Equations 4 and 5, respectively (*25, 26*):

$$MTTD = \begin{cases} 0.75 \text{ } if \text{ } MP>6\% \\ -61.126 + 64.181*e^{-\left(\frac{V}{C}*MP\right)} if \text{ } MP \leq 6\% \end{cases} \quad (4)$$

where, MTTD is mean time to detect (min), MP is the CV market penetration (%), V is the volume in vehicle per hour, and C is the capacity in vehicle per hour:

$$MTTD = 4.375 - \left(6.427*\frac{V}{C}\right) + (0.003*D) \quad (5)$$

where, MTTD is mean time to detect (min), V is volume (vph), C is capacity (vph) and D is the distance between the closest upstream point detector to the incident location (Ft).

Equations 1–5 include several stochastic variables that result in uncertainty in the results of the benefit analysis. The uncertainty in the incident rate, CV market penetration, traffic demand, incident location with respect to the point detectors, incident detection time without infrastructure detection, and the value of time were varied in the Monte Carlo analysis by expressing these variables as distributions rather than point estimates. The parameters of the distributions utilized in this study were borrowed from previous studies. The lane blockage incident rate was assumed to have a Poisson distribution with a mean of 0.616 per million VMT per hour per lane, based on a previous study by the authors for the same location of the study area (*27*). The variance of the Poisson distribution is equal to the mean. The utilized VOT stochastic distribution in the Monte Carlo simulation was calculated based on a distribution derived in a study by Hossan et al. (*28*). The travel demand is assumed to have a fixed coefficient of variation, calculated based on real-world data in Florida and with the mean demands changing according to a uniformly distributed random annual growth rate between 0.5% and 1.5%. The location of the incident is assumed to vary according to a uniform distribution between 20 Ft and 2,250 Ft from the upstream detector with a detector spacing of half of a mile.

The base case incident detection time (no detection technology) is assumed to follow a normal distribution with a mean of 6.856 and a standard deviation of 1.782 for detection based on service patrol and a mean of 4.63 minutes and a standard deviation of 3.273 minute for detection using other methods (CCTV, FHP, and other manual methods), as estimated by Qiang et al. for the study area (*29*). These distributions are used in the Monte Carlo simulation to determine the detection time with no detection technology. The connected vehicle market penetration is assumed to follow a lognormal distribution, as derived in a study by Iqbal et al. (*30*).

Benefits due to travel time accuracy: The benefits of travel time accuracy in terms of the impact on diversion decisions made by travelers during incidents were also included in the return-on-investment analysis. The Standard Deviation of Percentage Error (SDPE) was used a measure to estimate the accuracy. The $SPDE_i$ with point traffic detection was calculated by assuming a random value between 10% and 20% based on a previous assessment of this parameter for estimating travel time during incident conditions (*31*). The $\mathrm{SPDE_i}$ with CV data as a function of the stochastic CV market penetration was calculated using Equation 6 (*25*):

$$SPDE = 25.15 - 19.81*\cos\left(\frac{V}{C}\right) - 0.75*\log(MP) - 11.56*\frac{V}{C} + 0.02*(\mathrm{MP}) \quad (6)$$

where, *SPDE* is the standard deviation of percentage error, V is volume (vph), C is capacity (vph) and MP is the CV proportion.

To convert the travel time estimation error impact on traveler diversion decisions to dollar values, the relationship developed by Toppen et al. (*32*) which converts the SDPE to dollar values were used, as shown in Equation 7. The yearly benefits of travel time accuracy were then calculated by multiplying the dollar value obtained from Equation 7 by the yearly number of trips passing the incident location in the presence of queue and are willing to divert, as shown in Equation 8:

$$S = 1.2 - 0.003918*SPDE_i^2 \quad (7)$$

where, S is the monetary value of travel time accuracy per trip (\$) and $SPDE_i$ is the ith year travel time standard deviation of percentage error:

$$Travel\ Time\ Accuarcy\ Benefit_i = \left(1.2 - 0.003918*SPDE_i^2\right)*tQ_i*D_i*IF_i*DR_i \quad (8)$$

where $SPDE_i$ is the ith year travel time error, tQ_i is the ith year queue duration calculated using queuing theory as presented in Equation 3, D_i is the ith year travel demand, and IF_i is the ith year incident frequency. DR_i is the ith year diversion rate, which is assumed to be a random value between 20% and 40%. It should be noted that as with the calculation of the incident detection benefits, as reported in the previous section, variables in the above equation are considered random variables in the Monte Carlo analysis.

Traffic Data Collection/Monitoring Deployment Costs. Based on the Florida Department of Transportation (FDOT) District 6 cost database, the point detector cost at each location is estimated to be $21,800. This cost includes capital, replacement, calibration, maintenance, design, and mobilization costs. It is assumed that these detectors are installed at half-mile intervals, resulting in a total capital cost of $665,200 for the 15-mile segment of study. The connected vehicle deployment costs were extracted from "The Life Cycle Cost Model (LCCM)" developed by the Federal Highway Administration (FHWA) (*33*). With this tool, the user is asked to specify the CV Vehicle-to-Infrastructure (V2I) application of interest as well as other questions to determine the quantities of the components required for the application deployment. If necessary, other types of costs such as backhaul communication upgrade can be added by the user. The total cost of CV deployment for a 15-mile corridor was estimated to be $609,100, with a $23,100 maintenance cost. Due to the uncertainty in the CV costs, it is assumed that the cost varies within 15% of these values according to a uniform distribution.

Analytical Hierarchy Process Application

As mentioned earlier, the AHP method was selected to perform the multi-criteria decision analysis of this study. The AHP method decomposes the decision criteria into hierarchy by reducing the number of criteria for comparison at each level, thus simplifying the structure of the comparison of the relative importance of these criteria considering the number of criteria that may need to be considered. Both quantitative and qualitative criteria can be assessed using this method. The pairwise comparison by the decision-makers facilitates the estimation of the weights of the criteria, and the consistency in the judgment between different comparisons can be checked. The method is also generally easy to apply using a spreadsheet or commercial off-the-shelve tools that have more advanced analysis and visualization capabilities of the decision-making elements and results. The AHP method has three steps: calculating the weights for each criterion (vector of criteria weights), calculating the rating for each alternative with respect to each criterion (matrix of alternative scores), and calculating the weighted average rating for each alternative and ranking them. These steps are briefly described below.

Calculating the Vector of Criteria Weights. In the AHP, a pairwise comparison matrix A (m*m) is created where m in number of evaluation criteria. Each element a_{ij} of the matrix indicates the importance of ith criterion relative to the jth criterion. If $a_{ij} > 1$, criterion i is more important than criterion j. The values to measure the relative importance between two criteria are selected from a numerical scale from 1 to 9 and reciprocals are used when the second criterion is more important (1/2, 1/3...1/8, 1/9). After creating the pairwise comparisons matrix, a normalized pairwise matrix A_{norm} is created and each element of the matrix is calculated as shown in Equation 9:

$$\bar{a}_{ij} = \frac{a_{ij}}{\sum_{i=1}^{m} a_{ij}} \quad (9)$$

Finally, each criterion's weight (w_j) is calculated by averaging the elements on each row of the normalized matrix using Equation 10:

$$w_j = \frac{\sum_{j=1}^{m} \bar{a}_{ij}}{m} \quad (10)$$

Calculating the Matrix of Alternatives Scores. The score matrix S is then calculated. S is a n*m matrix where n is the number of alternatives and m is the number of criteria. To derive the matrix S, a pair-wise matrix $B^{(i)}$ needs to be created for each of the m criterion (i = 1,..., m). Matrix $B^{(i)}$ is a n*n matrix and each element a^{i}_{kh} of the matrix $B^{(i)}$ indicates the importance of kth alternative compared with h^{th} alternative with respect to the ith criterion. If $a^{i}_{kh} > 1$ the kth alternative is better than h^{th} alternative with respect to the ith criterion. Then, the same procedure, applied to matrix A, is applied to each matrix $B^{(i)}$. Each element of the matrix is divided by the sum of entries in the same column and then the elements of each row are averaged to calculate the score vectors $s^{(i)}$, i = 1,...,m. In other words, $s^{(i)}$ represents the costs of the scores of different alternatives with respect to the ith criterion. Finally, the score matrix S is defined as shown in below:

$$S = \left[s^{(1)} \dots s^{(m)}\right] \quad (11)$$

Ranking the Alternatives. Having the weight vector w and the score matrix S, the global score for each alternative is calculated by multiplying S and w, as shown below:

$$v = S.w \quad (12)$$

A consistency ratio (CR) needs to be calculated to make sure that the rankings given by different decision makers and used as inputs to the AHP application are consistent. CR is calculated using Equation 13:

$$CR = \frac{CI}{RI} \quad (13)$$

Where CI is consistency index calculated as shown in Equation 14 and RI is the random index obtained from Reference (22):

$$CI = \frac{\lambda_{max} - n}{n - 1} \quad (14)$$

where, λ_{max} is the maximum Eigen value of the comparison matrix and n is number of criteria being compared. If the consistency ratio is less than 0.1, the rankings are consistent otherwise they have to be revised by the decision-makers.

Case Study and Results

This section illustrates the application of the methodology that combines the stochastic return-on-investment analysis and the AHP, as described earlier, to a case study in Southeast Florida. The case study addresses an investment decision for a 15-mile segment of a freeway corridor with two assumed volume demand levels: moderate and light traffic (V/C ratio = 0.8 and V/C ratio = 0.4). The different volume levels affect the number of CV vehicles available to provide vehicles in the traffic stream and thus the quality of the data, as shown in Equation 6. In addition, the volume levels affect the MMTD for both CV and point detectors, as shown in Equations 4 and 5.

The results from the Monte Carlo analysis are used to generate the NPV distribution for each alternative. Then, the results from the Monte Carlo analysis are used as one of the inputs to the AHP analysis and the examined alternatives are ranked based on their final scores, as described below.

Monte Carlo Simulation Results

One thousand Monte Carlo analysis runs were performed with the input variables generated randomly from the distributions specified for each stochastic variable in the return on investment analysis. The NPV estimated based on each run was saved, and the results from the 1,000 runs were used to generate the NPV distribution. The analysis was done for 6 hours of operations of the corridor with the moderate traffic (V/C ratio of 0.8) and 6 hours of operations with the light traffic (V/C ration of 0.4), separately. Table 2 shows the generated performance measures from the Monte Carlo runs for each of the alternatives. As can be seen from this table, the quality of the CV data improves through the years due to the increase in the CV market penetration.

The Incremental NPV of the CV utilization for traffic monitoring is defined as the difference in the NPV between the utilization of CV and traffic detectors for traffic monitoring, calculated as described in the methodology section. The incremental NPV distribution utilizing Monte Carlo simulation outputs was generated for each year starting with the first year of CV deployment, assumed to be the year when the CV technology becomes mandated on all new vehicles by the United States Department of Transportation. The results, shown in

Table 2. Alternatives Performance Measures Based on Monte Carlo Runs

			Year 1	Year 3	Year 5	Year 7	Year 9
V/C = 0.4	50th percentile (median) results						
	SPDE	CV	10.76	8.38	7.05	6.02	5.12
		Point detector	15.14	15.14	15.14	15.14	15.14
	MTTD	CV	2.20	0.75	0.75	0.75	0.75
		Point detector	5.03	4.96	4.90	4.84	4.76
	85th Percentile Results						
	SPDE	CV	11.59	8.92	7.48	6.40	5.48
		Point detector	18.48	18.48	18.48	18.48	18.48
	MTTD	CV	2.57	0.75	0.75	0.75	0.75
		Point detector	7.70	7.66	7.61	7.50	7.47
V/C = 0.8	50th percentile (median) results						
	SPDE	CV	10.23	7.81	6.44	5.44	4.51
		Point detector	14.91	14.91	14.91	14.91	14.91
	MTTD	CV	1.66	0.75	0.75	0.75	0.75
		Point detector	2.94	2.81	2.72	2.61	2.54
	85th percentile results						
	SPDE	CV	11.02	8.34	6.89	5.84	4.91
		Point detector	18.30	18.30	18.30	18.30	18.30
	MTTD	CV	2.24	0.75	0.75	0.75	0.75
		Point detector	5.83	5.70	5.62	5.54	5.42

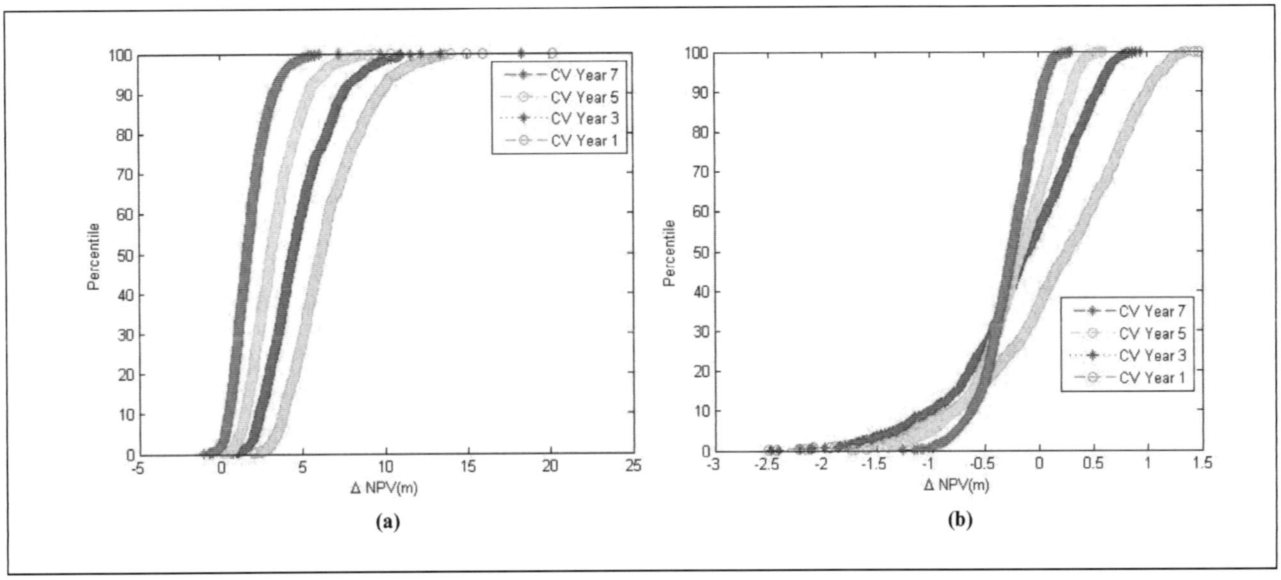

Figure 1. Incremental NPV distributions based on the CV deployment vs. point detector for each year of starting utilizing CV data after CV technology becomes mandated: (*a*) corridor with moderate traffic (V/C = 0.8); (*b*) corridor with light traffic (V/C = 0.4).
Note: ΔNPV = the difference in the NPV between the utilization of CV and traffic detectors for traffic monitoring. Unit = millions of dollars.

Table 3. Decision Objectives Pairwise Comparison by the Decision Makers

Decision Maker	Objectives	Provision of the required functions	Achieving the required performance	Minimizing the risks and constraints	Maximizing the return on investment
Decision Maker 1	Provision of the required functions	1	1	2	2
	Achieving the required performance	1	1	2	2
	Minimizing the risks and constraints	0.5	0.5	1	2
	Maximizing the return on investment	0.5	0.5	0.5	1
Decision Maker 2	Provision of the required functions	1	1	0.5	0.5
	Achieving the required performance	1	1	0.5	0.5
	Minimizing the risks and constraints	2	2	1	1
	Maximizing the return on investment	2	2	1	1

Figure 1, indicate that utilizing CV data for freeway segments is significantly more cost-effective than point detectors in detecting incidents and providing travel time estimates about one year after the CV mandated on all new vehicles becomes effective, particularly for the corridor with moderate traffic. However, for the corridor with light traffic (V/C = 0.4), there is a probability of the CV deployment not being effective in the first few years of the CV deployment due to low measurement reliability of travel time and high latency of incident detection.

AHP Application Results

In this study, managers in two transportation system management agencies in Florida were asked to fill the rating tables that are required as inputs to the AHP methodology. The AHP analysis was done separately using each of the two inputs provided by the managers, as their priorities are different. Table 3 shows the highest level ratings that involve comparing the high-level objectives given by the first and second decision makers, referred to as Decision Maker 1 and 2. The colored cells are the only ones needed to be ranked as the uncolored ones are the reciprocal of the colored ones. Decision Maker 1 put the highest emphasis on achieving the required functions and achieving the required performance and gave these two objectives twice the importance of minimizing the risk and constraints and maximizing the return on investment. This rating favors new technologies like utilizing CV data compared with existing technologies. Unlike Decision Maker 1, minimizing the risks and constraints and maximizing the return on investment are the two most important factors to Decision Maker 2. This decision-maker appears to be more conservative than Decision Maker 1 in accepting new technologies that provide additional functions and better performance, if there are additional risks associated with the technology.

The rating at the sub-criteria level is shown in Table 4. The sub-criteria comparisons given by Decision Maker 1 are only shown in this table due to space limit. The following can be concluded based on Table 4:

- With regard to the functions provided, Decision Maker 1 rated the measurement of travel time as important as measuring volume and density, and twice as important as the ability to measure new measures that cannot be obtained using current detection technologies such as acceleration noise, stops, breaks, wiper on/off, lights on/off, emission, potential for crashes, etc.
- With regard to achieving the required performance; the estimation accuracy, as measured by the mean absolute error, and the estimation reliability, as measured by standard deviation of error, were ranked as equally important and twice as important as the timeliness of incident detection, probably reflecting the current use of the agency of their detector data.
- With regard to the risks and constraints, Decision Maker 1 gave the highest rating to technology uncertainty, standard stability, and uncertainty is the business model. Data archiving and processing and associated technical skill requirements, funding, and data availability and sharing, and security and privacy concerns were not ranked high; reflecting the current availability of good technical support and resources for the agency.
- With regard to the return on investment, Decision Maker 1 rated the 50% percentile NPV twice as important as the 15% NPV, reflecting that the median value of the return on investment is more important than the uncertainty or stochasticity associated with the return-on-investment analysis.

The consistency index of the AHP was calculated for each of the tables to make sure that Decision Maker 1's rating of criteria relative to each other was consistent. After confirming that the consistency index was acceptable, the AHP analysis was conducted for different demand levels and analysis years. Table 5 shows examples of the AHP analysis results. As can be seen from this table, in all of the cases, utilizing CV data has a higher final score than point detectors. This is because CV has a significantly better performance in incident detection and travel time estimation. The partial score for each of the criteria is also shown in the table. Utilizing CV data has lower scores in the third criterion which is minimizing the risk and constraints. However, the score improved in future years, as the quality of data improves due to increased market penetration and the uncertainty and risks associated with the CV deployment decreases.

Table 6 shows the AHP results based on Decision Makers 2's ratings. Comparing Tables 5 and 6, it is clear that the CV data utilization alternative has a higher relative score based on Decision Maker 1's inputs than that based on Decision Maker 2's inputs, reflecting the more conservative view of Decision Maker 2, although utilizing CV data is still the preferred alternative to collect data for segments with moderate to heavy volumes.

Conclusions and Recommendations

This study has presented a multi-criteria decision-making methodology with consideration of uncertainties involved in the alternative assessment. An AHP approach was used to select between utilizing CV-based alternatives versus legacy technology alternatives. The four objectives specified in the AHP analysis are providing the required functions, providing the required performance, minimizing the risks and constraints, and maximizing the return on investment. A stochastic return-on-investment analysis using Monte Carlo simulation was used in combination with the AHP method. The method was applied to the selection between CV deployment and legacy detection technology to support the freeway traffic data collection and monitoring service which includes incident detection and travel time estimation. However, the methodology can be applied to any other ITS service by identifying the functions, the required performance, the associated risks, and the return-on-investment method associated with the service. Two alternatives were evaluated in the case study: utilizing CV data versus utilizing point detector data. The results of the stochastic return-on-investment analysis indicate that utilizing CV data for freeway segments is significantly more cost-effective than using point detectors in detecting incidents and providing travel time estimates about one year after the CV technology becomes mandated on all new vehicles, for corridors with moderate to heavy traffic. However, for corridors with light traffic (V/C = 0.4), there is a probability of the CV deployment not being effective in the first few years of the CV deployment due to low measurement reliability of travel time and high latency of incident detection, associated with the smaller sample size of the collected data. The AHP analysis results indicate that the scores

Table 4. Decision Sub-Criteria Pairwise Comparison by the First Decision Maker

Provision of the required functions

	Measuring travel time	Measuring volume and/or density	Measuring new measures
Measuring travel time	1	1	2
Measuring volume and/or density	1	1	2
Measuring new measures	0.5	0.5	1

Achieving the required performance

	Mean absolute error	Std. dev. of error	Detection timelines
Mean absolute error	1	1	2
Std. dev. of error	1	1	2
Detection timelines	0.5	0.5	1

Minimizing the risks and constraints

	Technology uncertainty	Standard stability	Uncertainty in business model	Data archiving and processing and technical skill concerns	Funding	Data availability and sharing concern	Security and privacy concerns
Technology uncertainty	1	1	1	9	9	9	9
Standard stability	1	1	0.5	9	9	9	9
Uncertainty in business model	1	2	1	9	9	9	9
Data archiving and processing and technical skill concerns	0.11	0.11	0.11	1	1	1	1
Funding	0.11	0.11	0.11	1	1	1	1
Data availability and sharing concern	0.11	0.11	0.11	1	1	1	1
Security and privacy concerns	0.11	0.11	0.11	1	1	1	1

Maximizing the return on investment

	Median present worth	15% present worth
Median present worth	1	2
15% present worth	0.5	1

of the evaluated alternatives vary depending on stakeholder priorities. When the provision of the required function and achieving the required performance are more important than minimizing the risk for an agency, the CV-based alternative has a higher score than when minimizing the risk has a high weight in the MCDA. For practical applications of the tool, it is recommended that a tool be developed that incorporates and automates the calculations involved in the Monte Carlo return-on-investment and AHP process to make it easier for agencies to use the method developed in this study in the decision making processes.

Table 5. AHP Results Based on Decision Maker 1 Ratings

CV deployment start from year 1		Provision of the required functions	Achieving the required performance	Minimizing the risks and constraints	Maximizing the return on investment based on NPV	Final score
CV	V/C = 0.8	0.58	0.66	0.21	0.84	0.56
Point detector		0.42	0.34	0.79	0.16	0.44
CV	V/C = 0.4	0.58	0.64	0.21	0.54	0.52
Point detector		0.42	0.36	0.79	0.46	0.48

CV deployment start from year 5		Provision of the required functions	Achieving the required performance	Minimizing the risks and constraints	Maximizing the return on investment based on NPV	Final score
CV	V/C = 0.8	0.58	0.76	0.34	0.74	0.60
Point detector		0.42	0.24	0.66	0.26	0.40
CV	V/C = 0.4	0.58	0.71	0.34	0.45	0.55
Point detector		0.42	0.29	0.66	0.55	0.45

Table 6. AHP Results Based on Decision Maker 2 Ratings

CV deployment start from year 1		Provision of the required functions	Achieving the required performance	Minimizing the risks and constraints	Maximizing the return on investment based on NPV	Final score
CV	V/C = 0.8	0.58	0.60	0.23	0.85	0.56
Point detector		0.42	0.40	0.86	0.15	0.47
CV	V/C = 0.4	0.58	0.55	0.23	0.54	0.45
Point detector		0.42	0.45	0.86	0.46	0.58

CV deployment start from year 5		Provision of the required functions	Achieving the required performance	Minimizing the risks and constraints	Maximizing the return on investment based on NPV	Final score
CV	V/C = 0.8	0.58	0.73	0.37	0.75	0.59
Point detector		0.42	0.27	0.72	0.25	0.44
CV	V/C = 0.4	0.58	0.64	0.37	0.45	0.48
Point detector		0.42	0.36	0.72	0.55	0.55

Acknowledgments

The authors of this paper are grateful for the financial support of the sponsor of this research, the National Cooperative Highway Research Program (NCHRP) of the National Academics of Sciences, Engineering, and Medicine.

Author Contributions

The authors confirm contribution to the paper as follows: study conception and design—Samaneh Khazraeian (50%), Mohammed Hadi (50%); data collection—Samaneh Khazraeian (70%), Mohammed Hadi (30%); analysis and interpretation of results—Samaneh Khazraeian (50%), Mohammed Hadi (50%); draft manuscript preparation—Samaneh Khazraeian (70%), Mohammed Hadi (30%). All authors reviewed the results and approved the final version of the manuscript.

References

1. Intelligent Transportation Systems, Joint Program Office. *Knowledge Resources: 2016-2017 Deployment Tracking*

Survey Results [Online]. U.S. Department of Transportation, 2013. http://www.itsdeployment.its.dot.gov/.
2. Gordon, S., and J. Trombly. *Deployment of Intelligent Transportation Systems: A Summary of the 2013 National Survey Results*. Final Report. FHWA-JPO-14-146. FHWA, U.S. Department of Transportation, Washington, D.C., 2014.
3. Wright, J., J. Garrett, C. Hill, G. Krueger, J. Evans, S. Andrews, C. Wilson, B. Rajbhandari, and B. Burkhard. *National Connected Vehicle Field Infrastructure Footprint Analysis*. Final Report. FHWA-JPO-14-125. FHWA, U.S. Department of Transportation, Washington, D.C., 2014.
4. Black, F., and M. Scholes. The Pricing of Options and Corporate Liabilities. *Journal of Political Economy*, Vol. 31, No. 3, 1979, pp. 637–659.
5. Cox, J., S. Ross, and M. Rubinstein. Option Pricing: A Simplified Approach. *Journal of Financial Economics*, Vol. 7, No. 3, 1979, pp. 229–263.
6. Gilbert, E. Investment Basics XLIX. An Introduction to Real Options. *Investment Analysts Journal*, Vol. 33, No. 60, 2005, pp. 49–52.
7. Boyle, P. P. Options: A Monte Carlo Approach. *Journal of Financial Economics*, Vol. 4, No. 3, 1977, pp. 323–338.
8. Yang, S., S. Luou, and M. Hadi. Risk Analysis to Account for Uncertainty in Benefit-Cost Evaluations of Intelligent Transportation Systems. *Transportation Research Record: Journal of the Transportation Research Board*, 2007. 2035: 187–194.
9. Neufville, R., and S. Scholtes. *Flexibilty in Engineering Design*. MIT Press, Cambridge, Mass., 2011.
10. Sullivan, W., and R. Orr. Monte Carlo Simulation Analyzes Alternatives in Uncertain Economy. *Industrial Engineering*, Vol. 14, No. 11, 1982, pp. 42–49.
11. Cox, A., R. Fenton, and P. Carlock. Incorporating Contingency Risk into Project Cost and Benefit Baselines: A Way to Enhance Realism. Presented at 9th Annual Symposium of the International Council of Systems Engineering (INCOSE), Brighton, England, 1999.
12. Sullivan, W., and S. LeClair. Justification of Flexible Manufacturing Systems. *Proc., Autofact 85' Conference*, Society of Manufacturing Engineers, Detroit, MI, 1985.
13. Fishburn, P. C. Additive Utilities with Incomplete Product Set: Applications to Priorities and Assignments. *Operations Research*, Vol. 15, No. 3, 1967, pp. 537–542.
14. Annette, J., A. Banu, and P. Chandran. Comparison of Multi Criteria Decision Making Algorithms for Ranking Cloud Renderfarm Services. *Indian Journal of Science and Technology*, Vol. 9, No. 31, 2016, pp. 1–5.
15. Hwang, C., and K. Yoon. *Multiple Attribute Decision Making: Methods and Applications*. Springer, New York, 1981.
16. Brans, J., and P. Vincke. A Preference Ranking Organization Method (The PROMETHEE Method for Multiple Criteria Decision Making). *Management Science*, Vol. 31, No. 6, 1985, pp. 647–656.
17. San Cristobal, J. Critical Path Definition Using Multicriteria Decision Making: PROMETHEE Method. *Journal of Construction Engineering and Management*, Vol. 29, No. 2, 2013, pp. 158–163.
18. Roy, B. Classement et Choix en Présence de Points de vue Multiples (la méthode ELECTRE). *La Revue d'Informatique et de Recherche Opérationelle (RIRO)*, Vol. 2, No. 8, 1968, pp. 57–75.
19. Benayoun, R., B. Roy, and N. Sussman. Manual de Reference du Programme Electre, Mote de Synthese et Formaton. Presented at Direction Scientifique SEMA, Paris, France, 1966.
20. Hokkanen, J., and P. Salminen. Choosing a Solid Waste Management System Using Multicriteria Decision Analysis. *European Journal of Operational Research*, Vol. 98, No. 1, 1997, pp. 19–36.
21. Jato-Espino, D., E. Castillo-Lopez, J. Rodriguez-Hernandez, and J. Canteras-Jordana. A Review of Application of Multi-Criteria Decision Making Methods in Construction. *Automation in Construction*, Vol. 45, 2014, pp. 151–162.
22. Saaty, T. *The Analytic Hierarchy Process*. McGraw-Hill, New York, 1980.
23. Macharis, C., and A. Bernardini. Reviewing the Use of Multi-Criteria Decision Analysis for the Evaluation of Transport Projects: Time for a Multi-Actor Approach. *Transport Policy*, Vol. 37, 2015, pp. 177–186.
24. May, A. *Traffic Flow Fundamentals*. Prentice Hall Inc., Englewood Cliffs, N.J., 1990.
25. Hadi, M. *NCHRP Report 20-07/376: Framework to Support Transportation Agency ITS Infrastructure and ITS Legacy Decisions with Consideration of Connected Vehicle Deployment and Automated Vehicle Initiatives*. Transportation Research Board of the National Academies, Wahington, D.C., 2017.
26. Iqbal, S., S. Khazraeian, and M. Hadi. A Methodology to Assess the Quality of Travel Time Estimation and Incident Detection based on Connected Vehicle Data. *Transportation Research Record: Journal of the Transportation Research Board*, 2018. Advance online publication. https://doi.org/10.1177/0361198118773199
27. Hadi, M., Y. Xiao, H. Ozen, and P. Alvarez. *Evaluation Tools to Support ITS Planning Process*. Prepared for FDOT Research Center by Florida International University, Miami, 2008.
28. Hossan, S., H. Asgari, and X. Jin. Investigating Preference Heterogeneity in Value of Time (VOT) and Value of Reliability (VOR) Estimation for Managed Lanes. *Transportation Research Part A: Policy and Practice*, Vol. 94, 2016, pp. 638–649.
29. Qiang, H., M. Hadi, Y. Xiao, and C. Zhan. Estimating Incident Detection Time based on Incident Management and Traffic Detector Data. Presented at 89th Annual Meeting of the Transportation Board, Washington, D.C., 2010.
30. Iqbal, S., M. Hadi, and Y. Xiao. Assessment of Link Level Variation of Connected Vehicle Market Penetration. Presented at 96th Annual Meeting of the Transportation Research Board, Washington, D.C., 2017.
31. Xiao, Y. *Hybrid Approaches to Estimating Freeway Travel Times Using Point Traffic Detector Data*. Florida International University, Miami, 2011.
32. Toppen, A., and K. Wunderlich. *Travel Time Data Collection for Measurement of Advanced Traveler Information*

Systems Accuracy. Project No. 0900610-D1. FHWA, U.S. Department of Transportation, Washington, D.C., 2003.

33. Federal Highway Administration. *Near-Term V2I Transition and Ohasing Analysis, Life Cycle Cost Model User's Guide*. Federal Highway Administration, Washington, D.C., 2015.

The Standing Committee on Intelligent Transportation Systems (AHB15) peer-reviewed this paper (18-05453).

This study is a part of the NCHRP 20-07/376 project and the finding of the study reflects the authors' opinions not the NCHRP and its constitution members.

Article

TRR
JOURNAL OF THE TRANSPORTATION RESEARCH BOARD

Transportation Research Record
2018, Vol. 2672(19) 13–24

DOI: 10.1177/0361198118791668
journals.sagepub.com/home/trr
SAGE

Predictive Dynamic Speed Limit in a Connected Environment for a Weather Affected Traffic Network: A Case Study of Chicago

Archak Mittal[1], Eunhye Kim[1], Hani S. Mahmassani[1], and Zihan Hong[1]

Abstract

Dynamic speed limits (DSLs) are used to improve safety and mobility on freeways in unfavorable traffic conditions due to recurring congestion, roadworks, incidents, or adverse weather. The evaluation of in-field deployment reveals that the effectiveness of DSLs can be hampered by low compliance rates or lack of inherent capacity. With the emergence of vehicle-to-vehicle (V2V) or vehicle-to-infrastructure (V2I) communication, it is believed that the operation of DSLs will be able to take advantage of vehicle connectivity. In this paper, the effectiveness of the predictive DSL operation in a connected environment is investigated on the weather affected traffic network of Chicago city under different operational conditions. For the sensitivity test, different market penetration rates of connected vehicles are tested in microsimulation. Microscopic models are used to simulate information exchange by V2V or V2I communication. However, such an application over a large network with mixed traffic can be computationally expensive. A mesoscopic or macroscopic tool is needed that can scale and be computationally economical at the network level. This study integrates the microscopic aspect of V2V communication and the macroscopic for dynamic traffic assignment at a network level. The evaluation of effectiveness at network level is conducted by the Traffic Estimation and Prediction System (TREPS), which is a mesoscopic simulator. The results show, depending on the strategy applied, meaningful increases in both throughput and prevailing speed.

Dynamic speed limit (DSL) control is an active traffic management strategy that adjusts speed limits based on prevailing or predicted traffic states, events (roadworks or accidents), or weather conditions. This strategy aims to reduce incident-related risk and to keep operational performance of the traffic network at or above the desirable level. Adjusted speed limits are communicated to drivers through dynamic message signs or vehicle communication. The success of the DSL strategy is achieved through reducing the variability of vehicle speeds that may result in shock waves and hence perturbations in the traffic flow. Speed harmonization helps to smooth the traffic state transition and maintain stable traffic operations, resulting in improved safety and mobility on the road (*1*).

As found through in-field deployment, speed harmonization can help to enhance road safety, sometimes at the cost of reduced throughput or increased travel time (*2–5*). Speed harmonization was introduced in Germany in 1970, the Netherlands in 1981, the United Kingdom in 2001, and has been implemented in several states of the U.S.A, including Washington and Michigan. Safety assessments have validated the positive effects of DSL. A successful and efficient implementation of DSL depends very much on the sensor coverage of intelligent transportation systems, the enforcement strategies to achieve higher driver compliance with the posted speed limit (*6*), speed limit selection criteria, and the selection of road segments for implementing the corresponding logic (*7*).

Recent studies have demonstrated the impact of speed harmonization on operations and safety in connected traffic environments, where greater flow is achievable. In this study, we investigate mixed traffic behavior with DSL control at the corridor level within several scenarios involving varying levels of connectivity and weather conditions. The results show, depending on the strategy applied, meaningful increases in both throughput and prevailing speed.

[1]Transportation Center, Northwestern University, Evanston, IL

Corresponding Author:
Address correspondence to Hani S. Mahmassani: masmah@northwestern.edu

This paper is organized as follows. The second section reviews the relevant literature on DSL. The third section explains the DSL strategy and the experimental settings. There follows the evaluation of the impact of the DSL strategy within different traffic scenarios. Conclusions and recommendations for future studies conclude the paper.

Literature Review and Previous Work

Since the late 2000s, empirical and simulation-based approaches have been applied to investigate the potential benefits DSL control can offer. While empirical studies have demonstrated the positive effects of DSL on safety (*8–10*), there is no consensus on its operational benefits. Meanwhile, various control methodologies have been tested to investigate the validity of DSL in simulation-based studies. Among them, optimal control studies used various objective functions such as total travel time (*11, 12*), mean speed, throughput (*13*), or a combination of them (*14*). Alessandria et al. (*13*) conducted a macro-simulation study investigating the dynamic optimal control problem near congestion areas, which is nonlinear and nondifferentiable. Extended Kalman Filter was adopted for state estimation and the cost function measures expected total throughput in discrete time intervals. Simulation results show traffic under DSL control maintains traffic stream density under its critical value and high mean speeds in stark contrast to the uncontrolled case. Yet, the mean travel time was only minimally improved. One of the main criticisms of the traditional DSL strategy is that it is too late to activate the updated speed limits after a traffic stream disruption occurs. Therefore, traffic state prediction within the DSL procedure is required. Hegyi et al. (*12*) tested a model predictive control (MPC) in a rolling horizon scheme to minimize the number of shock waves and their adverse effect on the upstream traffic stream. The results showed a 17.3% decrease in total time spent after deploying MPC.

These macro-simulation approaches enable traffic control and evaluation of its effectiveness at a network-wide scale (*14*). However, they are not fully capable of capturing drivers' behavior, which is the fundamental prerequisite of DSL control. Micro-simulators such as VISSIM (*15*), PARAMICS (*4*), and MITSIM (*16*) are well adapted to emulate individual vehicle movements on the road and evaluate drivers' reactions to controller indications. One of the benefits of micro-level approaches is the high resolution of the input data when estimating traffic states which is crucial for the successful implementation of dynamic traffic management strategies. Talebpour et al. (*17*) used information on acceleration/deceleration of individual vehicles to detect the early stage of a shock wave. With the rise of vehicle communication technology, microsimulation studies have been preferred when studying traffic behavior in a mixed environment with human-driving, connected, and autonomous vehicles (*15, 18–20*). With connectivity, information is collected and distributed between connected vehicles in real time. This enables more accurate traffic state estimation, prediction, and real-time traffic control at individual vehicle level (*15*). The other advantage of connectivity, which this study aims to demonstrate, is that a connected traffic stream can utilize the same facility at a higher capacity. In other words, connected vehicles would modify the fundamental diagram of the network by increasing the maximum flow and critical density, especially at lower speed limits (*21*). Hence, the same demand within the connected vehicle environment would be served with less severe or even no breakdown compared with the case without connectivity with the assistance of DSL.

Microscopic models are used to simulate information exchange through vehicle-to-vehicle (V2V) or vehicle-to-infrastructure (V2I) communication. However, such an application over a large network with mixed traffic can be computationally expensive. Hence, a mesoscopic or macroscopic tool is needed that can scale and be computationally economical at the network level. This study integrates the microscopic aspect of V2V communication and the macroscopic aspect for dynamic traffic assignment at a network level. In the following sections, the methodology is presented followed by the results and recommendations.

Methodology

The overall method consists of three parts; (i) calibration of the traffic flow models that identifies speed-density relationships of a given network with a mixed population of vehicles (human-driving vehicles and connected vehicles); (ii) a Traffic Estimation and Prediction System (TREPS) model that estimates and predicts the traffic demand and dynamic traffic conditions; and (iii) a decision support system that activates traffic management operations according to the prevailing traffic states. The effectiveness of the strategy is evaluated in various weather conditions. The integration of microsimulation and mesosimulation models is presented in Figure 1. This approach has been developed by the authors in an earlier study and has been discussed in detail in previous publications (*21, 22*).

The following section provides details on the modeling approach adopted for speed harmonization.

The traffic behavior at different market penetration rates (MPRs) of connected vehicles is obtained from a micro-simulator. Converged speed-density curves are

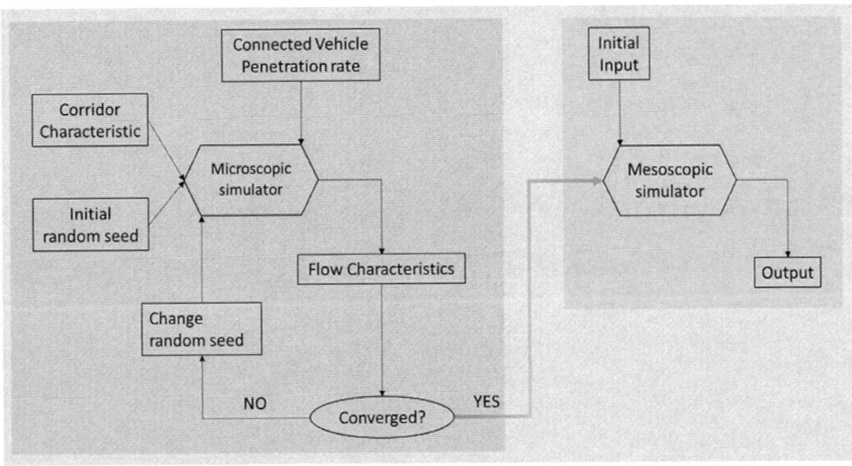

Figure 1. Schematic of the calibration and simulation frameworks.

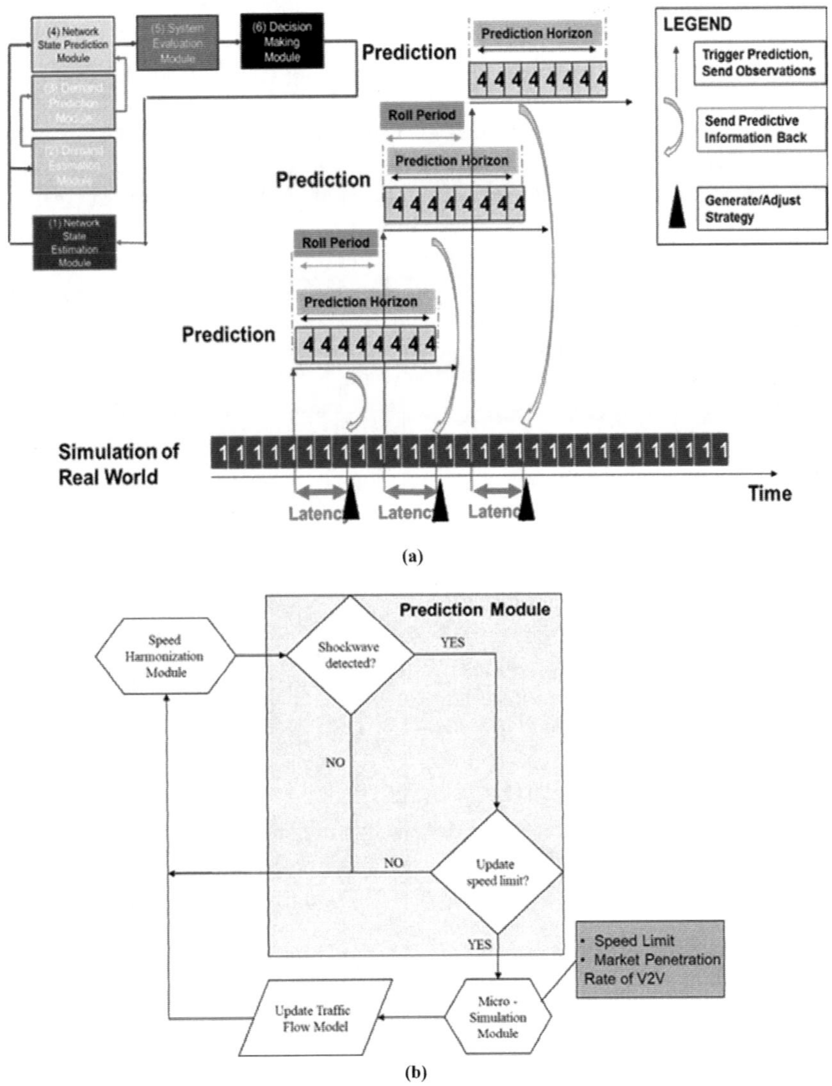

Figure 2. Flowchart of the (*a*) modeling framework and (*b*) prediction module for variable speed limits.

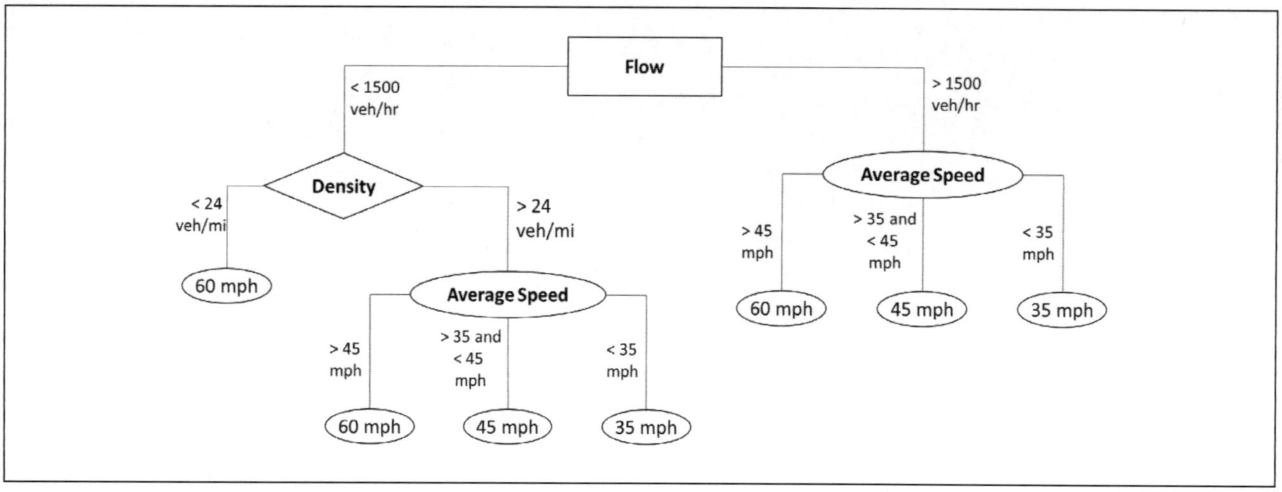

Figure 3. Speed harmonization decision tree (veh = vehicle).

calibrated based on multiple simulation runs. As for the mesoscopic simulation, the modified dual-regime Greenshields model is used in which constant free-flow speed is specified for the free-flow conditions (1st regime) and a modified Greenshields model is specified for congested-flow conditions (2nd regime). This step is followed by the prediction of the traffic state (shock wave) and the implementation of DSL within the mesoscopic simulation. Readers should refer to reference (*23*) for further information on the integration of microscopic and mesoscopic models.

In the TREPS model, the time and location of the traffic shock wave are predicted and if it is expected to occur then speed harmonization is activated to prevent or mitigate the shock wave effects in the near future. Figure 2 presents the schematic diagram of the prediction module of the speed harmonization strategy. The optimal interval between a predicted breakdown location and DSL application is obtained by conducting a sensitivity analysis. Each traffic prediction iteration is executed every 10 minutes, with a 30-minute prediction horizon, and corresponding speed limits are introduced 5 minutes before the occurrence of the predicted shock wave. Figure 3 shows the decision tree for activating the variable speed limit operations, which is adapted from the work of Allaby et al. (*24*). The decision is based on the traffic conditions (i.e., speed, flow, and density) at a point where a shock wave might be predicted.

The network-level effect of speed harmonization with various levels of connectivity (including no connectivity) is evaluated through the methodology presented above. The DSL messages are sent to all the vehicles through dynamic message signs to human drivers and through V2V and V2I communications to connected vehicles. Furthermore, to investigate the effectiveness of the strategy in different weather conditions, three different operational conditions (OCs) were tested: OC1 (clear/no rain, no snow), OC2 (moderate/heavy rain changing to moderate snow), and OC3 (moderate and heavy snow).

In summary, three OCs were tested to evaluate the impact of speed harmonization with and without connected vehicles. Three MPRs were tested in conjunction with speed harmonization. The first 12 hours, midnight to noon, were simulated and examined. The following sections present and discuss the results obtained from the simulation runs. The impact of the DSL strategy is presented and discussed in comparison with the base case,

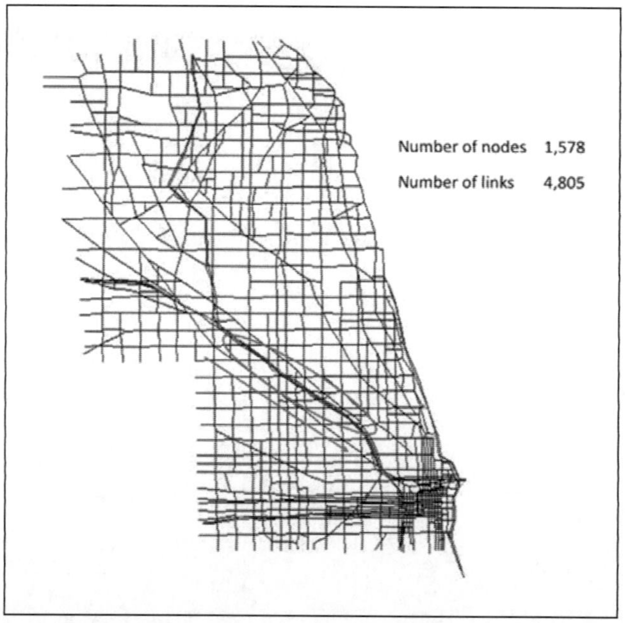

Figure 4. Schematic diagram of Chicago city with the selected corridor marked in red.

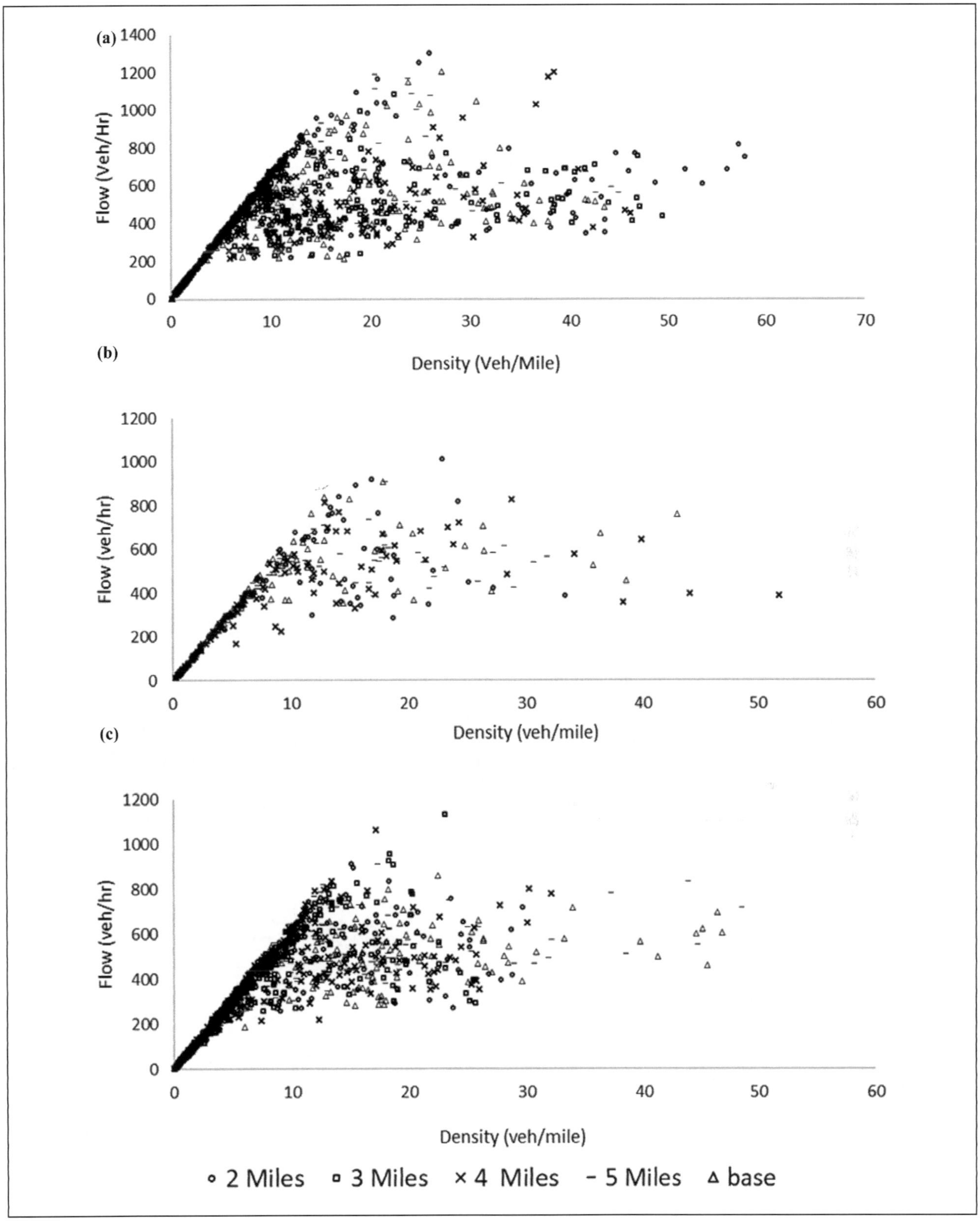

Figure 5. Speed harmonization: optimal upstream location selection: (a) OC1, (b) OC2, (c) OC3.

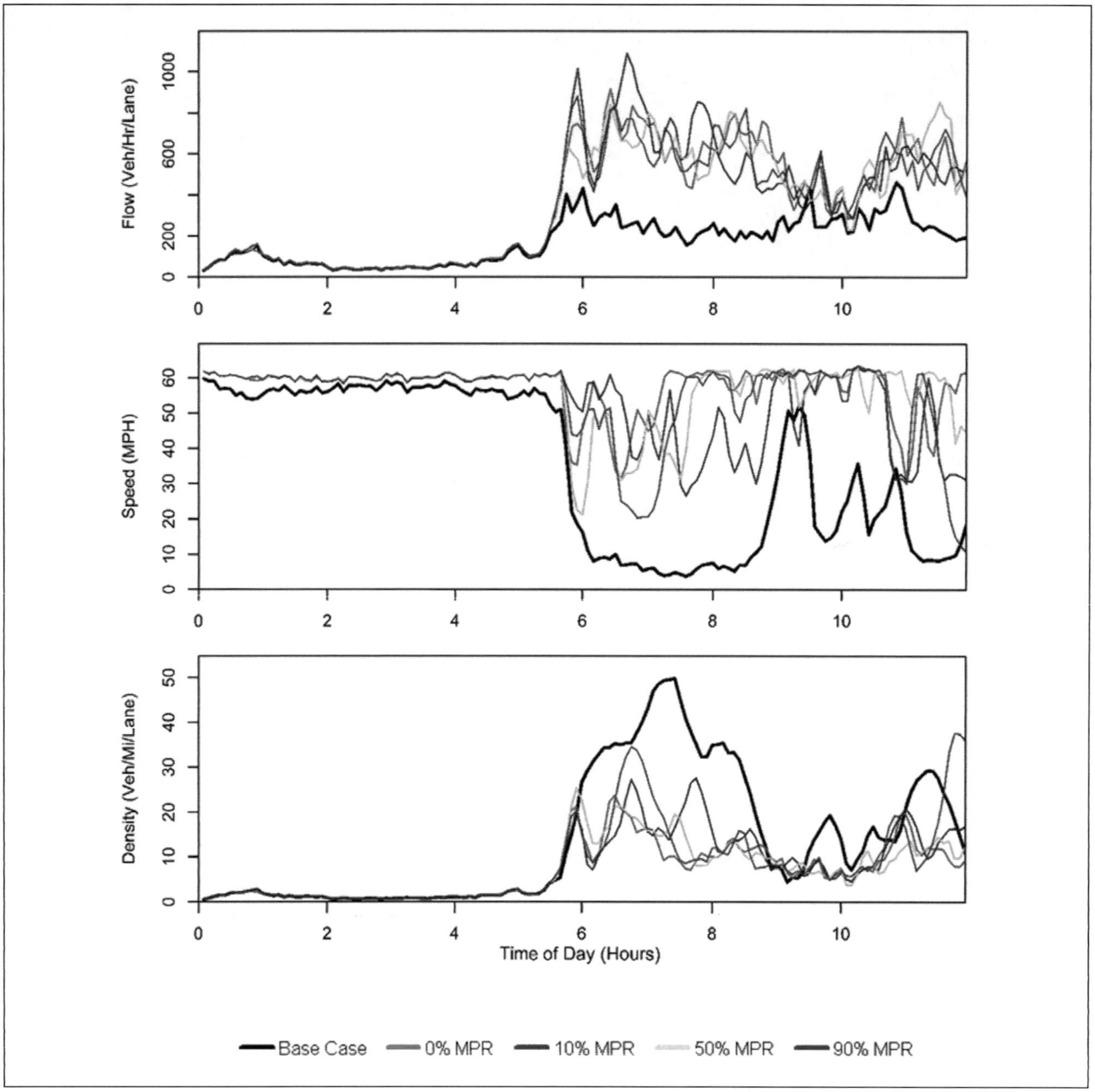

Figure 6. Comparison to base case conditions for OC1.

that is, each OC's environmental condition with no speed harmonization and no V2V connectivity.

Results

Speed harmonization was tested on a five-lane, seven-mile stretch of the I-90 freeway going southeast in Chicago, Illinois (Figure 4).

Sensitivity analysis, to identify the optimal distance interval between a predicted breakdown location and DSL, was conducted through changing speed limits upstream at distances ranging between two and five miles from the breakdown locations. From the flow-density diagrams it was observed that the extremes, that is, the highest flow and the highest densities, were affected by the upstream position where the speed limit adjustment took place. The optimal interval allows the highest flow at a low density (Figure 5). Legends in Figure 5 refer to the upstream distance in miles at which the speed limit was changed. Based on the results, the selected optimal upstream locations for OC1, OC2, and OC3 were two, three, and three miles respectively.

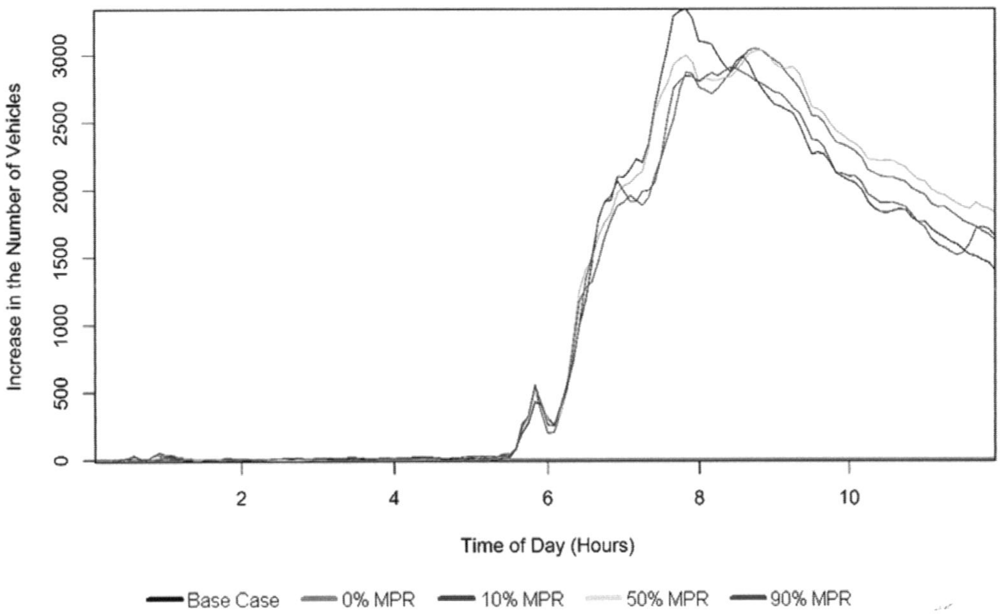

MPR	Flow (Veh/Hr/Lane)	Speed (MPH)	Density (Veh/Mi/Lane)	Travel Time Rate (Minute/Mile)
0%	172	39	-6.78	-3.46
10%	171	32	-5.71	-3.27
50%	169	37	-6.52	-3.4
90%	164	28	-5.26	-3.16

Figure 7. Increase in the parameters for OC1 at different MPRs of connected vehicles.

Figure 6 through Figure 11 present the impact of speed harmonization with different MPRs of connected vehicle in 3 OCs. Results shown are compared to the base case where no strategy is applied for the respective OCs.

The base case of OC1, which is a clear day with no precipitation, observed an average throughput of 175 veh/h/lane, speed of 13 mph, density of 13.44 veh/mile/lane and travel time rate of 4.61 minutes per mile. A total of 20,905 vehicles traversed the corridor within the selected time frame. The impact of the strategy in OC1 is presented and compared with the base case in Figures 6 and 7.

From the results, it can be observed that speed harmonization contributes to improved traffic conditions by increasing the throughput and speed and decreasing the density and travel time rate. An increase of 170 vehicles per hour and speed increase of 30 to 40 mph was observed with speed harmonization in a connected environment. However, the impact of connectivity was not found to be monotonous and was not different from that of speed harmonization alone, with no connected vehicles present. Hence, on a clear day, V2V communication does not provide greater benefits than those already achieved through speed harmonization.

The base case of OC2, which represents conditions of heavy rain with instances of light snow, experienced an average throughput of 165 veh/h/lane, speed of 24 mph, density of 6.84 veh/mile/lane and travel time of 2.49 minutes per mile. Under the base case of OC2, traffic stream density was half of that in the base OC1 scenario. The total number of vehicles that traversed the facility was 15,776. It is thus established that, under OC2, the vehicles were moving sparsely at a higher speed compared with OC1. The impact of the strategy in OC2 is presented in Figures 8 and 9. Comparable to the OC1 results, the speed harmonization technique improves the service rate of the facility and overall traffic conditions. Specifically, the throughput and speed increased while the density and travel time rate reduced. An increase of 109 vehicles per hour and speed increase of 25 mph was observed with

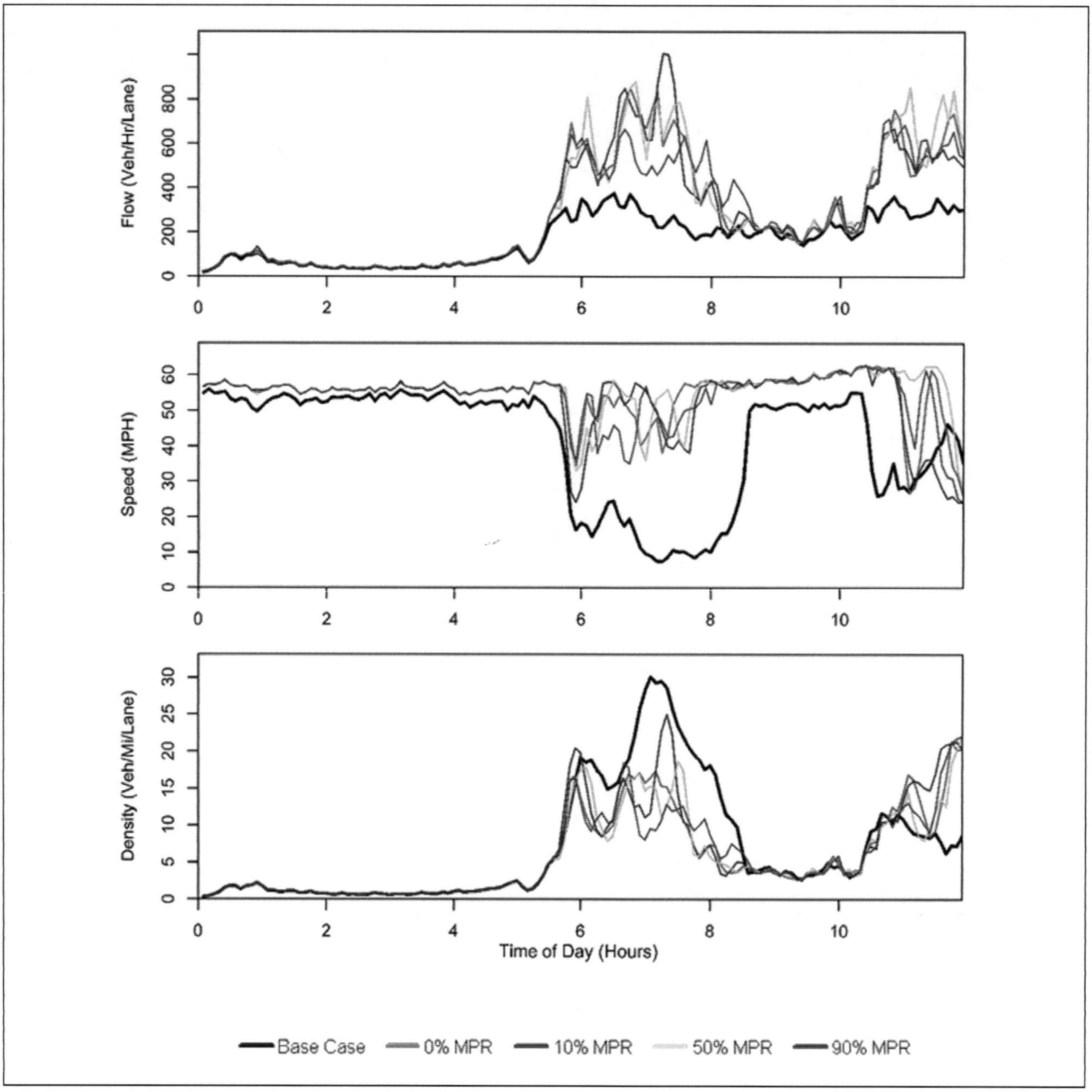

Figure 8. Comparison with base case conditions for OC2.

speed harmonization in a non-connected environment. Here, the beneficial impact of connectivity was found to be monotonously increasing with the increase in connectivity until reaching 50% MPR and plateauing thereafter. Hence, under lower visibility, that is, during adverse weather conditions, V2V connectivity complements the benefits of speed harmonization.

The base case of OC3, which represents a heavy snow day with poor visibility, observed an average throughput of 265 veh/h/lane, speed of 21 mph, density of 12.5 veh/mile/lane and travel time of 2.84 minutes per mile. The total number of vehicles traveling along the facility reached 16,157. OC3 is characterized by a lower demand level compared with OC1 and is comparable to OC2, however, its density is twice that of OC2. The impact of the strategy in OC3 is presented in Figures 10 and 11. In heavy snow conditions, similar trends with respect to the other two cases were observed. Traffic operational characteristics were enhanced through throughput and speed increases and

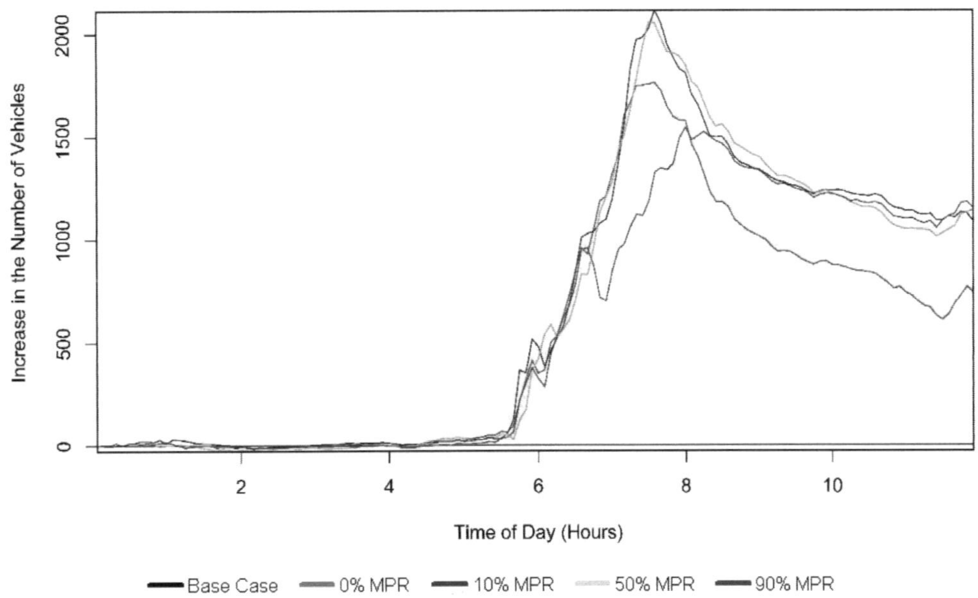

MPR	Flow (Veh/Hr/Lane)	Speed (MPH)	Density (Veh/Mi/Lane)	Travel Time Rate (Minute/Mile)
0%	109	25	-1.31	-1.28
10%	107	21	-0.88	-1.17
50%	119	28	-1.38	-1.34
90%	101	25	-1.37	-1.26

Figure 9. Increase in the parameters for OC2 at different MPRs of connected vehicles.

the corresponding decrease in density and travel time rate. An increase of 172 vehicles per hour and speed increase of 39 mph was observed with speed harmonization in a non-connected environment. Here, the beneficial impact of connectivity was found to be monotonously increasing with the increase in MPR. Hence, under lower visibility and reduced capacity due to lane closures resulting from snow accumulation, in other words, inclement weather conditions, V2V connectivity complements the benefits of speed harmonization. V2V communication was found to be most beneficial in OC3. When the visibility decreases, that is, when a driver's visual inspection of the downstream surroundings no longer serves its purpose, V2V communication assists drivers in adjusting their speeds.

Connectivity among the vehicles in conjunction with speed harmonization helps the network to improve traffic flow conditions further. The impact of the connectivity was more prominent in adverse weather conditions.

Conclusion

In this study, flow characteristics in a connected environment were identified based on a microscopic simulation tool. The calibrated speed-density relations were utilized in a mesoscopic simulation tool to study the network-wide effects of connected vehicles.

A previous study based on simulated traffic data with only connected vehicles and no speed harmonization showed that, with increase in MPR of connected vehicles, the network exhibited a lowered maximum density and an increased flow rate for the same density level (*23*). In this paper, we tested the effectiveness of the speed harmonization strategy within a connected environment. The experiments were conducted on a corridor of weather affected traffic network in the city of Chicago, IL. Speed harmonization was found to help the facility achieve higher throughput in each of the evaluated scenarios. Connectivity, in addition to speed harmonization, proved even more beneficial with respect to overall traffic OCs.

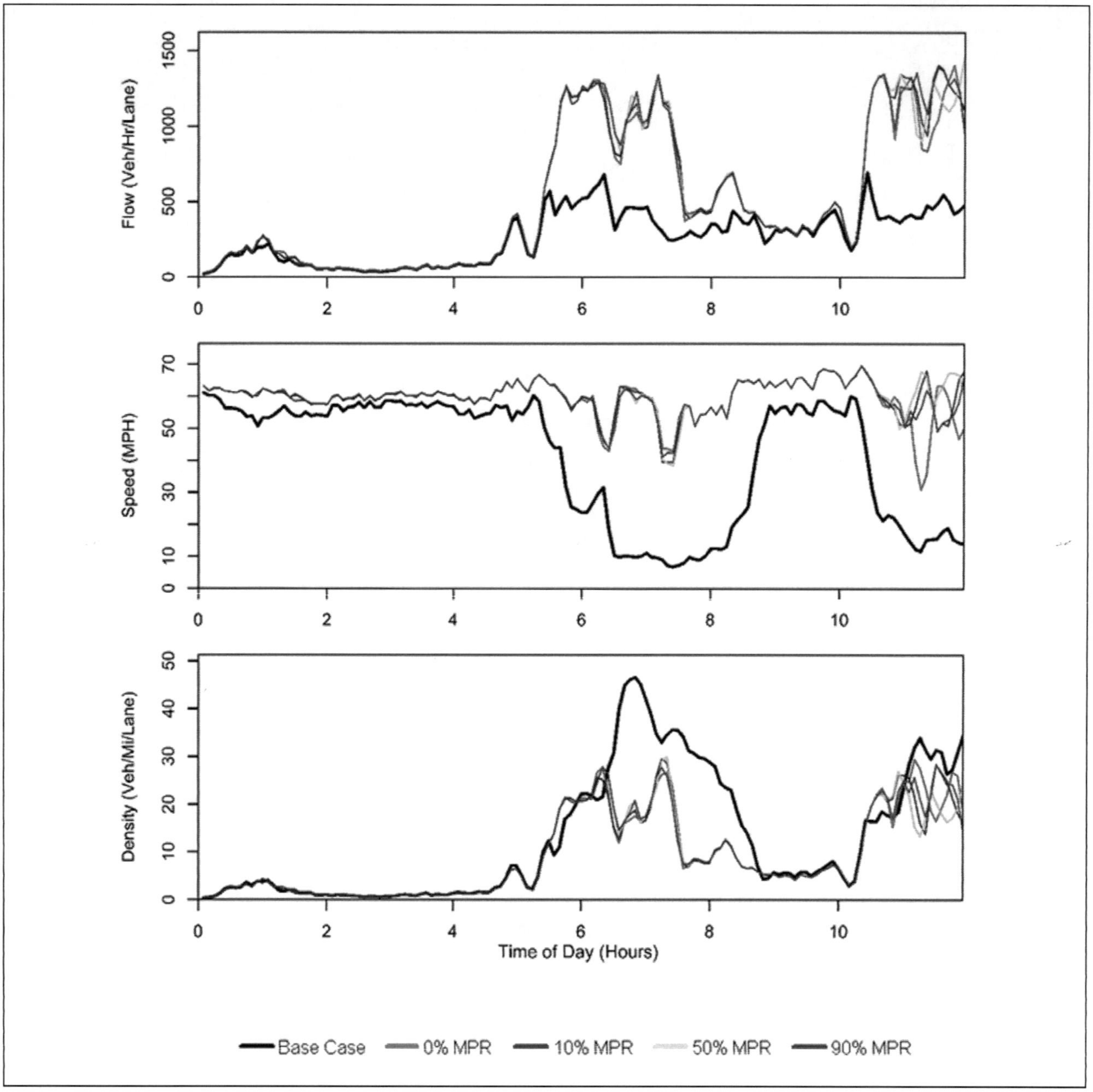

Figure 10. Comparison with base case conditions for OC3.

The impact of the connectivity was more prominent in adverse weather conditions.

Note that, in the study, V2I communication was not modeled in the mesoscopic simulation framework. The effect of connectivity was determined through the microscopic framework and then transferred into a mesoscopic model. Real-time traffic information was assumed to be collected from sensors covering the entire corridor, that is, full coverage. In future work, it is worthwhile to investigate the impact of the strategy in conditions in which V2I communication is enabled with no infrastructure to infrastructure communication. Speed harmonization in a connected environment provides an improvement in the throughput, especially under adverse weather conditions. In this study, we found the strategy was helpful in improving the overall, corridor-level traffic operations. With the strategy, an increase of up to 242 veh/h/lane in the flow rate and 39 mph in speed was achieved, though it is not possible to generalize such values to other contexts.

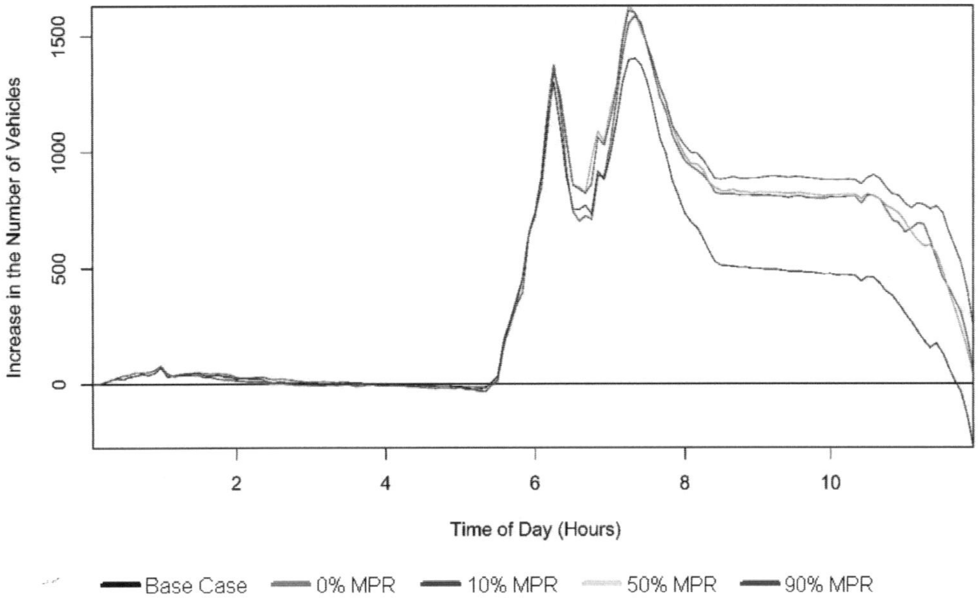

MPR	Flow (Veh/Hr/Lane)	Speed (MPH)	Density (Veh/Mi/Lane)	Travel Time Rate (Minute/Mile)
0%	227	35	-3.73	-1.76
10%	239	36	-3.72	-1.79
50%	240	37	-3.83	-1.8
90%	242	36	-3.64	-1.79

Figure 11. Increase in the parameters for OC3 at different MPRs of connected vehicles.

Further investigation of V2V communication in a mesoscopic model needs to be conducted, by incorporating both connected and automated vehicle types.

Author Contributions

All authors contributed to all aspects of the study from conception and design, to data collection, to analysis and interpretation of results, and manuscript preparation. All authors reviewed the results and approved the final version of the manuscript.

References

1. Mahmassani, H., H. Rakha, E. Hubbard, and D. Lukasik. *Concept Development and Needs Identification for Intelligent Network Flow Optimization (INFLO)–Assessment of Relevant Prior and Ongoing Research*. U.S. Department of Transportation ITS Joint Program Office Research and Innovative Technology Administration (RITA), 2012.
2. Lu, X. Y., and S. E. Shladover. Review of Variable Speed Limits and Advisories Theory, Algorithms, and Practice. *Transportation Research Record: Journal of the Transportation Research Board*, 2014. 2423: 15–23.
3. Juan, Z. C., X. X. Zhang, and H. W. Yao. Simulation Research and Implemented Effect Analysis of Variable Speed Limits on Freeway. *Proc., The 7th International IEEE Conference on Intelligent Transportation Systems*, Washington, IEEE, New York, 2004, pp. 894–898.
4. Lee, C., B. Hellinga, and F. Saccomanno. Evaluation of Variable Speed Limits to Improve Traffic Safety. *Transportation Research Part C: Emerging Technologies*, Vol. 14, No. 3, 2006, pp. 213–228.
5. Olia, A., H. Abdelgawad, B. Abdulhai, and S. N. Razavi. Assessing the Potential Impacts of Connected Vehicles: Mobility, Environmental, and Safety Perspectives. *Journal of Intelligent Transportation Systems*, Vol. 20, No. 3, 2016, pp. 229–243.
6. Soriguera, F., and M. Sala. Freeway Lab: Testing Dynamic Speed Limits. *Procedia-Social and Behavioral Sciences*, Vol. 160, 2014, pp. 35–44.

7. Waller, S. T., M. Ng, E. Ferguson, N. Nezamuddin, and D. Sun. *Speed Harmonization and Peak-Period Shoulder Use To Manage Urban Freeway Congestion.* Report FHWA/TX-10/0-5913-1. Center for Transportation Research, Texas A&M University-Kingsville, 2009.
8. Abdel-Aty, M. A., and A. Dhindsa. Coordinated Use of Variable Speed Limits and Ramp Metering for Improving Safety on Congested Freeways. Presented at 86th Annual Meeting of the Transportation Research Board, Washington, D.C., 2007.
9. Rämä, P. Effects of Weather-Controlled Variable Speed Limits and Warning Signs on Driver Behavior. *Transportation Research Record: Journal of the Transportation Research Board*, 1999. 1689: 53–59.
10. McMurtry, T., M. Saito, M. Riffkin, and S. Heath. Variable Speed Limits Signs: Effects on Speed and Speed Variation in Work Zones. Presented at 88th Annual Meeting of the Transportation Research Board, Washington, D.C., 2009.
11. Di Febbraro, A., T. Parisini, S. Sacone, and R. Zoppoli. Neural Approximations for Feedback Optimal Control of Freeway Systems. *IEEE Transactions on Vehicular Technology*, Vol. 50, No. 1, 2001, pp. 302–313.
12. Hegyi, A., B. De Schutter, and J. Hellendoorn. Optimal Coordination of Variable Speed Limits to Suppress Shock Waves. *IEEE Transactions on Intelligent Transportation Systems*, Vol. 6, No. 1, 2005, pp. 102–112.
13. Alessandri, A., A. Di Febbraro, A. Ferrara, and E. Punta. Nonlinear Optimization for Freeway Control Using Variable-Speed Signaling. *IEEE Transactions on Vehicular Technology*, Vol. 48, No. 6, 1999, pp. 2042–2052.
14. Hadiuzzaman, M., and T. Z. Qiu. Cell Transmission Model Based Variable Speed Limit Control for Freeways. *Canadian Journal of Civil Engineering*, Vol. 40, No. 1, 2013, pp. 46–56.
15. Khondaker, B., and L. Kattan. Variable Speed Limit: A Microscopic Analysis in a Connected Vehicle Environment. *Transportation Research Part C: Emerging Technologies*, Vol. 58, 2015, pp. 146–159.
16. Yang, Q., and H. N. Koutsopoulos. A Microscopic Traffic Simulator for Evaluation of Dynamic Traffic Management Systems. *Transportation Research Part C: Emerging Technologies*, Vol. 4, No. 3, 1996, pp. 113–129.
17. Talebpour, A., H. Mahmassani, and S. Hamdar. Speed Harmonization: Evaluation of Effectiveness Under Congested Conditions. *Transportation Research Record: Journal of the Transportation Research Board*, 2013. 2391: 69–79.
18. Lee, J., and B. Park. Evaluation of Variable Speed Limit under Connected Vehicle Environment. *Proc., 2013 International Conference on Connected Vehicles and Expo (ICCVE)*, Las Vegas, Nev., IEEE, New York, 2013, pp. 966–967.
19. Ma, J., X. Li, S. Shladover, H. A. Rakha, X. Y. Lu, R. Jagannathan, and D. J. Dailey. Freeway Speed Harmonization. *IEEE Transactions on Intelligent Vehicles*, Vol. 1, No. 1, 2016, pp. 78–89.
20. Ramezani, H., and R. F. Benekohal. Optimized Speed Harmonization with Connected Vehicles for Work Zones. *Proc., 2015 IEEE 18th International Conference on Intelligent Transportation Systems*, Las Palmas, Spain, IEEE, New York, 2015, pp. 1081–1086.
21. Hong, Z., H. S. Mahmassani, X. Xu, A. Mittal, Y. Chen, H. Halat, and A. Roemer. Effectiveness of Predictive Weather-Related Active Transportation and Demand Management Strategies for Network Management. *Transportation Research Record: Journal of the Transportation Research Board*, 2017. 2667: 71–87.
22. Mahmassani, H. S., Z. Hong, X. Xu, A. Mittal, B. Yelchuru, and R. Kamalanathsharma. *Analysis, Modeling, and Simulation (AMS) Testbed Development and Evaluation to Support Dynamic Mobility Applications (DMA) and Active Transportation and Demand Management (ATDM) Programs — Evaluation Report for the Chicago Testbed.* Research Report FHWA-JPO-16-387. Office of Operations, Federal Highway Administration, U.S. Department of Transportation, 2017.
23. Mittal, A., A. Talebpour, and H. Mahmassani. Network Flow Relations and Travel Time Reliability in a Connected Environment. *Transportation Research Record: Journal of the Transportation Research Board*, 2017. 2622: 24–37.
24. Allaby, P., B. Hellinga, and M. Bullock. Variable Speed Limits: Safety and Operational Impacts of a Candidate Control Strategy for Freeway Applications. *IEEE Transactions on Intelligent Transportation Systems*, Vol. 8, No. 4, 2007, pp. 671–680.

The Standing Committee on Intelligent Transportation Systems (AHB15) peer-reviewed this paper (18-06414).

Article

TRR
JOURNAL OF THE TRANSPORTATION RESEARCH BOARD

Transportation Research Record
2018, Vol. 2672(19) 25–33

DOI: 10.1177/0361198118775870
journals.sagepub.com/home/trr
SAGE

Framework for Estimating the Impact of Camera-Based Intelligent Transportation Systems (ITS) Technology on Incident Duration

Emily Moylan[1], **Sai Chand**[1], **and S. Travis Waller**[1]

Abstract

Safety is a major motivator of intelligent transportation systems (ITS) projects, and most efforts have addressed the potential to avoid incidents. Managing and reducing the duration of incidents is another key application for ITS despite challenges in distinguishing the true versus the reported duration of an incident. This paper presents a framework for modeling the impact of camera-based (closed-circuit television or CCTV) ITS technology on incident duration including an increase in the reported duration and a reduction in the true duration. The framework is validated against a data set of 121,793 accidents in New South Wales, Australia, covering 4.5 years. The results demonstrate that the use of CCTVs for incident duration contributes a 4.5 min reduction in average duration (as earlier detection can lead to more efficient clearance) and a 9% reduction in variance in the duration (as a uniform detection method supports standardized response procedures). These impacts are only visible when the 8.5 min median detection delay (the difference between the recorded duration and the true duration) is modeled and accounted for. These results offer a quantitative support tool for decision makers wishing to assess the value of incident-detection ITS projects.

Since the beginning, safety has been a key motivator in the adoption of intelligent transportation systems (ITS). While many ITS technologies focus on crash avoidance (smart signs, variable speed limits, and driver fatigue detection), ITS also plays a role in managing the impact of safety-related events through traveler information signs and rapid incident detection and clearance. Incident detection builds on various solutions including freeway service patrols, informants to emergency numbers, and video surveillance. Video surveillance cameras provide the traffic management center with instant information about collisions, breakdowns, and hazards, which facilitates potential travel time savings (economic benefits) and follow-on crash reductions (safety benefits).

Measuring, modeling, and managing incident duration are essential in evaluating the impact of safety-based ITS technologies. Significant effort has been put into modeling incident durations (*1–5*) to assess the value of incident-management ITS technologies (*6–10*). One of the challenges is accounting for the time between the occurrence of the incident and its detection and verification. A common ITS application is to deploy technologies (freeway patrols, call boxes, and video surveillance) that reduce this interval through rapid detection. For locations where the gap between occurrence and detection remains, it is helpful to quantify the contribution of this phase to the overall duration to compare with the benefit–cost ratios of reductions to each phase of an incident (response, clearance, and recovery) and reduce the risk of secondary incidents (*11,12*) and fatalities (*13*).

In this paper, we focus on video surveillance cameras (also referred to as closed-circuit television or CCTV) as an ITS technology for rapid incident detection. Their advantages as a data source for incidents include spatial coverage over a stretch of road rather than a single point, archival footage that can be used to pinpoint the exact start time of the incident even when the video is not monitored continuously (as described by Bertini et al. [*8*]), and the option to lower the detection threshold to include near-misses. Additionally, CCTV footage can be paired with video processing for automatic incident

[1]Research Centre for Integrated Transport Innovation, School of Civil and Environmental Engineering, UNSW Sydney, Australia

Corresponding Author:
Address correspondence to Emily Moylan: e.moylan@unsw.edu.au

detection (AID), which provides accurate times for both the start and end of the incident (*14*). Compared to traditional AID algorithms (*15–17*), CCTV-based AID can focus on identifying the start of the incident (e.g. the swerve causing the crash or the appearance of a hazard) rather than the start of the impact on the traffic flow. Future applications of video surveillance might include aerial drones that provide widespread, adaptive, spatial coverage.

It is difficult to quantify how much CCTV cameras reduce the duration of incidents as the start times of many incidents are ambiguous in the pre-CCTV baseline. Non-CCTV-detected incidents may appear to be shorter because the incident was detected (a) many minutes after it started, or (b) was resolved by the same on-site person who reported it, or (c) both. In contrast, an incident detected by CCTV may be detected earlier, before the queue has time to build. This can result in the reported duration of the incident being longer, whereas the true duration of the incident may be shorter in comparison to non-CCTV-detected incidents.

This paper addresses these components of incident duration using a large (approximately 325,000) data set of incidents over 4.5 years. Comparing the durations of incidents detected by CCTV versus those detected by other means, this paper verifies a statistical difference in the distributions and identifies several mechanisms that affect the difference. The immediate value of this work is to provide an estimate on the reduction in incident duration caused by using CCTV. A future application is to develop methods for correcting the measured durations from non-CCTV-detected incidents for use in a wide range of ITS program evaluations.

Data

The data set includes 327,298 unplanned incidents from across the Australian state of New South Wales in the period from January 2012 to June 2016. A map of the incidents is shown in Figure 1. The data set includes breakdowns, hazards, towing, crashes, and police activity. Approximately two thirds of the data are recorded through the InterCAD Emergency Messaging System (ICEMS). Less than 1% are sourced through clearway towing logs, road occupancy logs, police reports and manual entries. The remaining incidents are reported by the facility's operator. Overall, 3.6% of the incidents include a reference to a CCTV device. As new ITS infrastructure is added to the network, more incidents are reported through CCTV (Figure 2). Therefore, the impact of this ITS technology on incident detection, duration, and management is the primary focus of this paper.

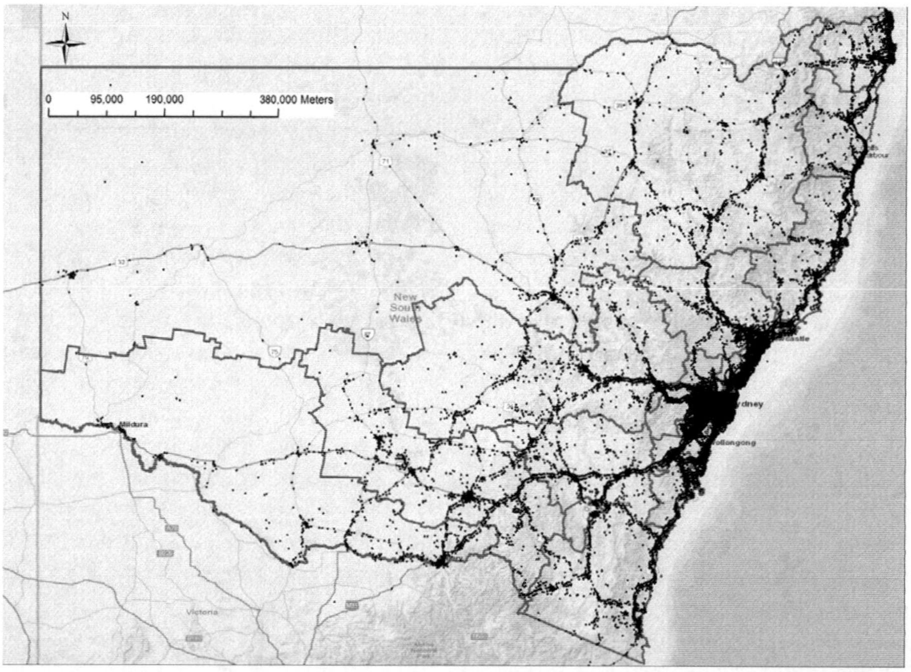

Figure 1. Map of New South Wales with green lines showing the boundaries of Statistical Area 4s (SA4s) and black dots showing locations of incidents between January 2012 and June 2016.

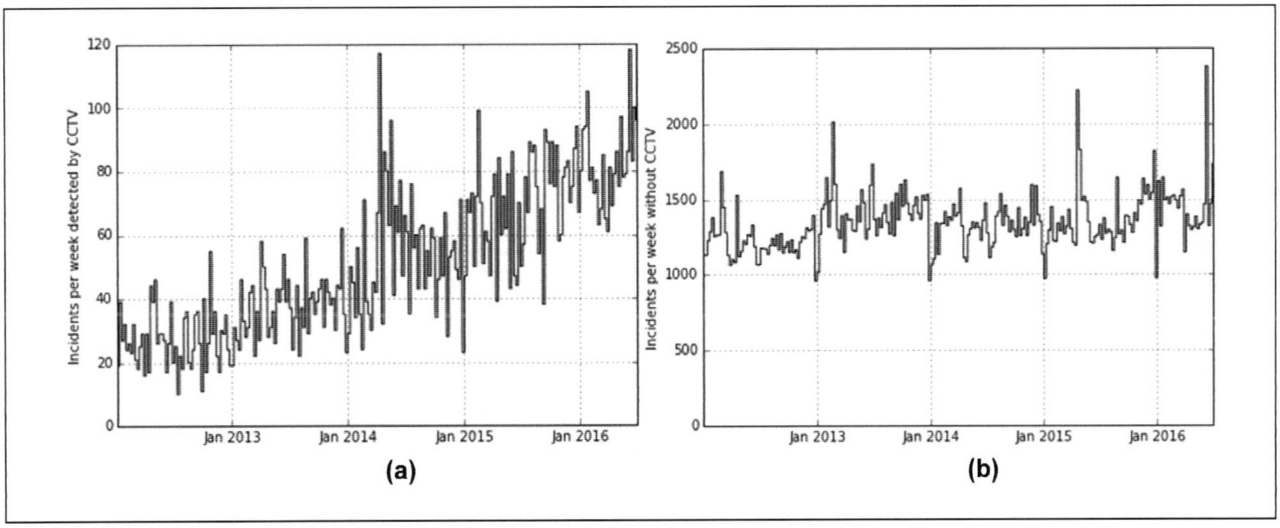

Figure 2. The number of CCTV-detected incidents (*a*) is increasing dramatically compared to incidents detected by other means (*b*).

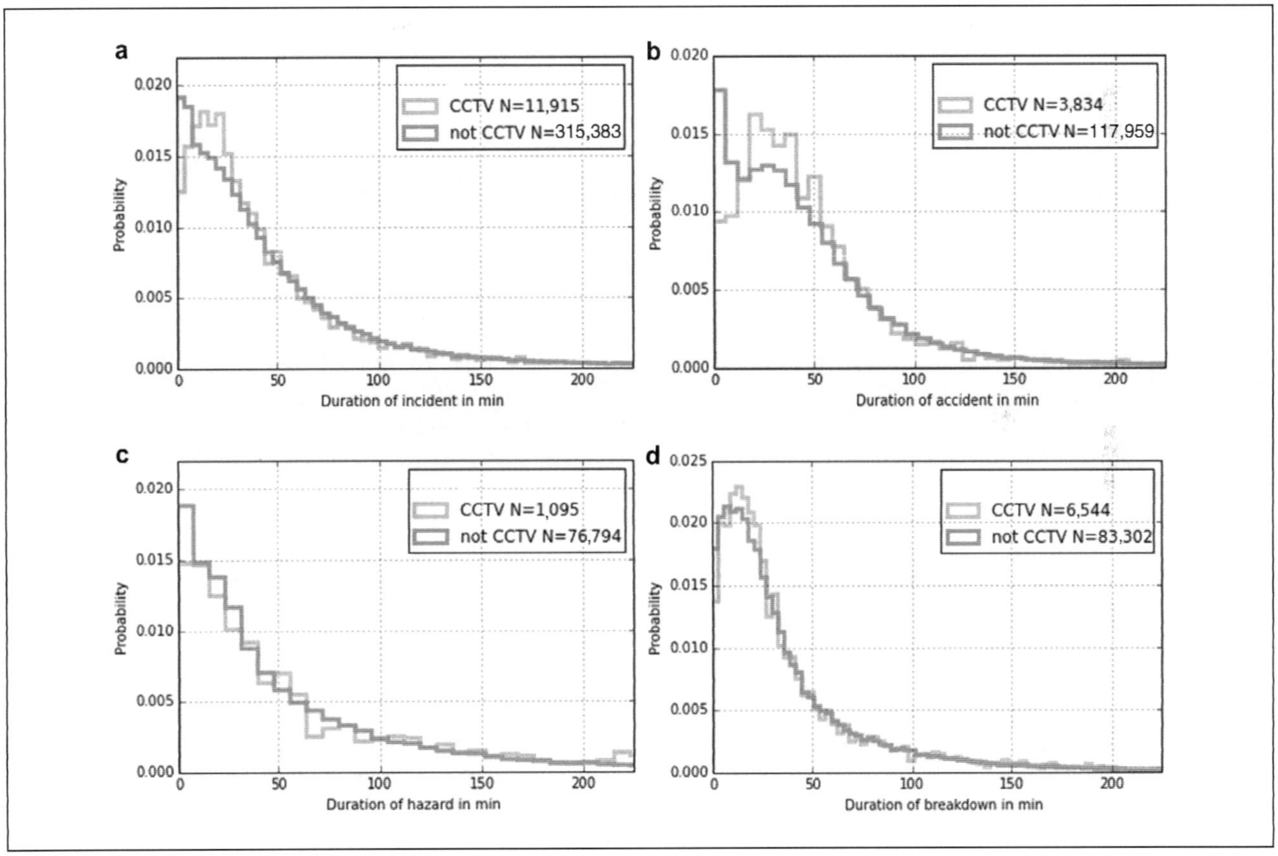

Figure 3. Incident duration by type of incident: (*a*) all, (*b*) accidents, (*c*) hazards, (*d*) breakdowns.

Each incident record includes a recorded duration. As shown in Figure 3, the distributions of CCTV-detected and non-CCTV-detected incident durations take different shapes. Most striking is the overabundance of extremely short incidents in the non-CCTV sample, particularly for the accidents. This distinction dominates over more subtle differences such as changes in the width or in the location of the distribution. The methodology presented below identifies the components of the difference between the CCTV and non-CCTV durations and

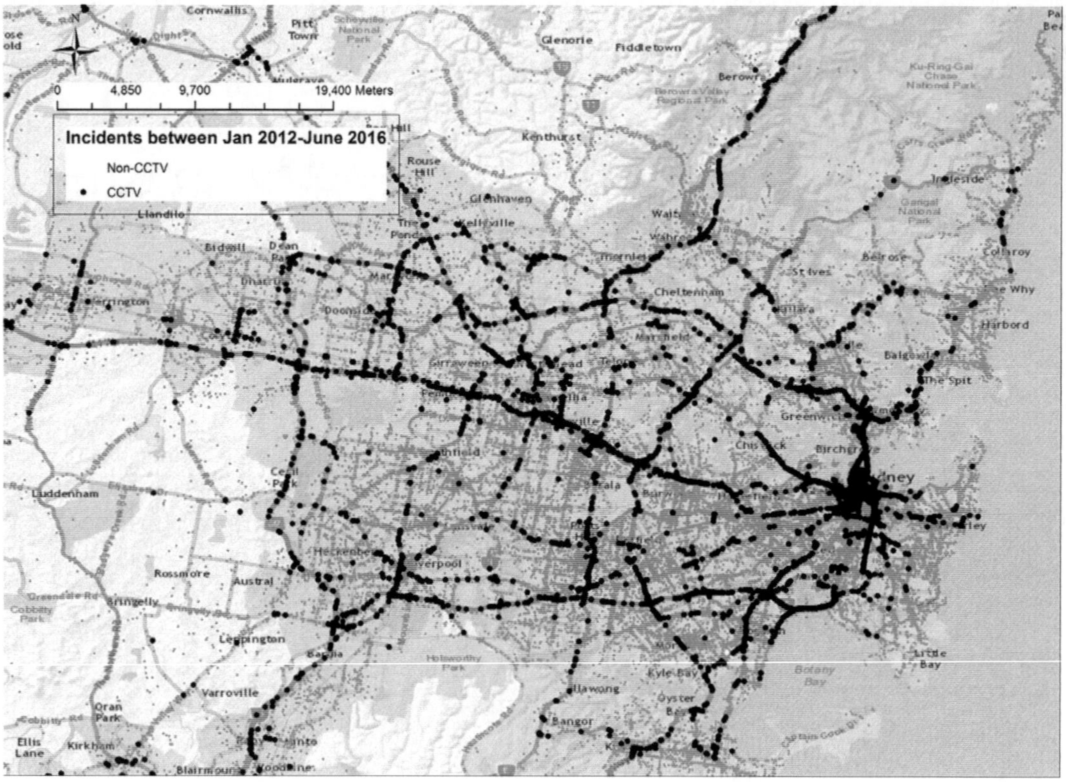

Figure 4. Incidents in Sydney showing that CCTV-detected incidents occur mainly on motorways and major arterials.

suggests a framework for quantifying the impact of CCTV detection on incident duration.

Assumptions and Attributes

The data set records the general category of the detection source for each incident (predominately ICEMS or operator), but the specific type of detection can only be inferred from the incident description (e.g. "informant" implies that the incident was called in to an emergency number, a reported CCTV ID number implies that it was either detected or verified on CCTV). These sources are non-exclusive, so a single incident might be linked to only one of its several detection sources. In this way, the distinction between CCTV- and non-CCTV-detected incidents is not strictly unambiguous, which may introduce some bias in the results.

One of the assumptions behind the methodology discussed below is that the samples of incidents detected with and without CCTV are identical except for the method of detection. This paper hypothesizes that the method of detection may affect the duration of the incident, but to isolate this effect it is necessary to assume that the samples are otherwise identical. Some reasons to question this assumption are the preferential placement of CCTVs on certain kinds of roads (e.g. motorways with high speeds and therefore high-speed crashes) or on accident-prone facilities (due to a historical prevalence of incidents at that location). In Figure 4, the preponderance of CCTV-detected incidents on motorways and major arterials illustrates this attribute. Therefore, some systematic differences are expected in the samples, including an overabundance of high-speed, serious collisions, a paucity of pedestrian involvement, and a high rate of incidents per location in the CCTV sample. Modeling these impacts is reserved for future work.

The shape of the distribution of non-CCTV incident durations implies that the recorded durations are sometimes shorter than the true distributions. One aspect of the methodology proposed below is to quantify the delay in detection that causes the preponderance of very short durations in the non-CCTV sample. To support this methodology, it is necessary to assume that the measured durations of CCTV-detected incidents are the true durations. This implies that the CCTVs are monitored continuously or that footage is reviewed to identify true start times for the database. The impact of these biases on the results will be revisited in the discussion section.

Methodology

The methodology is designed to establish the statistical significance of the difference between the durations of incidents detected with and without CCTV technology and to attribute the differences to the introduction of the technology. The difference between the CCTV and non-CCTV samples is established using Kolmogorov-Smirnoff (KS) tests to assign a probability to the two samples being drawn from the same distribution. The KS test is a non-parametric test for the equivalence of two samples. It is based on the supremum difference between the empirical distribution functions, so it is sensitive to differences in both shape and scale.

Building on the significance of the difference, the goal is to attribute the difference in the distributions to the ITS intervention. It is necessary to begin with the assumption that the only difference in the two samples is the method of detection. As discussed in the data section, the method of detection is hypothesized to affect the duration of the incident, and no other attributes contribute to differences in the duration.

We propose a framework for the impact of CCTV detection on the duration of the incident with three components. First, the time of detection is affected by the method of detection in such a way that the start time of the incident is recorded some interval after the true start time, resulting in a shorter measured duration for incidents detected without CCTV. The interval between occurrence and detection in non-CCTV incidents can be modeled with a random variable, t, drawn from an exponential decay function such that

$$f(t;\lambda) = \lambda \exp(-t\lambda) \quad (1)$$

where $\frac{1}{\lambda}$ is the half-life and $f(t;\lambda)$ is the probability of an incident being detected t minutes after it occurred. An illustration of this impact is the scenario where there is a hazard (e.g. a traffic cone in the driving lane). Without CCTV, the hazard is easy to remove, it might be addressed by the very person (police, maintenance crew, or passer-by) who reports it, and so the recorded duration is very short. With CCTV, the hazard can be recorded remotely when there is no one on site to address the problem, and the recorded duration will include the time until a crew can arrive on site as well as the (possibly negligible) time to fix the hazard. As this scenario demonstrates, the interval between occurrence and detection will lead to overall shorter recorded durations and an overabundance of extremely short durations in the non-CCTV sample.

Second, the earlier detection of the incident allows the incident to be managed earlier resulting in shorter durations on average. One illustration of this impact is that a tow truck might be dispatched earlier to a breakdown detected with CCTV. As a result of the earlier detection, there is less opportunity for queue formation, and the tow truck can arrive at the broken-down vehicle sooner. This impact can be visualized as a left shift in the distribution of CCTV-detected durations by a shift parameter, β. It is expected that this impact might not be noticeable unless the detection delay is controlled. This contribution also benefits from the assumption that the measured durations of CCTV incidents are correct, although it is only necessary to assume that the delay between occurrence and detection is shorter on average for the CCTV sample than for the non-CCTV sample.

The final impact of CCTV technology on the duration of incidents is a reduction of the variance in duration. Considering the diversity of detection methods included in the non-CCTV sample, it is reasonable that the duration of an incident will vary considerably depending on whether it was phoned in by an informant, contributed by on-site emergency services personnel, or discovered using other ITS technologies such as AID on loop detector data. The heterogeneity in detection source serves as a convenient proxy for variation in the type of incident, the incident's context, and the severity of incident. Compared to the single detection source in the CCTV sample, these incidents will contribute a broad variation of durations shown as a stretching of the distribution by some scaling parameter, α. As discussed above, we consider only changes in the duration that are directly caused by the detection method, which means that all of the underlying diversity is captured in the detection source.

To quantify the contribution from the detection delay, the decrease in average duration, and the decrease in variance, we employ a numerical fitting technique to identify the best-fit values for the decay parameter, λ, the shift parameter, β, and the scaling parameter, α. As an initial application, this methodology is applied to the duration of 121,793 accidents (Figure 3*b*) since the difference between the CCTV and non-CCTV durations is self-evident. The sample of CCTV-detected accident durations is used as the baseline measure of true durations, and the goal is to reproduce the distribution of durations observed with the non-CCTV sample. To reduce noise, each true duration is sampled 100 times and scaled by α. A random detection delay is then subtracted from each scaled duration and the mean shift, β, is added. Any negative duration values are corrected to 1 min to reflect how they would be recorded in the database. The resulting distribution is termed the adjusted distribution, and this is compared to measured non-CCTV durations for goodness of fit. Exploring the range of reasonable values for λ, α and β, the best set of parameters to quantify the impact of CCTV on incident duration results in the best match between the adjusted and measured distributions. Goodness of fit is quantified with an χ^2-style metric on the number of observations in each 5-min interval in the probability mass function:

$$\text{Goodness of fit} = \frac{\sum_{\text{bins}}(\text{adjusted} - \text{measured})^2}{n_{\text{bins}}} \quad (2)$$

The search of the parameter space considers values for λ in the range [1,11], α in the range [0.5,1.5], and β in the range [–2.0,10]. The ranges reflect realistic adjustments to the observed duration distribution. Using the goodness of fit measure and a KS test-style supremum in the difference between the cumulative mass functions, it is possible to narrow and refine the parameter space under consideration. The parameters that produce the best match are selected.

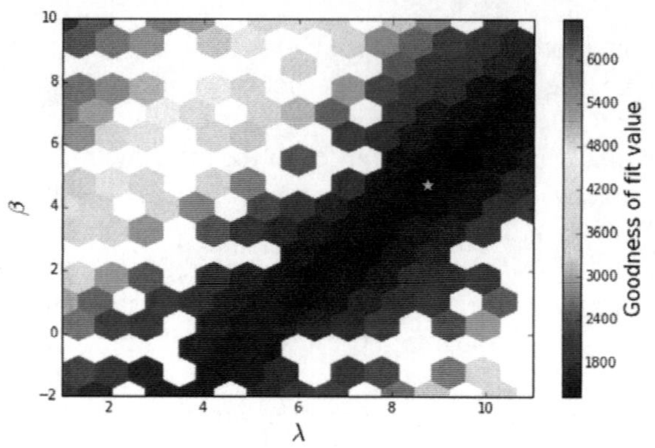

Figure 5. Goodness of fit (GoF) measures for varying values of β and λ. Each cell corresponds to the α value resulting in the best GoF for that combination of β and λ. There is strong correlation between β and λ values that result in good fits. The star shows the best GoF value.

Results

Significance of the Difference between CCTV and Non-CCTV Incidents

The distribution of the full sample of incident durations differs significantly by source of detection (CCTV vs non-CCTV), as shown in Figure 3*a*. Notably, the overrepresentation of very short incidents is evident in the duration distribution of accidents (Figure 3*b*) and hazards (Figure 3*c*), but it is weaker in breakdowns (Figure 3*d*). A two-sample KS test rejects the null hypothesis that the CCTV and non-CCTV duration distributions are the same, with greater than 99% confidence for each of the categories (Table 1).

Although the KS tests establish a statistically significant difference in the reported duration of all types of incidents, they do not demonstrate a reduction in true duration because of ITS applications of CCTV. In fact, the predominant difference between the distributions is an overabundance of short incidents in the non-CCTV sample. Because the KS test relies on the supremum of the distance between the two empirical distribution functions, it will be sensitive to this localized discrepancy regardless of whether the distribution on the whole is shifted to shorter durations or compressed to less variability in the duration.

Quantification of the Impact of CCTV on Accident Duration

Initially, the parameter space was explored using generous ranges for the reasonable values for λ, α and *β*. The goodness of fit measure quantifies how well the probability mass functions of the measured non-CCTV durations and the adjusted CCTV durations match, and this metric is supported by using the supremum distance between the two cumulative mass functions (CMFs).

In Figure 5, the two-dimensional projection of the three-dimensional parameter space shows how the best values for goodness of fit lie on a diagonal line, indicating a trade-off between average shift in duration, β, and the magnitude of the detection delay, λ. This makes sense as λ represents the half-life of the decay function—a larger typical delay between occurrence and detection can be offset by a systematic increase in the duration of the accident. However, the two parameters are not strictly degenerate because of interaction between the stochasticity of the decay value and the requirement that duration values be greater than or equal to 1. A large value for λ will result in a larger typical delay with a larger variance in the delays. Although a corresponding average shift, β, will somewhat compensate for this change, it will still result in more negative durations that are reassigned to 1 min. As a result of this subtlety, it is possible to identify a minimum along the valley shown in Figure 5.

In Figure 6, the remaining projections of the three-dimensional parameter space show the sensitivity of goodness of fit to changes in α. The scaling parameter is stable with a factor of 1.05 consistently performing well for combinations of β and λ.

The three best combinations of parameters are shown in Table 2. All three sets fall in the dark blue region that has been explored minutely in Figure 5. The difference in the goodness of fit is less than 1%, indicating that they each offer similar explanatory power. Likewise, the maximum distance between the observed and adjusted CMFs is roughly 4% for each set of parameters. Figure 7 shows the comparison of the probability mass functions for the parameter combinations yielding the best goodness of fit measure and the smallest supremum distance in the CMF (3.4%). Goodness of fit is the preferred measure as

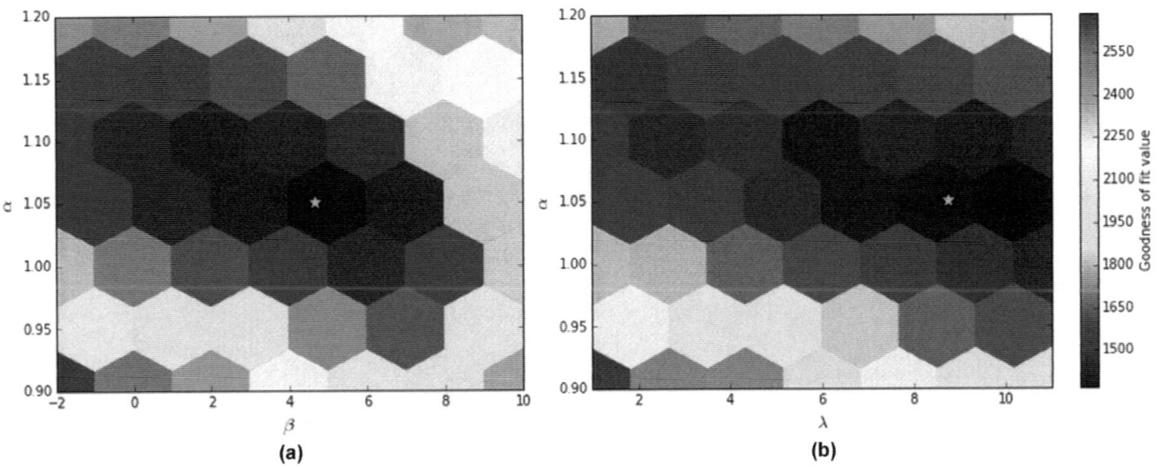

Figure 6. GoF sensitivity to α over a range of β (*a*) and λ (*b*) values. Each cell corresponds to the **λ** or **β** value respectively that results in the best GoF for that combination of parameters. The star marks the best GoF value.

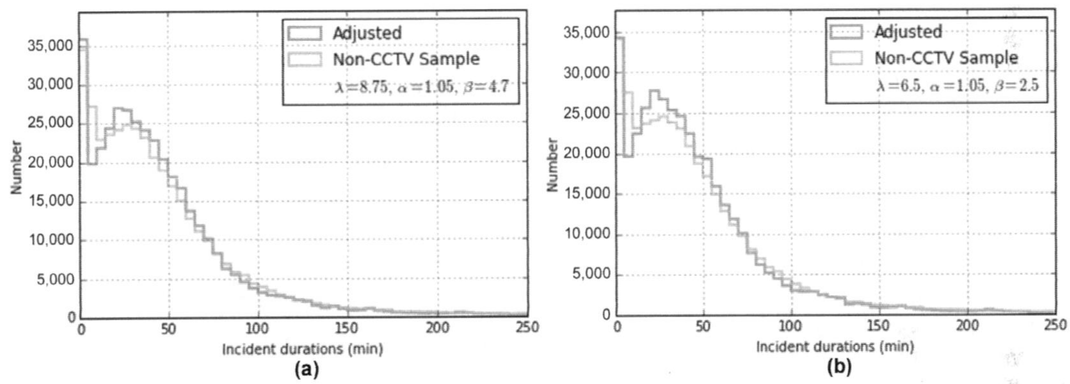

Figure 7. Probability mass function comparisons for the parameter combinations that yield the best goodness of fit (*a*) and the smallest supremum distance (*b*).

Table 1. KS Test Results for Each Category of Incident

	D	*p*-value
All incidents	0.037	<0.01
Accidents	0.075	<0.01
Hazards	0.103	<0.01
Breakdowns	0.021	<0.01

it reduces sensitivity to the correction for reported versus true durations. Figure 7 shows that the CMF distance metric results in parameters that minimize the gap between adjusted and non-CCTV observations in the dip between very short durations and the main peak (i.e. the second bin of the histogram), whereas the goodness of fit metric results in overall better agreement through the peak of the curve. Since the parameters presented in Table 2 are tightly grouped in value as well as performance, we offer a generic interpretation in the discussion section.

Discussion and Conclusions

The fitted parameter values describe the transformation that converts the distribution of CCTV accident durations into non-CCTV accident durations. To describe the impact of CCTV on incident duration, we simply invert the transformation.

This paper demonstrates that CCTV detection decreases accident duration by approximately 5 min. It

Table 2. Three Best Combinations of Parameters According to Goodness of Fit Measure

λ	α	β	CMF distance	Goodness of fit
8.75	1.05	4.7	3.98%	1323.7
9.50	1.05	5.5	4.15%	1326.4
8.75	1.05	4.6	4.07%	1332.0

also reduces the variation in the duration by a factor of $\frac{1}{1.05} \cong 0.95$. For a decision maker wishing to evaluate the value of CCTV, it may be difficult to identify these benefits in light of the increase of 8.75–9.5 min in the median reported duration. A key contribution of this methodology is to control for the detection delay to separate the impact of the reported duration from the impact on the true duration.

Revisiting the assumptions introduced in the data section, we consider the interpretation of the parameter values in light of concerns about biases in the data. Ambiguities in the detection source have the potential to contribute noise to the results. However, the KS tests clearly show that the CCTV and non-CCTV samples, as they are defined, are distinct. Better information about the detection source would yield an even clearer benefit of CCTV.

Selection biases in the locations of CCTVs indicate that CCTV-detected incidents should include long-duration, high-speed crashes, as well as locations with recurring, serious incidents. The positive values for both the scale, α, and location, β, parameters indicate that non-CCTV incidents are longer and more variable in duration. Therefore, this bias does not affect the interpretation. More specifically, it implies the benefits from CCTV are, if anything, underestimated.

Likewise, the assumption that the measured durations of CCTV accidents are the true duration is supported by the shape of the duration distribution (a log-normal style distribution with only one inflection point). If this assumption is false, we can conclude that the difference between the true durations of CCTV and non-CCTV is even more pronounced.

Quantification of the impacts of CCTV on incident duration clearly shows that CCTV offers benefits for both mean duration and variation in duration. The framework presented here also underscores the importance of accounting for detection delay.

It is expected that video footage of traffic flow will become more widespread, especially with the growth of aerial drones which is supported by the incident-management benefits foretold in these results. A future extension to this work is the development of a correction factor that adjusts the recorded duration of a non-CCTV incident into a true duration. True durations have the potential to improve accuracy in planning (cost–benefit analysis) and operations (incident response plans) decisions.

Author Contributions

The authors confirm contribution to the paper as follows: study conception and design: E. Moylan; data collection: E. Moylan, S. Waller; analysis and interpretation of results: E. Moylan, S. Chand; draft manuscript preparation: E. Moylan. All authors reviewed the results and approved the final version of the manuscript.

References

1. Giuliano, G. Incident Characteristics, Frequency, and Duration on a High Volume Urban Freeway. *Transportation Research Part A: General*, Vol. 23, No. 5, 1989, pp. 387–396.
2. Garib, A., A. E. Radwan, and H. Al-Deek. Estimating Magnitude and Duration of Incident Delays. *Journal of Transportation Engineering*, Vol. 123, No. 6, 1997, pp. 459–466.
3. Nam, D., and F. Mannering. An Exploratory Hazard-based Analysis of Highway Incident Duration. *Transportation Research Part A: Policy and Practice*, Vol. 34, No. 2, 2000, pp. 85–102.
4. Boyles, S., D. Fajardo, and S. T. Waller. A Naive Bayesian Classifier for Incident Duration Prediction. Presented at 86th Annual Meeting of the Transportation Research Board, Washington, D.C., 2007.
5. Hojati, A. T., L. Ferreira, S. Washington, and P. Charles. Hazard Based Models for Freeway Traffic Incident Duration. *Accident Analysis & Prevention*, Vol. 52, 2013, pp. 171–181.
6. Chu, L., H. Liu, and W. Recker. Using Microscopic Simulation to Evaluate Potential Intelligent Transportation System Strategies Under Nonrecurrent Congestion. *Transportation Research Record: Journal of the Transportation Research Board*, 2004. 1886: 76–84.
7. Thill, J. C., G. Rogova, and J. Yan. Evaluating Benefits and Costs of Intelligent Transportation Systems Elements from a Planning Perspective. *Research in Transportation Economics*, Vol. 8, 2004, pp. 571–603.
8. Bertini, R. L., M. W. Rose, and A. El-Geneidy. Using Archived ITS Data Sources to Measure the Effectiveness of a Freeway Incident Response Program. Presented at 84th Annual Meeting of the Transportation Research Board, Washington, D.C., 2005.
9. Guin, A., C. Porter, B. Smith, and C. Holmes. Benefits Analysis for Incident Management Program Integrated with Intelligent Transportation Systems Operations: Case Study. *Transportation Research Record: Journal of the Transportation Research Board*, 2007. 2000: 78–87.
10. Özbay, K. M. A., W. Xiao, G. Jaiswal, B. Bartin, P. Kachroo, and M. Baykal-Gursoy. Evaluation of Incident Management Strategies and Technologies Using an Integrated Traffic/Incident management simulation. *World*

Review of Intermodal Transportation Research, Vol. 2, Nos. 2/3, 2009, pp.155–186.
11. Karlaftis, M. G., S. P. Latoski, N. J. Richards, and K. C. Sinha. ITS Impacts on Safety and Traffic Management: An Investigation of Secondary Crash Causes. *Journal of Intelligent Transportation Systems* Vol. 5, No. 1, 1999, pp. 39–52.
12. Khattak, A., X. Wang, and H. Zhang. Are Incident Durations and Secondary Incidents Interdependent? *Transportation Research Record: Journal of the Transportation Research Board*, 2009. 2099: 39–49.
13. Evanco, W. M. The Impact of Rapid Incident Detection on Freeway Accident Fatalities. Report No. WN 96W0000071, Mitretek, McLean, VA, 1996.
14. Jarašūniene, A. Research into Intelligent Transport Systems (ITS) Technologies and Efficiency. *Transport*, Vol. 22, No. 2, 2007, pp. 61–67.
15. Stephanedes, Y., and J. Hourdakis. Transferability of Freeway Incident Detection Algorithms. *Transportation Research Record: Journal of the Transportation Research Board*, 1996. 1554: 184–195.
16. Abdulhai, B., and S. G. Ritchie. Enhancing the Universality and Transferability of freeway Incident Detection Using a Bayesian-based Neural Network. *Transportation Research Part C: Emerging Technologies*, Vol. 7, No. 5, 1999, pp. 261–280.
17. Yuan, F., and R. L. Cheu. Incident Detection Using Support Vector Machines. *Transportation Research Part C: Emerging Technologies*, Vol. 11, No. 3, 2003, 309–328.

The Standing Committee on Intelligent Transportation Systems (AHB15) peer-reviewed this paper (18-00545).

Article

TRR
JOURNAL OF THE TRANSPORTATION RESEARCH BOARD

Transportation Research Record
2018, Vol. 2672(19) 34–44

DOI: 10.1177/0361198118787083
journals.sagepub.com/home/trr
SAGE

Video Analytics Usage in Transportation Agencies: Nationwide Survey and Maryland Feasibility Study

Zachary Vander Laan[1], Kaveh Farokhi Sadabadi[1], and Thomas Jacobs[1]

Abstract

This paper summarizes the findings from a study conducted by the University of Maryland Center for Advanced Transportation Technology (CATT) regarding the potential to extract meaningful traffic information from existing Maryland Department of Transportation-State Highway Administration (MDOT-SHA) closed-circuit television (CCTV) video feeds via image processing. It describes the survey that was developed and administered by CATT between March 10 and June 8, 2017, focusing on transportation agencies' current use cases and future plans for video analytics. Feedback from 21 agencies (9 of which are using video analytics) indicates a general excitement about the technology, although a persistent theme throughout the responses is that most current solutions are not yet able to provide satisfactory quantitative results (e.g., traffic counts, speeds, turning movements), especially in low light and high glare scenarios. Additionally, the report summarizes the vendor evaluation procedure that was undertaken, which includes identifying suitable video analytics vendors, developing a testbed of representative video clips from MDOT-SHA cameras under various conditions, asking the vendors to demonstrate their product capabilities on the testbed, and analyzing results. With proper camera positioning and calibration, the two vendors who participated achieved results within 5% to 15% of manual counts, and 4% to 7% of probe speeds in the northbound travel direction during a one-hour test clip. However, only one vendor produced results for the more challenging southbound direction, and the estimates were far less accurate (within 25% of manual counts and 20% of probe speeds). Accordingly, it appears that the estimation accuracy is highly sensitive to factors such as camera angle, resolution, and visibility.

The Maryland Department of Transportation-State Highway Administration (MDOT-SHA), in collaboration with the Center for Advanced Transportation Technology (CATT) at the University of Maryland, is exploring the potential use of video analytics for extracting traffic data (e.g., volume, speed, congestion, incident detection) from real-time video feed of existing closed-circuit television (CCTV) cameras. In an effort to understand how other agencies/jurisdictions use video analytics, MDOT-SHA requested that CATT design a questionnaire, reach out to appropriate agency contacts across the United States, and gather feedback on past, present, and planned video analytics pilots and/or deployments. Accordingly, this paper investigates the existing and planned use of video analytics in transportation agencies across the United States, using the results obtained from the nationwide survey to characterize the current state of practice. Furthermore, this paper describes the initial steps taken in an ongoing research effort to test the performance of existing video analytics solutions on a challenging, real-world testbed of video clips obtained from MDOT-SHA's CCTV video feeds.

Transportation agencies often use cameras to visually monitor traffic conditions and investigate potential incidents in real time. For example, operators in traffic management centers (TMC) use CCTV cameras in conjunction with traffic sensor data to identify accidents, mitigate congestion, and coordinate with law enforcement and emergency responders (*1*). Additionally, cameras are increasingly being used to enforce red light, stop sign, and speed violations, typically for the purpose of improving safety (*2–4*). While these systems are useful for specific use cases (e.g., taking a photograph of a vehicle that enters an intersection on a red light), they do not fully leverage the type of traffic information that can be extracted from video feeds. As image processing and machine learning technology continues to mature, transportation agencies may be able to augment traditional traffic

[1]Center for Advanced Transportation Technology, Department of Civil and Environmental Engineering, University of Maryland, College Park, MD

Corresponding Author:
Address correspondence to Zachary Vander Laan: zvanderl@umd.edu

data sources with this data, and leverage it for both planning purposes (e.g., to quantify historical performance, identify problematic areas, guide policy decisions) and real-time traffic management applications (e.g., incident detection, queue warnings).

Verma et al. (*5*) provide a thorough overview of video analytics, summarizing video data analysis techniques in terms of the following major steps: preprocessing images, segmenting objects, extracting features, classifying objects, and finally performing some type of relevant analysis (i.e., counting vehicles, calculating speeds, and detecting incidents in the case of transportation applications). Despite being a well-researched area with significant potential for transportation agencies that already have existing CCTV camera infrastructure, video analytics has well-documented challenges, which are described by Loce et al. (*6*). First, many results are generated based on ideal scenarios, and thus do not extend well to more realistic conditions that include inclement weather, camera positioning, lighting, and various traffic conditions. Additionally, the hardware required to implement these algorithms is expensive, although this issue is likely to become less significant in the long term as technology continues to evolve and prices decrease.

A significant body of literature has proposed various image processing solutions for transportation applications (e.g., (*7*, *8*)), while many others have addressed specific issues such as estimating vehicle speeds with uncalibrated cameras (*9*, *10*) or vehicle occlusion (*11*, *12*). However, this paper takes a different perspective, focusing specifically on how transportation agencies are currently using or plan to use such technology. The paper makes two main contributions:

- It synthesizes results from a nationwide video analytics survey, describing current and near-term future usage in transportation agencies across the United States. Based on agency feedback, it highlights the current shortcomings with the technology, and makes recommendations about best use cases based on current video analytics solutions and vendor performance. While other transportation surveys have considered specific aspects of video analytics (e.g., the Department of Highway Safety and Motor Vehicles survey summarizes use of red light cameras (*13*)), to the authors' knowledge none have performed a national survey synthesizing the existing state and usage of video analytics technology.
- It describes a feasibility study for the State of Maryland, focusing on the development of a representative testbed of MDOT-SHA CCTV video clips, and performing a vendor evaluation on a subset of them. Development of this detailed testbed helps facilitate a meaningful comparison of existing video analytics solutions under real-world conditions, which will be useful to transportation agencies considering this technology.

The remainder of this paper is organized in the following manner. First, it describes the survey methodology, after which it presents the survey results and discusses the agency responses from a number of perspectives. Next, it discusses a feasibility study for the State of Maryland, and describes the process of choosing representative Maryland CCTV video clips to form a testbed for a product evaluation. Finally, the paper concludes by analyzing the results of a vendor evaluation, and synthesizing the overall learning about the current and future state of video analytic technology for transportation agencies.

Survey Methodology

An online survey was developed to capture responses from relevant members of transportation agencies nationwide. The survey was first distributed via a ListServ to graduates of the I-95 Operations Academy (*14*), which includes leaders of many different U.S. transportation agencies who are heavily involved in traffic operations. Additionally, appropriate individuals were identified through publicly-available information on Transportation Research Board subcommittees (e.g., AHB-20, ABJ-35), AASHTO membership rosters, and attendance at the NATMEC 2016 conference. These additional prospective survey respondents were grouped by agency, and each agency was asked to select an appropriate individual to fill out the survey on behalf of the organization. In total, members of agencies from 49 states were contacted, either by ListServ or email.

The survey consisted of 19 multiple choice, 5 fill-in-the-blank, and 12 free-response questions, some of which were only applicable to agencies currently using video analytics. Broadly speaking, the survey was organized into the following categories:

- *Respondent Information*: Respondents were asked to provide their contact information, role in their organization, area of expertise, and willingness to be contacted for follow-up questions.
- *Camera Infrastructure*: Participants were asked about the number of cameras currently deployed, video coding format, resolution, and frame rate, as well as pan/tilt/zoom functionality, calibration requirements, and ability to archive footage.
- *Video Analytics*: Respondents were asked whether their agency is currently using or is planning to use video analytics in any capacity. If so, they were asked to describe their plans, and current users were asked to answer a series of questions about the inputs and outputs to their solution, how the output data is accessed, the frequency with which the data is made available, whether the results have been validated, and the reliability of the data in various conditions.
- *Vendors*: Respondents who answered that they are currently using video analytics in some capacity were

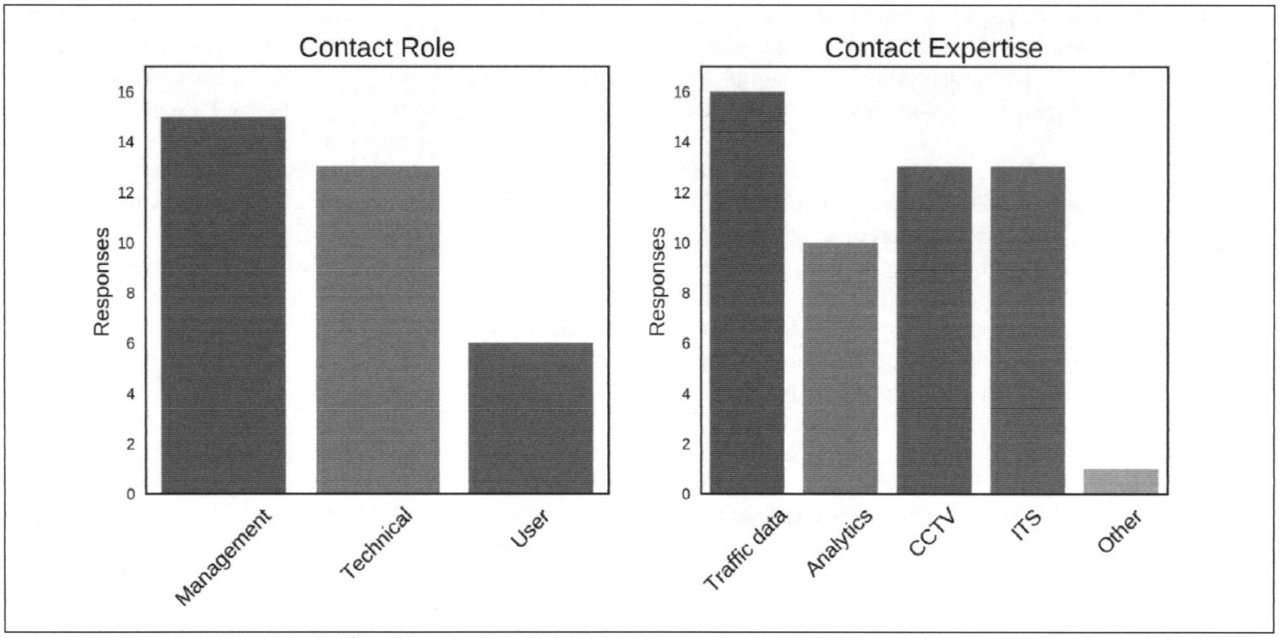

Figure 1. Survey contact roles and areas of expertise.

asked about the vendors(s) they worked with, including questions about how they selected the vendor(s), whether they received a product demonstration, and their perception of their customer service and overall capabilities.

- *General*: Respondents using video analytics were asked to assess overall strengths and weaknesses of their existing solutions, provide general comments, and rate their existing solution in the following categories: hardware, software, data type/quality/access/use, cost, vendor, and overall.
- *Ongoing/Future Plans*: Respondents who indicated they are currently using or plan to use video analytics in the near future were asked about their near-term plans regarding application areas, vendors, and data.

Survey Results and Discussion

In total, 21 agencies provided valid responses to the online survey between March 10 and June 8, 2017. The survey respondents represent a variety of roles within their respective agencies, including technicians, engineers, modelers, planners, managers, and directors. Each contact was asked to categorize his or her roles and areas of expertise from a predefined set of options, and an option was provided to write in additional areas of expertise. Figure 1 summarizes these results, and indicates that both management and technical personnel are well represented, and that respondents have a diverse set of interests related to video analytics. Because each contact was allowed to select multiple roles and areas of expertise, the total number of responses shown in Figure1 is much larger than the 21 valid surveys that were returned.

Figure 2 summarizes the number of agencies that reported that they are using or planning to use video analytics, showing that nine are currently using this technology in some fashion, with an additional four indicating that they plan to do so in the near future. The nine agencies currently using video analytics represent a diverse set of interests and application areas, including incident detection for real-time traffic management, traffic violation identification, intrusion surveillance, intersection management, and portable systems for performing temporary traffic counts. Here we compare the existing solutions from multiple perspectives.

Camera Infrastructure

Existing camera deployment varies significantly among the eight respondents using video analytics, and includes various camera types (CCTV, intersection, portable), frame rates (usually 15, 20, or 30 frames per second), video encoding types (MPEG-4, H.264, NTSC standard), and resolutions (most commonly 1CIF or 4CIF). The number of cameras deployed ranges from 20 to 800 with a mean of 270 and median of 133 cameras.

Figure 3 visualizes the following camera characteristics: (*a*) pan/tilt/zoom functionality, (*b*) calibration requirement, and (*c*) video archiving practice. Of the nine agencies that reported using video analytics in some capacity, five have cameras with pan/tilt/zoom functionality, and three require calibration to work properly. Furthermore, one agency

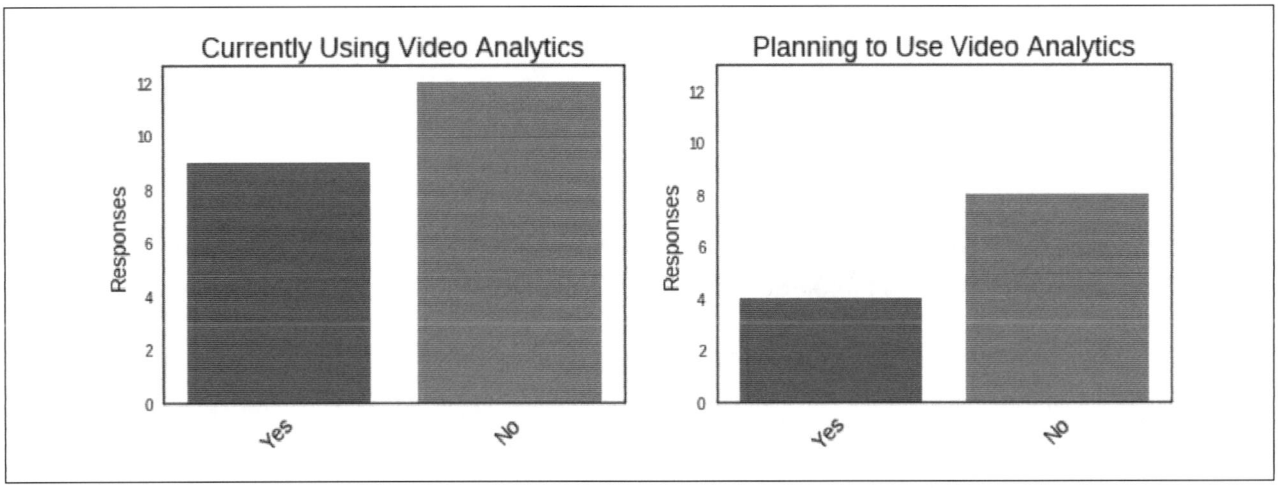

Figure 2. Current and future video analytics usage.

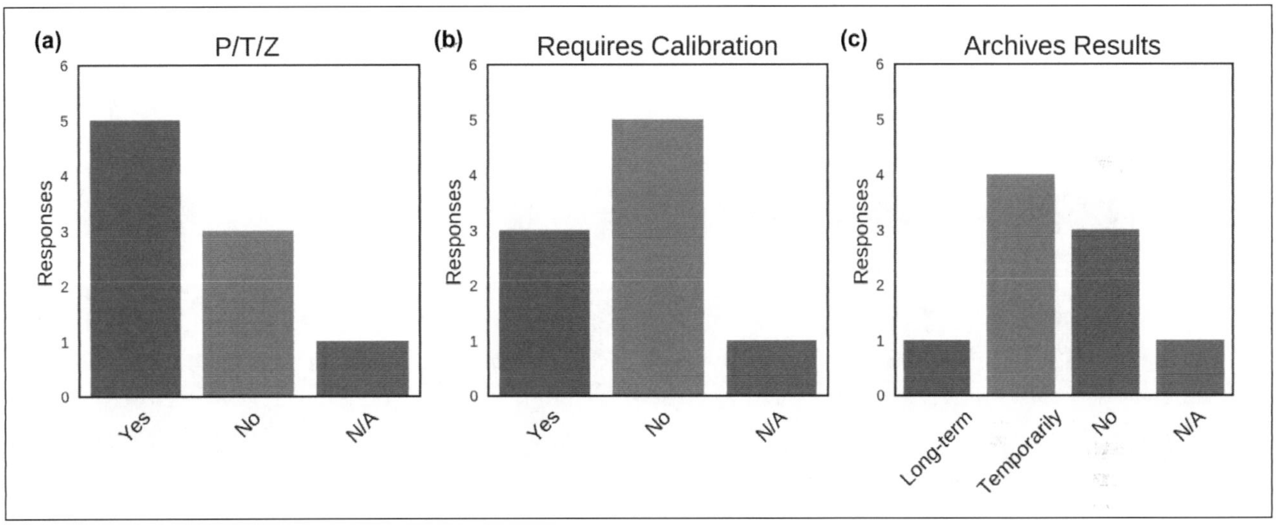

Figure 3. Camera characteristics.

archives video long term, four do so temporarily (less than a week), and three not at all.

Video Analytics Solutions

Figure 4 categorizes video analytics solutions based on whether they are integrated (i.e., all-in-one systems comprised of both cameras and software), or software-only systems, and also summarizes the extent to which data is accessible in real time. Figure 4*a* shows that six of the nine systems are software-only solutions, meaning the agencies purchased or developed software without also needing to buy integrated hardware. Of the two all-in-one systems, one purchased software and intersection cameras simultaneously, while the other used a portable system that bundled the camera and image processing software in a single unit. This portable unit also represents the only solution that did not provide real-time data, as shown in Figure 4*b*.

Figure 5 lists the reported outputs generated by the existing video analytics solutions. Note that, despite the various camera architectures and solution types, the following outputs were common to at least five of the nine agencies: speeds, counts, incidents, traffic violations, and disabled vehicle identification.

Respondents who reported using video analytics were asked to describe their current use cases in a free-response box. Interestingly, although most solutions appear to produce similar output metrics (as seen in Figure 5), each of the nine video analytics users described different use cases, many of which do not fully utilize all the available outputs. Although their responses do not fall neatly into discrete categories because of the open nature of the question, they can be

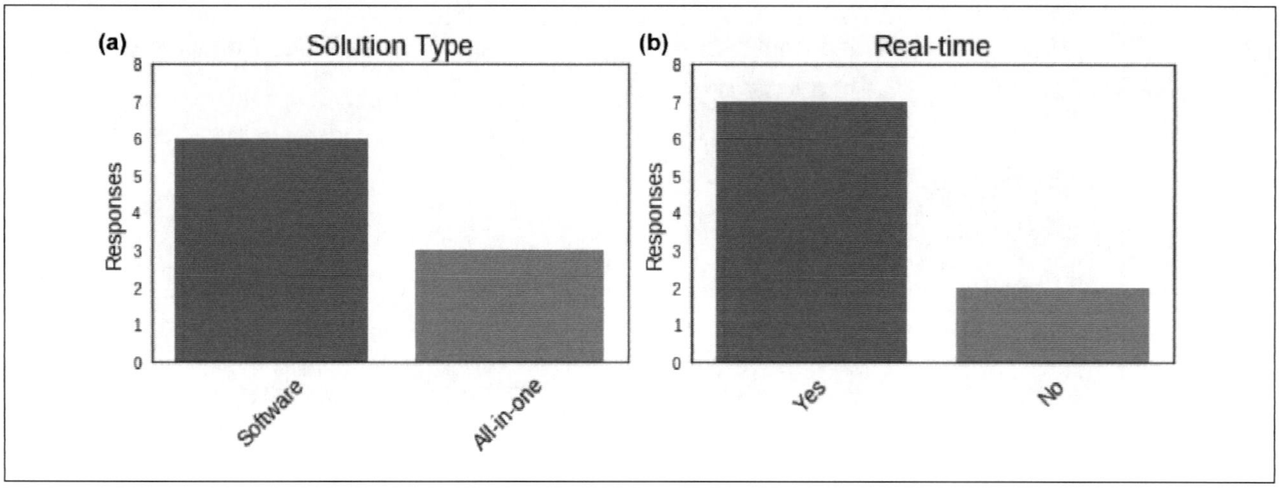

Figure 4. Video analytics solution characteristics.

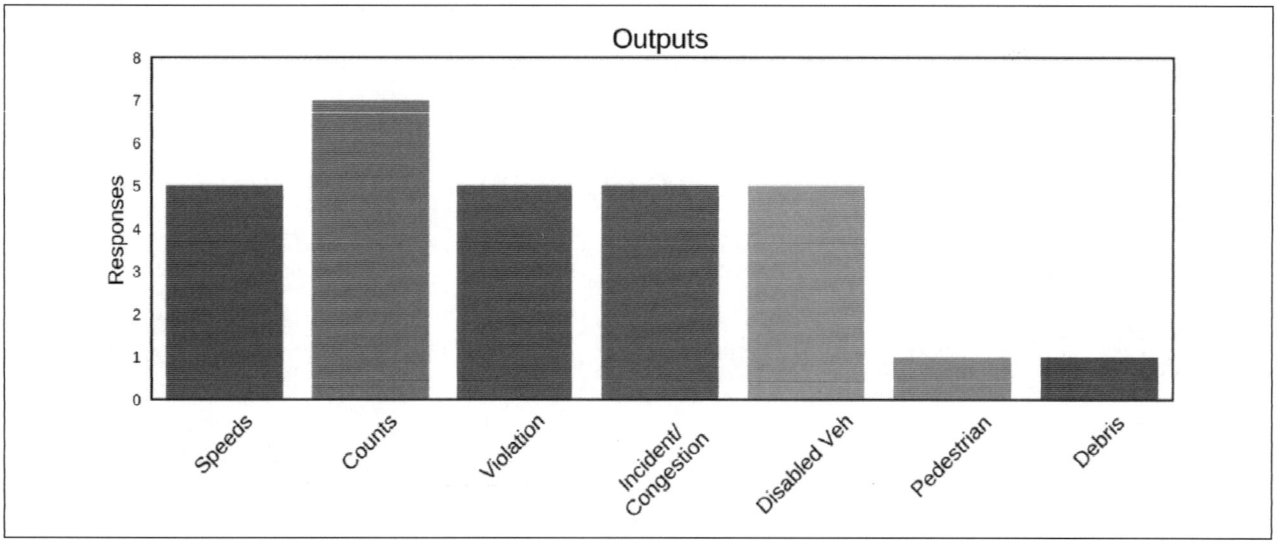

Figure 5. Video analytics solution outputs.

broadly separated into operational and planning solutions. Figure 6*a* summarizes the responses based on this classification, and shows that six applications are operations-oriented while two take a planning approach. Figure 6*b* further categorizes the solutions that deal with real-time operations, labeling each response with any of the following subcategories that apply: violations, incident/congestion detection, disabled vehicles, and intersection management.

Of the operations-based solutions, it is interesting to note that the use cases deal primarily with violations, incident/congestion detection, and disabled vehicles rather than speeds and vehicle counts. Thus it appears that, from an operational perspective, video analytics are being used primarily by operators (e.g., at TMCs) to identify and respond to congestion, incidents, and other anomalies. However, a few responders noted that in the future they hope to be able to utilize counts and speeds for data analytics, and one stated that they are evaluating their system for use with signal timing optimization.

Instead of using the technology for real-time operations, three agencies are currently taking advantage of video analytics for planning purposes. One noted that although the technology is urgently needed, existing solutions are not yet reliable enough to produce trustworthy results. Accordingly, they are currently using it for research and development, but hope to move to the operational level at some point in the future. Another agency uses a portable, all-in-one system to record temporary traffic counts at various locations. Rather than using the count data for real-time analytics, these temporary counts are used to calibrate permanent automatic traffic recorder stations. The final agency uses the data for 24-hour traffic counts, which are archived and will likely be

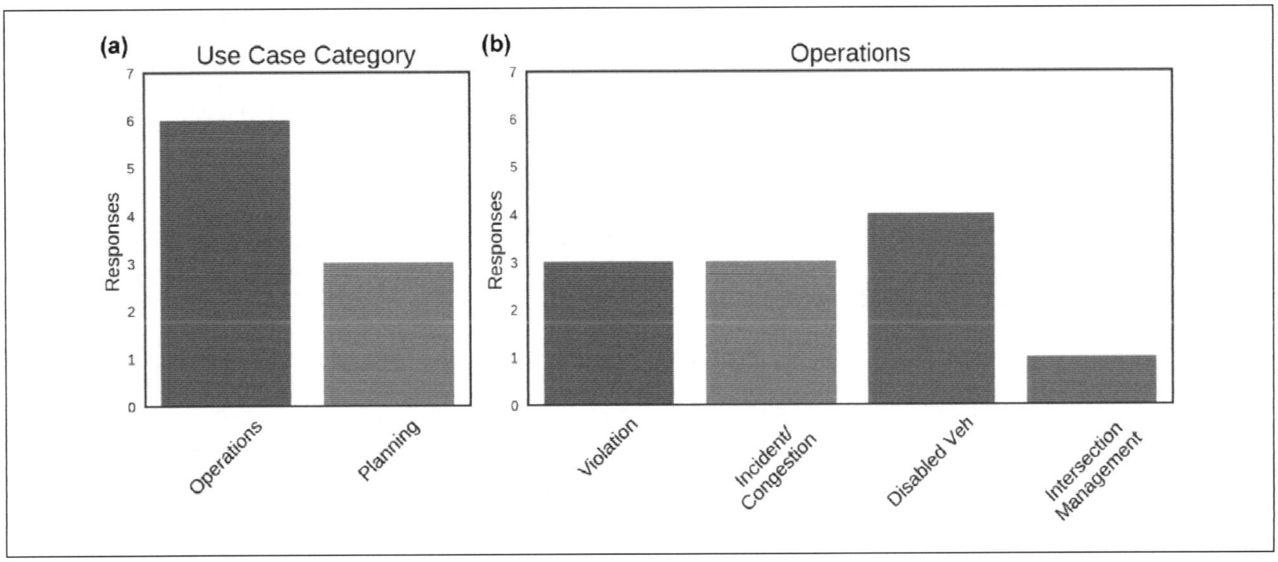

Figure 6. Video analytics use cases.

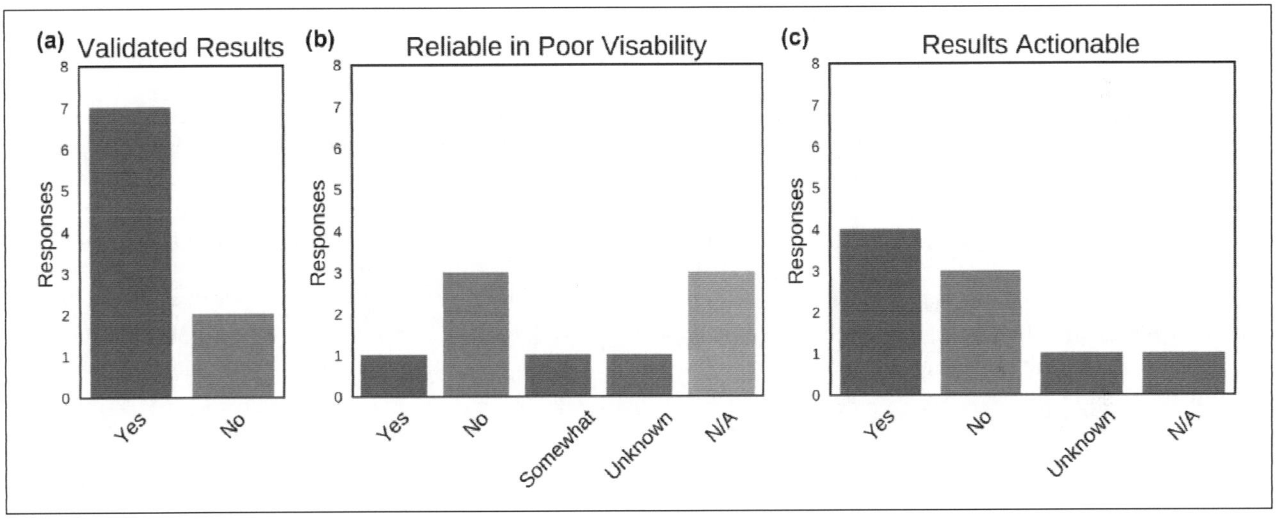

Figure 7. Data validity.

used for reporting, AADT calculations, and other future analysis.

Data Validity

Each respondent who reported using video analytics was asked to comment on the validity of the outputs generated, which is summarized in Figure 7. Figure 7*a* indicates that seven of the nine agencies performed some type of validation procedure, either manually or via a third party, while Figure 7*b* describes the extent to which the results are reliable in poor visibility conditions (e.g., low light, high glare). Interestingly, the most common responses were that the data was not trustworthy or had moderate problems, which indicates that vendors will need to improve performance in suboptimal conditions to generate confidence in their products. Overall, Figure 7*c* shows that the verdict is split; about half the respondents are willing to act on the results, while the rest are somewhat apprehensive. However, this likely depends on the desired use case and camera infrastructure, as those using video analytics for identifying anomalies or that have fixed intersection cameras tended to be more optimistic. Based on these responses, it appears that qualitative information produced by video analytics is more consistently reliable compared with quantitative data.

General Comments

Finally, respondents were asked to provide general feedback regarding their current video analytics solutions, and rate their system in the following categories: hardware, software, data type, data quality, data access, data use, cost, vendor, and overall. For each category, they were asked to indicate their satisfaction level among the following options: Extremely Dissatisfied, Dissatisfied, Somewhat Satisfied, Satisfied, and Extremely Satisfied (corresponding to a 1-to-5 scale, where 1 = Extremely Dissatisfied and 5 = Extremely Satisfied). The results indicate that in general most agencies are somewhat satisfied with their solution (overall average rating of 3.2), with data access and vendors receiving the highest average ratings (4.4 for each), and cost receiving the lowest (3.0).

It is important to remember, however, that these ratings depend on agencies' goals. Agencies that use video analytics to alert operators to potential anomalies appear to be more satisfied with the current state of technology, as the implications of a false positive are relatively insignificant. In contrast, those who hope to use video analytics for count data, real-time metrics and active traffic management appear to be more focused on the current shortcomings.

Summary. Based on the survey responses, it is clear that there is interest in video analytics in the transportation community. There are currently 13 agencies from 12 states that are using or planning to use video analytics in the near future. Accordingly, at least 24% of states in the U.S. have transportation agencies that are interested in exploring its potential application for operations or planning purposes. While this percentage may theoretically be higher, it should be noted that the current results may be biased towards agencies interested in video analytics (i.e., non-response bias), which would mean that additional responses from agencies in other states may not appreciably increase the percentage. Additionally, there is a general concern about the reliability of the results, particularly in low light, poor weather, and high glare scenarios. However, some agencies are only using the systems to alert traffic operators to abnormalities (i.e., are not relying on speed and count data to be reliable) or in tunnels where they do not deal with glare and weather issues. Accordingly, the suitability of existing solutions appears to depend heavily on the desired application area.

Maryland Feasibility Study

In order to evaluate the feasibility of using video analytics in Maryland, CATT developed a representative testbed of videos recorded with MDOT-SHA's existing CCTV cameras, and asked video analytics vendors to demonstrate their products' capabilities on clips from the testbed.

Camera Selection

Ten Maryland CCTV cameras were selected for recording, and are representative of the video quality and camera characteristics of Maryland's existing infrastructure. This selection contains a mixture of cameras that are conducive to video analytics, ones with known challenges (e.g., complex road geometry, challenging camera angle, glare or shadows at certain times of the day), and ones near existing fixed traffic sensors, which can be used for validation purposes.

Testbed Development

In order to develop a representative testbed of video samples, approximately two weeks of video feeds from March 12 to March 27, 2017, were recorded. The feeds were accessed through the Regional Integrated Transportation Information System (RITIS) (*15*), and a Python script was used to automate the process of continually recording one-hour clips at all locations. In total, 360 hours of video were obtained at each location, requiring about 320 GB of storage.

Based on this video database, 10 representative one-hour video clips were selected, capturing a variety of camera locations and angles, times of day, weather conditions, presence of incidents, traffic conditions, and other factors. Many of the clips represent scenarios that are known to be challenging for video analytics applications, including rain, snow, glare from headlights or the sun, camera jitter from the wind, complex road geometry, and camera movement by operators.

To facilitate selection of these scenarios, the video database was searched during specific hours of the day or during time periods when it had been snowing or raining. Furthermore, the locations of incident reports obtained through RITIS were visualized using QGIS software (*16*), and a spatial analysis was performed to identify incidents that may have been sufficiently close to the selected cameras.

Vendor Evaluation

Video analytics vendors were identified based on their representation at the Transportation Research Board's annual meeting, NATMEC 2016 conference, recommendations from survey respondents, and online presence. In total, seven companies were contacted, and the sample video clips were shared with six that were interested in demonstrating their products' capabilities.

Qualitative Results. Of the six vendors who received the video testbed, four responded with feedback, including two who provided a product demonstration (referred to as vendors A and B). The two who declined to participate in the evaluation stated that the testbed did not meet their camera positioning requirements, and deemed the CCTV footage inappropriate for performing proper video analytics, while the remaining two vendors abruptly stopped communicating after receiving the testbed.

Vendor A took a cautiously optimistic view of the testbed, stating that videos 1 to 4 were reasonable for video analytics and videos 5 to 7 were usable although noting that camera 6 was hampered by frequent jitter. The remaining videos were deemed poor for video analytics due to poor resolution and camera angle. In addition to agreeing to extract count and speed data from one clip (discussed in the next section), they provided four short annotated video clips, visually demonstrating their ability to count vehicles, automatically detect incidents and smoke, and explaining the negative impact of camera jitter.

Vendor B's assessment of the video clips was also positive; they indicated that their product can work on almost any camera pointed at traffic, including the types of samples provided in the testbed. Like vendor A, they qualitatively illustrated their capabilities by providing live annotations of various scenarios and conditions encountered in the clips. In particular, they demonstrated their product's ability to provide basic incident detection even without camera calibration, and then showed the additional capabilities available when the camera is calibrated and remains in the preset position (e.g., traffic metrics such as counts, speeds, flow rates and occupancy, as well as incident detection, stopped vehicles, pedestrians, slow speeds, and wrong-way vehicles).

Quantitative Results. Both vendors A and B agreed to extract raw data from one of the videos in the testbed (video 4), which is 58 minutes long and contains snowfall and poor visibility throughout the first half hour, representing moderately challenging conditions for performing video analytics. Furthermore, the camera is primarily pointed towards traffic on the near side of the image (traveling northbound), which makes the southbound traffic more difficult to measure.

Vendor A extracted data for the northbound direction only, and reported the count data in a comma separated value (csv) file and speed data for 10-minute aggregation time periods. Each record in the count csv file represents a vehicle passing through a predefined zone in the image, and contains two timestamps (representing the time entering and exiting the zone, respectively), a unique identifier, and confidence level that the identified count belongs to a vehicle. Raw speeds for each vehicle were not provided; the average speeds were instead reported for each 10-minute period.

Vendor B provided similar data, but extracted speeds and counts for both the northbound and southbound directions. Like vendor A, they shared the results in a csv format, with each record representing a vehicle count event. Additionally, they included the speed measurement for each record, and specified the specific lane in which the vehicle was counted. Data extracted from the video was manually validated and compared against other data sources in 10-minute periods to assess the vendors' performances. Ground truth count data was obtained via manual counts, while speeds were obtained from probe vehicle datasets (INRIX, HERE, and NPMRDS) via RITIS. It is worth pointing out that the probe speeds are spatially aggregated (i.e., over a TMC segment) for each period, which is slightly different than the temporally aggregated speeds calculated at a single location (i.e., the vendor speeds). However, although the probe speeds should not be viewed as a ground truth, they represent reasonable values for comparison with vendor data.

In addition to comparing absolute counts and speeds for each 10-minute period, it is instructive to express the estimation accuracy in terms of percentage error/deviation. Because probe speeds are not necessarily true speed values (i.e., they are a sample of overall traffic and are averaged differently), the term deviation will be used instead of error. The metric selected to capture how the vendor count/speed data differs from the manual count/probe speed data is called mean absolute percentage deviation (MAPD), and is defined as:

$$MAPD(\%) = \frac{100}{n}\sum_{i=1}^{n}\left|\frac{y_{validation} - y_{vendor}}{y_{validation}}\right| \quad (1)$$

where $y_{validation}$ is the manual count or probe vehicle speed, y_{vendor} is the count or speed estimate provided by the vendor, and n is the number of time period. Note that the absolute value ensures that positive and negative deviations are not canceled out over multiple time periods.

Figure 8 shows the validation results for traffic in the northbound direction, with the top chart comparing traffic counts, and the lower chart comparing speeds. In terms of counts, the results indicate that both vendors slightly underestimated the actual values, with vendor B closer to the ground truth than vendor A, corresponding to MAPD values of approximately 5% for B and 17% for A. With respect to speeds, both vendors performed quite well, with A closer to the measured probe speeds, and B consistently overestimating the values. INRIX probe speeds were chosen for y_{vendor} for MAPD calculations, which yielded values of 4% and 7% for A and B, respectively. Thus, both vendors estimated speeds that strongly agree with vehicle probe data. Figure 9 shows the validation results for southbound traffic, and only includes vendor B because A did not provide results for this direction. As compared to the northbound direction, B's predictions are significantly less accurate for both counts and speeds. For each 10-minute period, the traffic counts are about 75 to 150 vehicles fewer than the manually-counted value, which corresponds to a MAPD value of about 25%. Similarly, the speed estimates consistently underestimate the probe vehicle speeds by about 10 mph, corresponding to a MAPD value of about 20%.

Summary. Based on the vendor evaluation, it is apparent that traffic data extracted from video feeds is highly dependent on camera angle, resolution, and visibility. These requirements were mentioned by most vendors, and many deemed the representative MDOT-SHA samples to be too low-quality for accurate processing. In cases where ideal camera positioning can be guaranteed for a fixed period of time (i.e., good visibility, no pan/tilt/zoom after calibration), video analytics may be suitable for estimating short-term traffic volumes and speeds. With proper camera positioning and

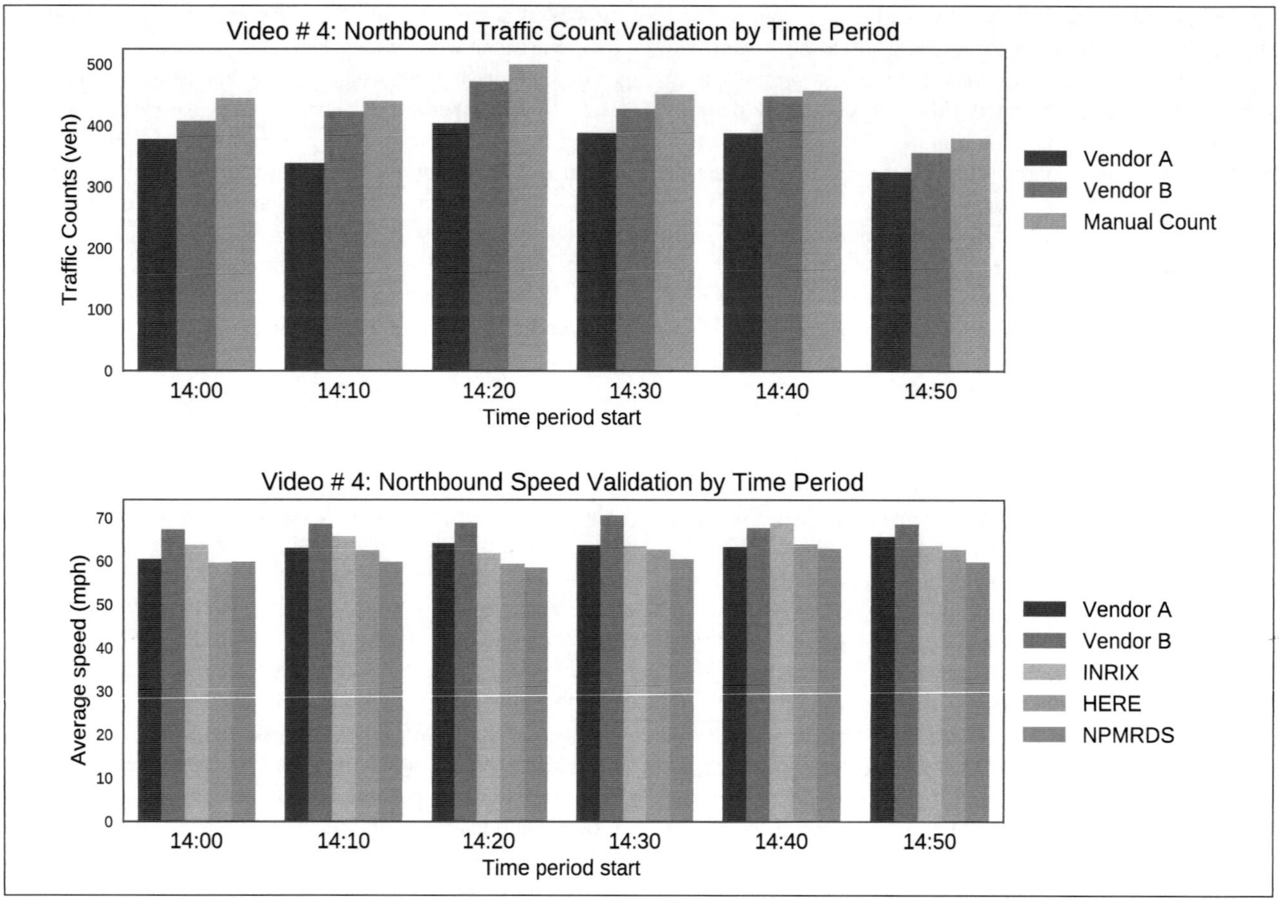

Figure 8. Northbound speed and count validation.

calibration, two vendors achieved results within 5% to 15% of manual counts, and 4% to 7% of probe speeds in the northbound travel direction during a one-hour test clip. However, results in the more challenging southbound direction were only provided by one vendor, and the estimates were only within 25% and 20% of manual counts and probe speeds, which reinforces the conclusion that accuracy is highly dependent on the positioning and video quality. Finally, even without proper calibration or ideal conditions, video analytics systems can often be used to detect incidents, which has important implications for real-time traffic management.

Conclusions

This paper investigates the current state-of-the-practice video analytics techniques being used by transportation agencies across the United States. To this end, we summarize results from a nationwide survey on transportation agencies' current and future plans to extract relevant traffic data from video feeds via image processing, and discuss the current status of a feasibility study in the State of Maryland.

The survey highlights a general optimism about the future of video analytics, although most respondents agreed that current solutions are not yet capable of providing satisfactory results in all circumstances, particularly in low light and high glare situations.

The results of the vendor evaluation corroborate the ideas communicated by transportation agencies through the survey; it does not appear that current video analytics solutions are capable of accurately extracting quantitative traffic data in all the varied conditions that characterize MDOT-SHA's CCTV cameras, particularly ones whose pan/tilt/zoom functionality is often employed. However, the results of this evaluation suggest that video analytics may be suitable for performing short-term measurements when properly-located/positioned cameras can be calibrated and held stationary for a fixed period. Additionally, because detailed count/speed data is not necessary for incident detection, it may currently be most useful for generating qualitative data; it can be used in real-time traffic management applications to help operators identify anomalies to investigate further.

Thus, based on the survey and vendor evaluation, it appears that video analytics may be most applicable for alerting operators to potential congestion patterns, incidents, and other abnormalities (i.e., qualitative results), with an additional possibility of occasionally collecting speed/count data

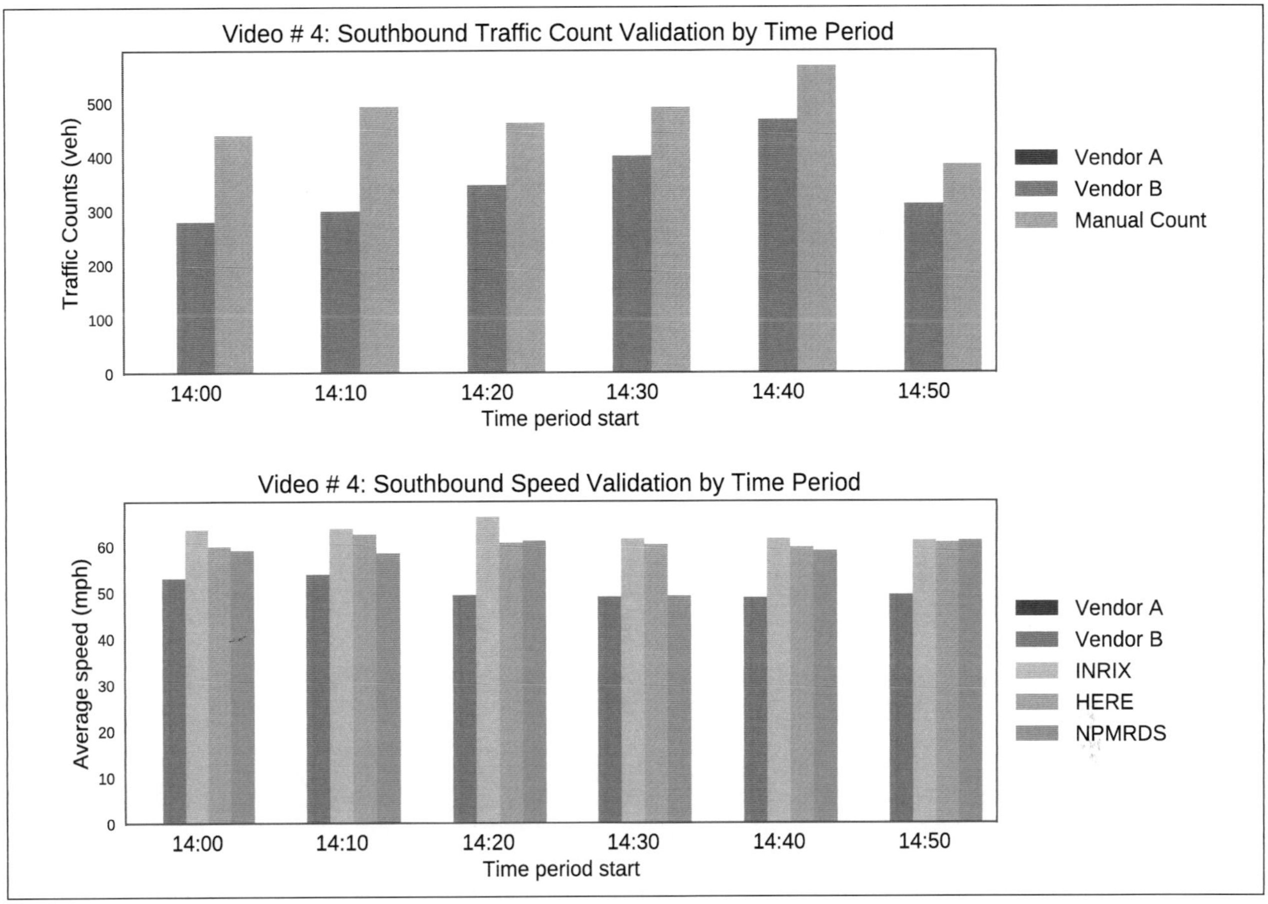

Figure 9. Southbound speed and count validation.

for specific studies (i.e., quantitative results). However, as more agencies adopt these solutions and apply pressure on vendors to improve, it is likely that solutions will become increasingly reliable.

Acknowledgments

This research was supported by the Maryland SHA office of Coordinated Highways Action Response Team (CHART). The authors gratefully acknowledge Maryland SHA for their support, and additionally thank the transportation agency representatives who participated in the video analytics survey.

Author Contributions

The authors confirm contribution to the paper as follows: study conception and design: ZV, KS, TJ; data collection: ZV, KS; analysis and interpretation of results: ZV, KS, TJ; draft manuscript preparation: ZV. All authors reviewed the results and approved the final version of the manuscript.

References

1. Williams, B. M., and A. Guin, Traffic Management Center Use of Incident Detection Algorithms: Findings of a Nationwide Survey. *IEEE Transactions on Intelligent Transportation Systems*, Vol. 8, No. 2, 2007, pp. 351–358.
2. Retting, R. A., S. A. Ferguson, and A. S. Hakkert. Effects of Red Light Cameras on Violations and Crashes: A Review of the International Literature. *Traffic Injury Prevention*, Vol. 4, No. 1, 2003, pp. 17–23.
3. Persaud, B., F. Council, C. Lyon, K. Eccles, and M. Griffith. Multijurisdictional Safety Evaluation of Red Light Cameras. *Transportation Research Record: Journal of the Transportation Research Board*, 2005. 1922: 29–37.
4. Shin, K., S. P. Washington, and I. Van Schalkwyk. Evaluation of the Scottsdale Loop 101 Automated Speed Enforcement Demonstration Program. *Accident Analysis & Prevention*, Vol. 41, No. 3, 2009, pp. 393–403.
5. Verma, B., L. Zhang, and D. Stockwell. Roadside Video Data Analysis: Deep Learning. *Studies in Computational Intelligence*, Vol. 711, No. 1, 2017, pp. 13–31.
6. Loce, R. P., E. A. Bernal, W. Wu, and R. Bala. Computer Vision in Roadway Transportation Systems: A Survey. *Journal of Electronic Imaging*, Vol. 22, No. 4, 2013, p. 041121.
7. Coifman, B., D. Beymer, P. McLauchlan, and J. Malik. A Real-time Computer Vision System for Vehicle Tracking and Traffic Surveillance. *Transportation Research Part C: Emerging Technologies*, Vol. 6, No. 4, 1998, pp. 271–288.

8. Zhang, G., R. Avery, and Y. Wang. Video-based Vehicle Detection and Classification System for Real-time Traffic Data Collection Using Uncalibrated Video Cameras. *Transportation Research Record: Journal of the Transportation Research Board*, 2007. 1993: 138–147.
9. Dailey, D. J., F. W. Cathey, and S. Pumrin. An Algorithm to Estimate Mean Traffic Speed Using Uncalibrated Cameras. *IEEE Transactions on Intelligent Transportation Systems*, Vol. 1, No. 2, 2000, pp. 98–107.
10 Schoepflin, T. N., and D. J. Dailey. Dynamic Camera Calibration of Roadside Traffic Management Cameras for Vehicle Speed Estimation. *IEEE Transactions on Intelligent Transportation Systems*, Vol. 4, No. 2, 2003, pp. 90–98.
11. Pang, C. C. C., W. W. L. Lam, and N. H. C. Yung. A Novel Method for Resolving Vehicle Occlusion in a Monocular Traffic-image Sequence. *IEEE Transactions on Intelligent Transportation Systems*, Vol. 5, No. 3, 2004, pp. 129–141.
12. Zhang, W., Q. J. Wu, X. Yang, and X. Fang. Multilevel Framework to Detect and Handle Vehicle Occlusion. *IEEE Transactions on Intelligent Transportation Systems*, Vol. 9, No. 1, 2008, pp. 161–174.
13. Florida Department of Highway Safety and Motor Vehicles, *Red Light Camera Summary Report*, 2013. https://www.flhsmv.gov/reports/redlightcameraanalysis2013.pdf
14. Operations Academy. *I-95 Corridor Coalition*. http://i95coalition.org/projects/operations-academy. Accessed July 26, 2017.
15. RITIS. *The Regional Integrated Transportation Information System (RITIS)*. http://www.cattlab.umd.edu/?portfolio=ritis. Accessed July 26, 2017.
16. QGIS Development Team. *QGIS Geographic Information System*. Open Source Geospatial Foundation, Chicago, IL, 2015.

The Standing Committee on Intelligent Transportation Systems (AHB15) peer-reviewed this paper (18-02857).

Article

TRR
JOURNAL OF THE TRANSPORTATION RESEARCH BOARD

Transportation Research Record
2018, Vol. 2672(19) 45–54

DOI: 10.1177/0361198118782798
journals.sagepub.com/home/trr
SAGE

Deployment and Testing of Optimized Autonomous and Connected Vehicle Trajectories at a Closed-Course Signalized Intersection

Aschkan Omidvar[1], Mahmoud Pourmehrab[1], Patrick Emami[2], Rebecca Kiriazes[1], John C. Esposito[3], Clark Letter[1], Lily Elefteriadou[4], Carl D. Crane III[3], and Sanjay Ranka[2]

Abstract

In this paper, we discuss the deployment and testing of an intelligent real-time isolated intersection traffic control system (IICS), designed to optimize simultaneously signal control and automated vehicle (AV) and connected vehicle (CV) trajectories for low demand condition. The work described here is part of an ongoing larger project (funded by the National Science Foundation and the Florida Department of Transportation, FDOT) to develop, test, and deploy the IICS. The focus of this paper is on the deployment and testing of the algorithm at the Traffic Engineering and Research Laboratory (TERL), FDOT's closed-course facility. The algorithm (described in more detail elsewhere) optimizes signal control and provides optimal AV and CV trajectories at isolated intersections. The algorithm is designed to handle AV, CV, and conventional vehicles in a mixed traffic and low demand condition. The paper provides an overview of the IICS framework, discusses the development and testing of the necessary software and hardware, and presents the scenarios tested at the TERL. The results from the field test confirm the feasibility of the IICS, and will be used to enhance it for future testing and ultimately for field deployment and in high demand condition.

Significant improvements in automated and connected vehicle technologies are expected to create a revolution in how we move and move things. It is highly likely that in the near future automated vehicles (AV) and connected vehicles (CV) will be operating side by side in large numbers, along with conventional vehicles. This creates many opportunities for improving the efficiency and safety of surface transportation. For example, the U.S. Department of Transportation Multimodal Intelligent Traffic Signal Systems (MMITSS) initiative aims to provide a comprehensive traffic information framework to service all modes of transportation in a connected vehicle environment (*1*, *2*).

Delays at traffic signals are estimated to cause 5% to 10% of all traffic delay on major roadways and to have contributed an estimated 25% to the increase in total highway traffic delays during the past 20 years (*3*). One source of delay at signals is inefficient utilization of green time in response to fluctuating demand. Another source is driver reaction-related delays, including delay in starting-up. Improvements in traffic signal timing, as well as the use of AV/CVs, have the potential to address those delays and significantly benefit the transportation system.

The objectives of this research are to develop, test, and deploy an intelligent real-time intersection traffic control system (IICS), designed to optimize simultaneously signal control and AV and CV trajectories. The focus of this paper is on the deployment and testing of the IICS at the Traffic Engineering and Research Laboratory (TERL), a Florida Department of Transportation (FDOT) closed-course facility. The algorithm (*4*) optimizes signal control and provides optimal AV and CV trajectories, and it is designed to handle AV, CV, and conventional vehicles.

The paper provides a brief literature review, and then an overview of the IICS framework. It discusses the development and testing of the necessary software and hardware, and presents the scenarios tested at the TERL, along with evaluation results.

[1]Department of Civil and Coastal Engineering, University of Florida, Gainesville, FL
[2]Department of Computer and Information Science and Engineering, University of Florida, Gainesville, FL
[3]Department of Mechanical & Aerospace Engineering, University of Florida, Gainesville, FL
[4]Department of Civil and Environmental Engineering, University of Florida, Gainesville, FL

Corresponding Author:
Address correspondence to Aschkan Omidvar: aschkan@ufl.edu

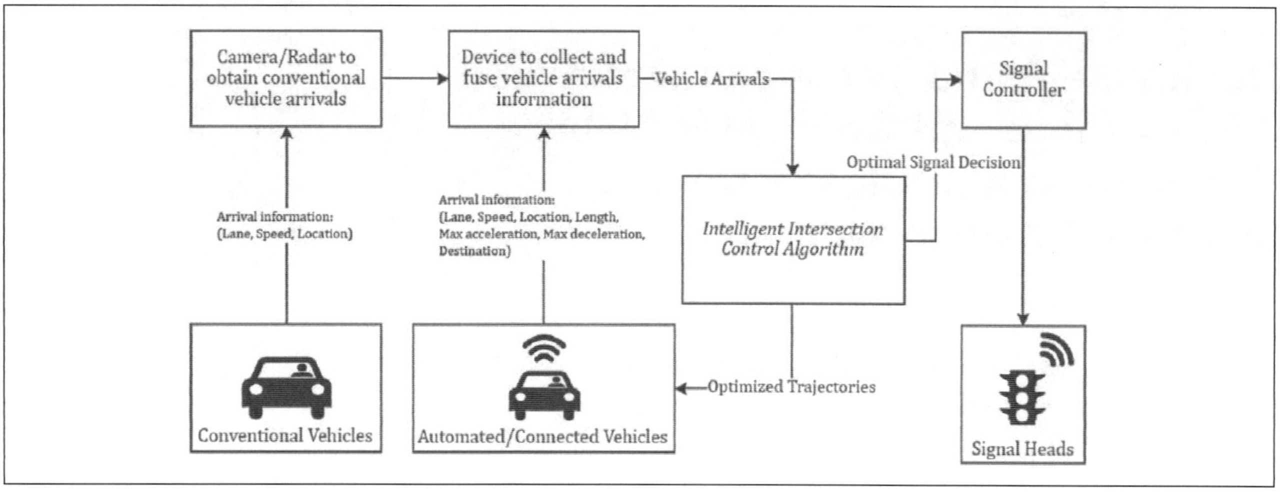

Figure 1. Intelligent intersection control system (IICS) (*4*).

Literature Review

In recent years, there has been a surge in the amount of research on transportation operations at signalized intersections under fully automated and/or connected conditions. Dresner and Stone (*5*) proposed an autonomous intersection management system which functions based on rejecting or granting AVs' requests for space and time within the conflict zone of an intersection. In general, when the traffic is fully composed of AVs and exchange of information is in real time and robust state, an algorithm can schedule the vehicle sequence without any need for signalization (*6–8*). This approach assumes the traffic flow is 100% AV. Several additional attempts have been made to eliminate the need for traditional signalization through novel methodologies (*9*, *10*).

Li et al. (*10*) designed an algorithm to optimize simultaneously signal control and AV trajectories at an isolated intersection with two conflicting through movements. The procedure takes arrival information of vehicles upon entering the detection/communication range as input and calculates the trajectory for each vehicle. Enumeration over a set of binary signal decisions is performed to make trajectory selection decisions, which results in the shortest average travel time delay. Several other papers subsequently took a similar approach by employing adaptive signal controlling logic for intersection optimization. One such approach is multimodal intelligent traffic signal systems (MMITSS) (*11–14*).

This paper reports on the results of a field test evaluating a real-time optimization algorithm for enhancing traffic signal control operations simultaneously with vehicle trajectories, when the traffic stream is composed of CVs, AVs, and conventional vehicles. The paper discusses the necessary hardware and interfaces for field implementation and reports on the test results of the integrated system with multiple vehicles in the field under different scenarios at FDOT's TERL facility.

Overview of the Intelligent Intersection Control System (IICS)

In this section, we briefly describe the IICS. Figure 1 depicts the overall structure of the IICS. As shown, arrival information is obtained through radar and cameras for both conventional vehicles and AVs as they enter the communication range. AVs and CVs communicate their position and other pertinent information by dedicated short-range communication (DSRC). The two sources of arrival information are fused, and the resulting arrivals are entered into the algorithm for processing. Optimized signal timings are sent to the signal controller, while optimized trajectories are sent to the AVs and CVs.

The IICS algorithm, which functions as the core logic of the IICS, uses arrival information of vehicles within the communication range (typically 500 ft) to optimize signalization and AVs' movement. The movement of an AV may be influenced by a leading vehicle, thus AV trajectories are optimized based on the predicted behavior and trajectories of conventional vehicles, assuming these follow the Gipps car-following model (*15*). It is assumed that there is no lane changing within the communication range, and that there are no cycle failures, so that vehicle trajectories are optimized only once upon their entrance to the detection zone. The algorithm receives and processes real-time lane-specific arrival data, and it can handle multi-approach and multi-vehicle input. The objective function is minimization of travel time delay.

The IICS algorithm receives vehicle arrival information through sensing and communication devices for each lane and for each vehicle. The trajectory optimization module of the algorithm has two sub-modules: trajectory generation (for AVs and CVs) and trajectory prediction, therefore location and speed estimation (for conventional vehicles). Optimized trajectory generation has two sub-models: trajectory optimization for lead AV and trajectory optimization for follower AV. Every time a new vehicle enters the detection range of an intersection the algorithm iterates over and generates a trajectory for each AV (*4*). The trajectory provides optimal speed components for AVs, from the detection point to the stop bar at the intersection. The same trajectory is translated into a recommended speed for CVs which is displayed through an onboard interface to CV drivers. The algorithm also develops regularly updated real-time optimal signal timings, which are transmitted to the controller.

In the core of the trajectory optimization module is a recursive model to compute the movement of detected vehicles in each lane. Depending on the type of vehicle: AV/CV or conventional, and the role of vehicle: lead or follower, the relevant sub-model to determine the trajectory is used. The sub-model for a lead conventional vehicle assumes constant speed all the way to the stop bar, for a follower conventional vehicle Gipps car-following speed profile prediction is implemented. For a lead AV the model solves a non-convex constrained mathematical program to minimize its travel time, for a follower AV it minimizes headway-to-lead to achieve earliest departure time at the stop bar. Once the set of optimal trajectories are computed, an adaptive signal control strategy allocates green intervals to serve newly detected vehicles. Under the worst case scenario, as many vehicles can enter the detection range as the number of incoming lanes which, due to *O(1)* complexity of proposed trajectory optimization algorithm, does not cause any computational difficulty for the central computer at the intersection (for more details on the algorithm refer to [*2*]).

We developed the IICS algorithm as a real-time control procedure, and implemented it in MATLAB 2017b for a four-leg intersection with six incoming lanes using a heuristic algorithm which obtains optimal solutions within 0.5 second. The algorithm is run in VISSIM simulation environment for validation (results and discussion available in [*4*]). For the field testing at TERL, we ran the algorithm on an Ubuntu machine with 2.4 GHz Inter Xeon Haswell processor and 16 GB Crucial DDR Ram.

Hardware and Software

Under this section, we describe the development of equipment and interface needs to deploy the IICS at the TERL. Equipment needs include:

- Signal control system which is compatible with the control cabinet at the TERL, along with connectivity software from the IICS algorithm to the signal controller
- Vehicle communication equipment for CV and AV
  - DSRC receiver on local server for forwarding messages to the IICS algorithm
  - Radio communication software to communicate between CVs/AVs and roadside equipment (RSE)
  - Hardware for the CVs to obtain exact location and speed, and to interface with the driver
- Sensing devices to identify vehicles at the detection range when approaching the intersection in order to emulate the CV broadcasting message for conventional vehicles
  - Roadside radar system
  - Sensor fusion system to detect and track all traffic approaching the intersection

Information on each of these components is provided below.

Signal Control System

The team developed an interface to control the actuated signal controller (ASC) remotely from the Ubuntu machine. This interface receives the optimal signalization schema from the IICS algorithm and translates it into signal phasing and timing (SPaT) legible by the ASC. Once the SPaT is sent to ASC, the cabinet implements the commands and changes traffic lights accordingly. Through the NTCIP 1202 protocol application, we establish a simple network management protocol communication between the local server and the ASC to take control of object identifiers (OIDs) available in the ASC in order to control the SPaT.

The ASC used in the field demonstration is a NEMA TS2 type 2 Econolite Cobalt ASC donated to the University of Florida by Econolite. Each data element has an OID that corresponds to its physical location within the global naming tree. The interface and all the necessary libraries and OID dictionaries were implemented in Python 2.7 on the local Ubuntu machine and tested and verified at UFTI-Transportation Signal Lab before field implementation.

Dedicated Short-Range Communication (DSRC) Equipment for CV and AV

Every CV during the testing had a Cohda Mk5 radio in order to be able to transmit and receive DSRC messages. The local server also has a radio (RSE) mounted inside, which is able to receive the basic safety message (BSM) sent with 10 Hz frequency by the vehicles and forward it to the IICS algorithm. Therefore, the local server functions as the core system, support system, and also roadside equipment to provide system connectivity. Then intersection radio receives another BSM from the IICS algorithm and forwards it to a designated CV.

The former BSM features the following data elements: message ID number followed by the vehicle ID, GPS coordinates (latitude and longitude), and time that the message was sent with millisecond precision. Additional information in the message includes the vehicle's current speed, length, max acceleration and deceleration capabilities, and the approach lane along with the direction it intends to follow as the vehicle passes through the intersection.

The latter BSM (Intersection Approach Message) lists a set of trajectory points (in terms of 2D GPS coordinates) along with a time for each point for the AV to follow with a

Figure 2. Connected vehicle communication hardware.

frequency of 10 Hz. This system works well for low demand conditions, which is the case in this research. For high demand, researchers may consider other system architectures. The development and implementation discussed here was made in the C environment on the same local server as discussed earlier.

We embedded the OBU DSRC radio unit inside an onboard module to relay information to the RSE for interaction with the CV drivers (Figure 2).

Sensing Devices and Sensor Fusion System

This study used a Doppler-based radar sensor developed by Smartmicro Systems to detect and classify all approaching vehicles. Only AVs/CVs are able to communicate directly with the intersection through DSRC. While this provides the arrival information for these two classes of vehicles, the controller relies on the radar sensor to obtain the necessary information from conventional vehicles that are not equipped with a radio transmitter.

The radar sensor generates a BSM for each vehicle inside the detection zone (500 ft in this demonstration) including location (distance from the stop bar and the lane ID) and speed of the vehicle. The BSM data obtained from the radar sensor are then fused with the DSRC radio BSMs that the intersection receives from AVs/CVs. We developed an algorithm informed by the fused data to classify incoming traffic as AV, CV, or conventional vehicle.

The sensor fusion component of our system is responsible for identifying the conventional vehicles, which cannot communicate via DSRC, and to obtain accurate estimates of the position and speed of all incoming vehicles. Later in the paper, we present the results of the system performance, which fused DSRC and radar data to classify and track all vehicles. The development was made in Python version 2.7 and run on the same local server discussed in the previous section (the fusion process and results are also reported in [*16*]).

Automated Vehicle System

This unit is composed of six components which altogether form the AV system (diagram shown in Figure 3). In this project, we updated the first five existing components, and also developed a vehicle-to-infrastructure (V2I) communication and system interoperability module which enables the autonomous system (first five components in Figure 3) to communicate with the traffic control system and execute appropriate commands. All these components, as well as additional interfaces and UDP/TCP/IP local communication bridges, work together in an integrated system setting to create an automated and connected environment.

Vehicle. This system was developed on a hybrid Toyota Highlander (Figure 3) for participation in the DARPA Urban Challenge competition. Actuators were added to control the steering and shifting.

System Architecture. GPS and IMU sensors are used for vehicle localization as well as vision, and Lidar sensors are used for obstacle detection and terrain mapping. Several situation assessment and behavior specialist components

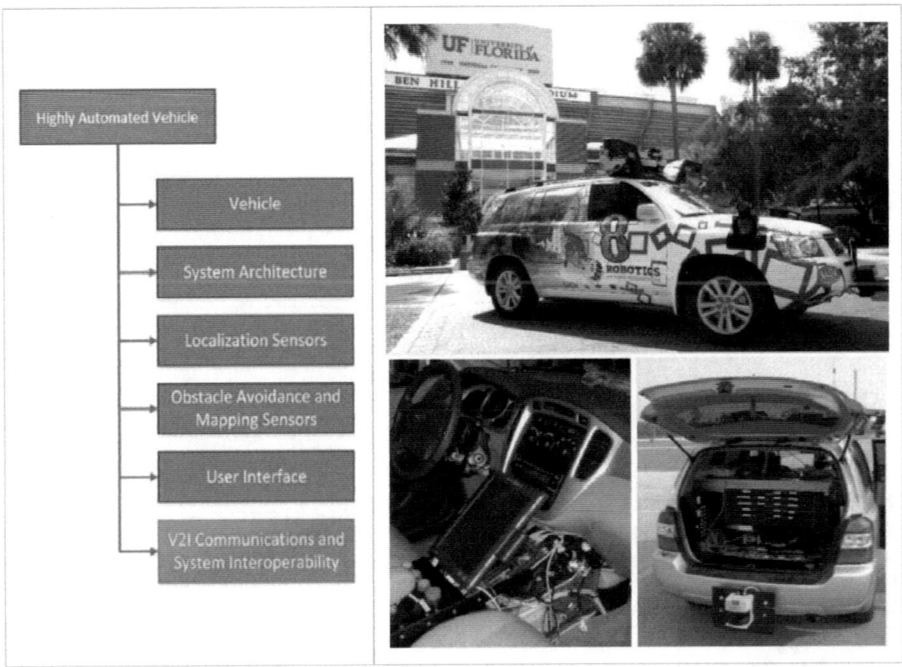

Figure 3. Automated vehicle (NaviGator) and its components.

are programmed to report findings from the surrounding environment and to control the vehicle's behavior.

Localization Sensors. For this project, a position and orientation accuracy of approximately 10 cm at high data rates is desired. Therefore, we install a GPS system, whose correction signals are received at no cost over the internet via FDOT's Florida Permanent Reference Network (FPRN). Centimeter-level accuracy was observed in the lab environment.

Obstacle Avoidance and Mapping Sensors. Six SickLMS 290 Lidar sensors are mounted to the vehicle. Two additional long-range Lidar sensors are mounted on the front quarter panels. These are augmented by two monocular vision sensors mounted at the top corners of the vehicle which map the location of lane markings.

User Interface. The touchscreen graphical user interface (GUI) allows the user to initialize the vehicle, so it can be driven manually or placed in autonomous mode.

V2I Connectivity and System Interoperability. Numerous hardware updates have been implemented to the AV, such as electrical interface, hardware interface for steering and shifting, and digital I/O among others, to provide the necessary platform for advanced communication capabilities. The V2I system in the vehicle is able to receive a trajectory sent from IICS algorithm at the roadside and use this trajectory to create the vehicle's path and desired speed to cross the signalized intersection.

Simulation and Optimization Algorithm Validation

We implemented the proposed optimization algorithm in MATLAB along with use of VISSIM COM interface. The simulation model was run over different scenarios to emulate the logic for a combination of arrival flows, AV ratios in traffic, detection ranges, and saturation headways. Table 1 represents the results for algorithm validation. An increase in AV penetration decreases average travel time delay. This is primarily because of ease of traffic management when all vehicles are automated. For all saturation headways under full AV penetration, the further up we detect the vehicles, the less delay we incur to the traffic at an intersection. However, for cases when conventional vehicles composed at least half of the traffic, the higher detection range may not benefit the system in terms of average travel time delay. We speculate that this behavior is due to error in the Gipps car-following model. Overall, higher traffic flow, higher saturation headway, and lower penetration rate of AV could lead to longer average travel time delay experienced by users.

Table 1. Average Travel Time Delay per 15-Minute Simulation Experiments of the Proposed Optimization Algorithm

		AV ratio/detection range					
Flow (veh/h)	Sat. HW[a] (sec)	0.3/150 m	0.3/900 m	0.5/150 m	0.5/900 m	1/150 m	1/900 m
60	1	3.9	4.6	3.9	2.6	3.8	1.0
60	1.5	4.0	4.7	4.0	2.7	3.8	1.0
60	2	4.0	4.4	4.0	2.9	3.8	1.8
100	1	5.5	7.8	5.3	5.5	5.1	1.5
100	1.5	5.6	8.0	5.4	5.7	5.1	1.5
100	2	6.1	8.5	5.5	6.2	5.3	1.8
450	1	16.8	26.0	14.3	18.5	11.4	2.0
450	1.5	51.2	65.8	28.8	36.0	14.5	4.5
450	2	156.8	175.2	130.7	126.4	31.3	22.6

[a]Saturation headway at the stop bar per lane.

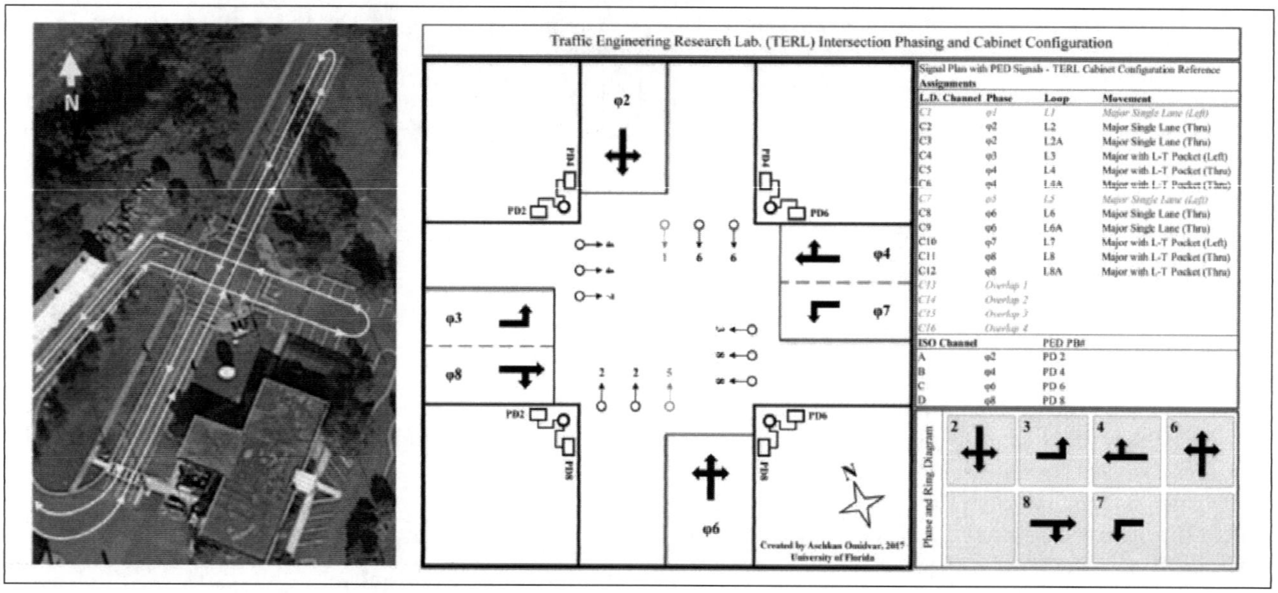

Figure 4. TERL intersection, testing route, and phasing.

Deployment and Field Testing at the TERL

Figure 4 shows an aerial photograph of the TERL and the test signalized intersection. Yellow and all-red intervals per phase are 3.0 and 1.5 seconds, respectively. Testing focused on through movements under low demand condition. There was no lane changing once a vehicle had arrived at the detection zone. During the two days of testing, a total of six vehicles were utilized: four CVs (one an AV with SAE level four and V2I DSRC connectivity), and two conventional vehicles. Scenarios vary according to the types of vehicles used (conventional, CV, AV) and vehicle inter-arrival times (i.e., level of demand). Each testing scenario includes approximately 15 minutes of setup time and 15 minutes of a continuous testing run with vehicles traveling along a predetermined path. The maximum speed for all scenarios and vehicles is 15 mph. The scenarios start from as simple as two CVs, up to mixed traffic of six CV/AV/conventional vehicles.

During the TERL tests, CVs were equipped with an application with a user-friendly GUI that received simplified trajectory messages and displayed recommended speeds for each approach to the drivers.

To evaluate the tested scenarios, and the performance of the vehicles and the effectiveness of the tested scenarios, several measures are tracked and analyzed. Data were recorded and stored on-site after each scenario from each vehicle's GPS system, the traffic signal controller, the radar, the communications components, intersection cameras, and a drone. We synchronized the local server's clock using the network time protocol to ensure that the various system components running from the server are time-aligned. The GPS systems measure the actual trajectory of each vehicle. We assessed how well the AVs/CVs were able to follow

Table 2. Root Mean Square Error (RMSE) between the Trajectories Assigned to the Vehicles and the Trajectory They Actually Followed

	Sample size	Min RMSE (ft)	Max RMSE (ft)	Mean RMSE (ft)
AV/CV	23	4.333 ± 3.507	64.856 ± 42.614	33.130 ± 18.587
Conventional	12	0.061 ± 0	75.499 ± 53.113	32.054 ± 20.923

the recommended trajectories. In addition, we evaluated the performance of our car-following model within the IICS by comparing the estimated trajectories with the actual trajectories based on data obtained from a traffic radar. Therefore, the performance measures for operations were:

- Mean travel time (vehicle information from GPS, DSRC communication hardware, and intersection cameras)
- Mean travel delay (vehicle information from GPS, DSRC communication hardware, and intersection cameras)
- Phasing and green duration (signal timing data from algorithm, intersection cameras)
- Vehicle trajectories (actual versus recommended trajectories by vehicle type)

Moreover, we collected data for each component separately and assessed the performance of each one. In detail, we analyzed run time, solution quality, recommended SPaT, and trajectories to evaluate the IICS, the communication latency and accuracy, the signal controller and cabinet along with signalization interface and setup, etc. Interested readers may refer to the project webpage at avian.essie.ufl.edu for additional information regarding equipment and for videos of the testing.

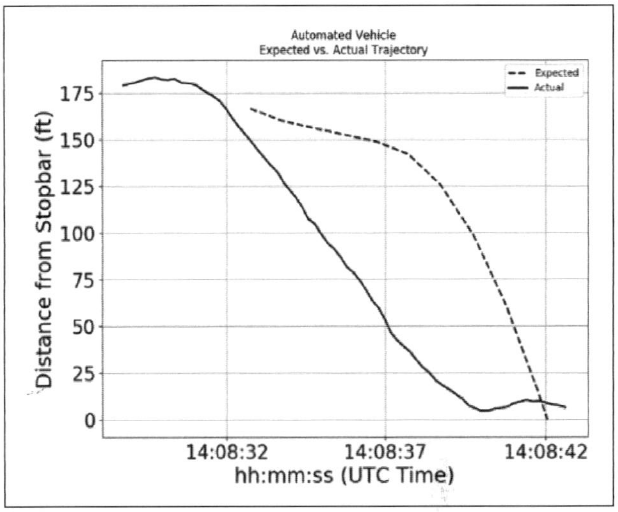

Figure 5. Example comparison of the expected trajectory versus the trajectory followed by an AV.

Results and Discussion

Two conventional vehicles and four CV/AVs looped around the facility using the configuration pattern depicted in Figure 4. Algorithm outputs captured run time, recommended trajectories, and recommended signal timings to evaluate the reliability and performance of the algorithm. Data on transmission times, errors in data transmission, and potential malfunctions of the communications components were collected to measure communication reliability. In addition, the performance of the radar was evaluated by comparing the actual arrivals to those detected by the radar. Delay in recognition and fusion of the data from the radar and the DSRC were also evaluated. Here we discuss some of the most notable results.

In order to assess how well a vehicle followed a recommended trajectory, we used the root mean square error (RMSE), which measures the mean absolute distance in feet between the vehicle's location at any given time and where it is supposed to be at that same time based on the recommended or estimated trajectory. All the vehicles were tracked from the GPS and speed data received from the DSRC and radar, and compared with the trajectories computed by the IICS. Table 2 shows the RMSE between the expected and actual trajectories of the vehicles averaged over a sample size of 23 trajectories. The minimum RMSE is the smallest observed at any point during the trajectory and the maximum RMSE is the largest observed. We provide uncertainty of one standard deviation with 95% confidence level.

The trajectories computed by the IICS demonstrated a much more nonlinear trend than the observed trajectories. We asked the drivers of the CVs to drive at a recommended speed and to attempt to follow the trajectory as closely as possible in a simplified setting. This approach was taken since it is challenging for the CV drivers to implement a nonlinear trajectory in real life. Furthermore, the RMSE could be reduced by improving the time synchronization of our real-time system.

With a clock synchronized with the DSRC communication hardware, the system is able to account for the 10 to 20 feet that a vehicle will move due to RSE to DSRC communication latency and synchronization between different components and platforms of the system. This offset effect is clearly illustrated in Figure 5 at the beginning of the trajectory. The solid curve shows the actual trajectory; we observed that drivers had difficulty matching the acceleration of the computed trajectories. Once vehicles start approaching the stop bar, expected and actual trajectories move in the same direction and the curves descend with similar slopes but with

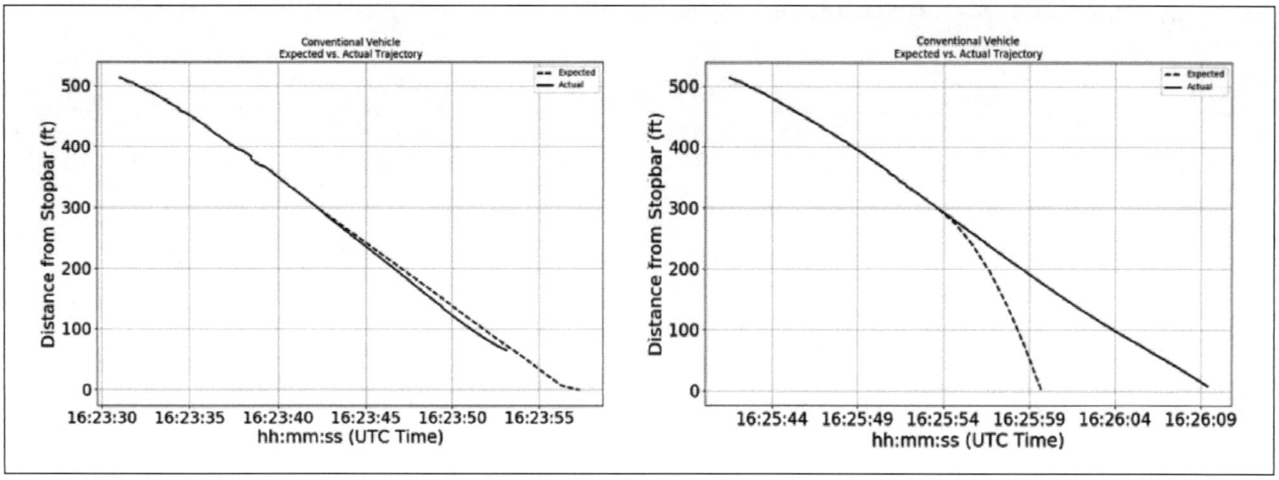

Figure 6. Comparisons of expected and actual trajectories for conventional vehicles.

a different time stamp. As is clear on Figure 5, the initial distance of a vehicle from the stop bar is 180 ft. This refers to the TERL configuration (approach 2 in Figure 4). As long as IICS has enough time to process and broadcast the BSM messages, the system is capable of handling the intersection. However, a typical recommended value in the literature (pilot deployment and simulation environments) is 500 ft. For further discussion please refer to (*17*).

In almost all instances during these tests, where a conventional vehicle approached the intersection, it was the lead vehicle in its lane. Depending on the current signalization, we observed that our algorithm estimated two different types of trajectories. If the lane already had green when the conventional vehicle arrived, the estimated trajectory would be linear. Otherwise, we estimated a nonlinear trajectory that would attempt to pass through the intersection at the start of the green time for the lane. We observed that the conventional vehicles maintained their speed consistently through the entire test (Figure 6). Table 1 shows the RMSE for the expected and actual trajectories, averaged over a sample of 12 conventional vehicle trajectories.

At the TERL the radar and DSRC fusion distinguished conventional vehicles from other types of vehicles with 100% classification accuracy. Details are provided in Figure 7, where the squares show the true and false positive rates. The color gradient on the right represents the number of vehicles used for each calculation.

We did not have access to ground-truth for evaluating the accuracy and precision of the system. Hence, we provide examples of how the DSRC and radar are fused to produce an accurate estimate of the true position and speed of the vehicle (Figure 8). Based on prior experimentation where ground-truth was available, we know that the radar produces more accurate estimates of the position; hence, the sensor fusion system "trusts" the data coming from the radar more, and the fused estimate weights the radar's estimate more

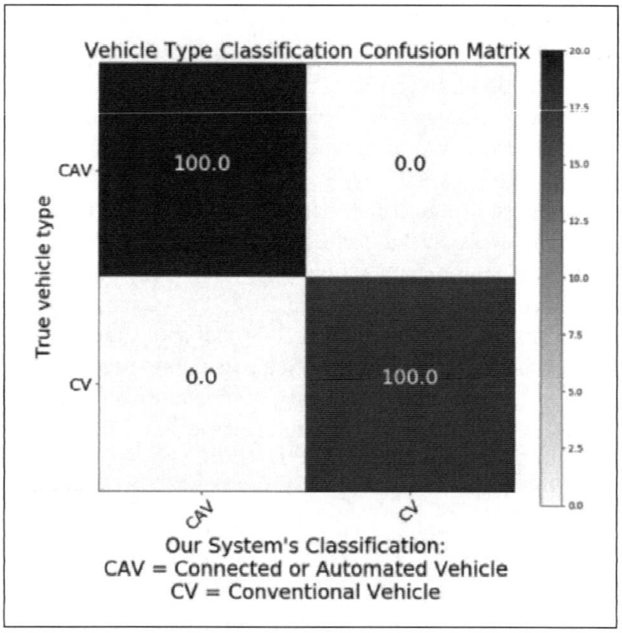

Figure 7. Vehicle type classification results for the sensor fusion system. Top left and bottom right show the true positive percentage.

heavily. However, for speed estimation, both radar and DSRC have high accuracy and precision, and therefore, the fusion system uses a balance of both sources. In total, the system was effective and capable of handling the mixed-traffic stream of vehicles.

Summary and Conclusions

This paper discusses the development and procurement of an intelligent real-time intersection traffic control system under mixed-traffic stream conditions for low demand to provide a

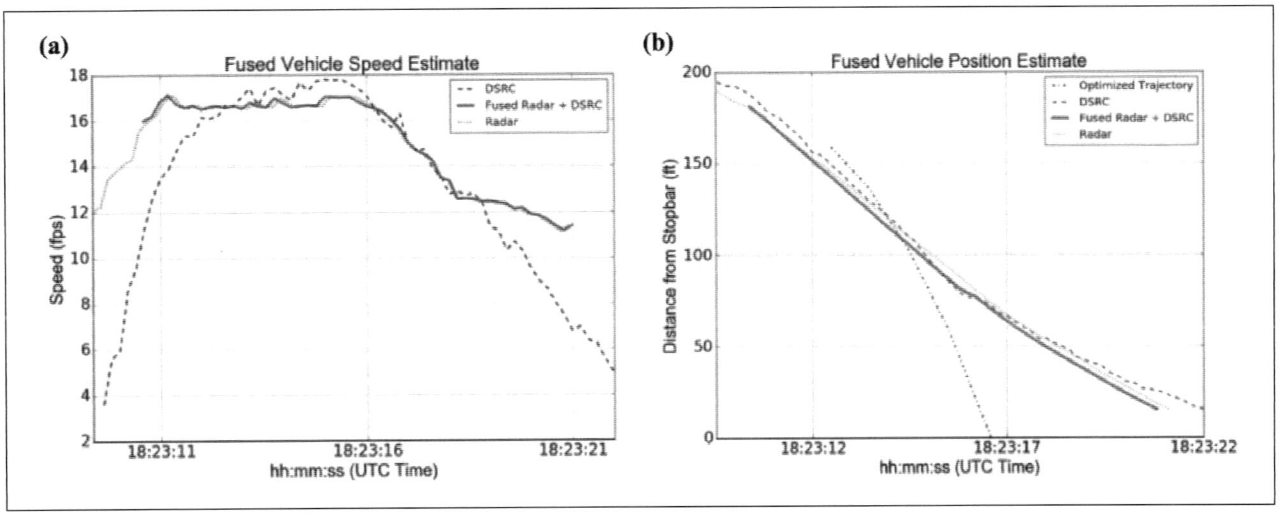

Figure 8. A trajectory showing the data fused position and speed estimate produced by the sensor fusion system. DSRC and radar are fused to produce an accurate estimate of the true position (*b*) and speed of the vehicle (*a*).

safe and efficient service to incoming vehicles at a given isolated intersection. We describe the development and procurement of the necessary software and hardware for field testing, namely: a local server, DSRC receiver on local server, interface to the signal controller and sensor fusion system on the roadside, and the radio communication software, the hardware for the AV/CVs to obtain exact location and speed and to interface with the driver, speed advisory applications, and AV system as onboard components.

The IICS was implemented at FDOT's TERL laboratory facility and tested under different scenarios over the course of a two-day experiment with six vehicles. The outputs and video footage from the test day showed promising direction for the next steps of the project. The system is capable of providing optimal trajectories instantly (usually <1 second) to CVs and AVs in order to reduce delay at isolated intersections due to unnecessary stops (visual results at avian.essie.ufl.edu). In future work, we will develop an improved driver-behavior model that allows more accurate representation of the behavior of lead conventional vehicles at traffic intersections. We are also extending the optimization algorithm to be able to handle high traffic volumes when the stream is composed of AVs, CVs, and conventional vehicles. Considering pedestrian and bicyclists in the model is another extension of the current system under consideration in the future.

Acknowledgments

This material is based upon work supported by NSF (Award 1446813), CPS: TTP Option: Synergy: Traffic Signal Control with Connected and Autonomous Vehicles in the Traffic Stream, and FDOT (BDV31 977-45, Development and Testing of Optimized Autonomous and Connected Vehicle Trajectories at Signalized Intersections). The authors are grateful to the officers and staff of TERL in Tallahassee, Florida, for facilitating the testing process. The authors also appreciate the assistance of the following individuals during the data collection: Marilo Martin Gasulla, Fabio Sasahara, Gustavo Riente, Pruthvi Manjunatha, Anuran Roychowdhury. We express our special gratitude to the City of Gainesville, Operations office, for their help in data collection and preliminary field testing in Gainesville, Florida. We also acknowledge our partners in industry, Econolite Group, Inc. and Image Sensing Systems, for their assistance and generous support.

Author Contributions

The authors confirm contribution to the paper as follows: study conception and design: Aschkan Omidvar, Mahmoud Pourmehrab, Patrick Emami, Rebecca Kiriazes, John C. Esposito, Clark Letter, Carl D. Crane III, Lily Elefteriadou, Sanjay Ranka; data collection: Aschkan Omidvar, Mahmoud Pourmehrab, Patrick Emami; analysis and interpretation of results: Aschkan Omidvar, Mahmoud Pourmehrab, Patrick Emami, Clark Letter, Lily Elefteriadou, Sanjay Ranka; draft manuscript preparation: Aschkan Omidvar. All authors reviewed the results and approved the final version of the manuscript.

References

1. Ahn, K., H. A. Rakha, K. Kang, and G. Vadakpat. Multimodal Intelligent Traffic Signal System Simulation Model Development and Assessment. *Transportation Research Record: Journal of the Transportation Research Board*, 2016. 2558: 92–102.
2. Zamanipour, M., K. L. Head, Y. Feng, and S. Khoshmagham. Efficient Priority Control Model for Multimodal Traffic Signals. *Transportation Research Record: Journal of the Transportation Research Board*, 2016. 2557: 86–99.
3. National Transportation Operations Coalition (NTOC). *National Traffic Signal Report Card*. NTOC, Washington, D.C., 2012.
4. Pourmehrab, M., L. Elefteriadou, S. Ranka, and M. Martin-Gasulla. Optimizing Signalized Intersections Performance under Conventional and Automated Vehicles Traffic, *Transportation Research, Part C: Emerging Technologies*. 2017 (forthcoming)

5. Dresner, K., and P. Stone. A Multiagent Approach to Autonomous Intersection Management. *Journal of Artificial Intelligence Research*, Vol. 31, 2008, pp. 591–656.
6. Tachet, R., P. Santi, S. Sobolevsky, L. I. Reyes-Castro, E. Frazzoli, D. Helbing, and C. Ratti. Revisiting Street Intersections Using Slot-Based Systems. *PLoS One*, Vol. 11, No. 3, 2016, p. e0149607.
7. Zhong, G., J. Zhang, T. Yin, and J. Nie. A Cooperative Management Strategy Designed for Unsignalized Intersections Under a Connected Vehicle Environment. In COTA International Conference of Transportation, 2015, pp. 233–245.
8. Elhenawy, M., A. A. Elbery, A. A. Hassan, and H. A. Rakha. An Intersection Game-Theory-Based Traffic Control Algorithm in a Connected Vehicle Environment. *Proc., 2015 IEEE 18th International Conference on Intelligent Transportation Systems (ITSC)*, IEEE, 2015, pp. 343–347.
9. He, Q., K. L. Head, and J. Ding. PAMSCOD: Platoon-Based Arterial Multi-Modal Signal Control with Online Data. *Transportation Research Part C: Emerging Technologies*, Vol. 20, No. 1, 2012, pp. 164–184.
10Li, Z., L. Elefteriadou, and S. Ranka. Signal Control Optimization for Automated Vehicles at Isolated Signalized Intersections. *Transportation Research Part C: Emerging Technologies*, Vol. 49, 2014, pp. 1–18.
11. Feng, Y., K. L. Head, S. Khoshmagham, and M. Zamanipour. A Real-Time Adaptive Signal Control in a Connected Vehicle Environment. *Transportation Research Part C: Emerging Technologies*, Vol. 55, 2015, pp. 460–473.
12. Feng, Y., M. Zamanipour, K. L. Head, and S. Khoshmagham. Connected Vehicle-Based Adaptive Signal Control and Applications. *Transportation Research Record: Journal of the Transportation Research Board*, 2016. 2558: 11–19.
13. Goodall, N., B. Smith, and B. Park. Traffic Signal Control with Connected Vehicles. *Transportation Research Record: Journal of the Transportation Research Board*, 2013. 2381: 65–72.
14.Lee, J., and B. Park. Development and Evaluation of a Cooperative Vehicle Intersection Control Algorithm under the Connected Vehicles Environment. *IEEE Transactions on Intelligent Transportation Systems*, Vol. 13, No. 1, 2012, pp. 81–90.
15. Gipps, P. G. A Behavioural Car-Following Model for Computer Simulation. *Transportation Research Part B: Methodological*, Vol. 15, No. 2, 1981, pp. 105–111.
16. Emami, P., L. Elefteriadou, and S. Ranka. Tracking Vehicles Equipped with Dedicated Short-Range Communication at Traffic Intersections. *Proc., 7th ACM International Symposium on Design and Analysis of Intelligent Vehicular Networks and Applications—2017 (DIVANet 2017)*, Miami Beach, FL, 2017.
17.Elefteriadou, L., M. Pourmehrab, P. Emami, A. Omidvar, C. Letter, P. Neal, R. Kiriazis, S. Ranka, C. Crane, and S. Ridgeway. Development and Testing of Optimized Autonomous and Connected Vehicle Trajectories at Signalized Intersections, 2017. http://www.fdot.gov/research/Completed_Proj/Summary_TE/FDOT-BDV31-977-45-rpt.pdf

The Standing Committee on Intelligent Transportation Systems (AHB15) peer-reviewed this paper (18-05710).

Article

TRR
JOURNAL OF THE TRANSPORTATION RESEARCH BOARD

Transportation Research Record
2018, Vol. 2672(19) 55–65

DOI: 10.1177/0361198118793257
journals.sagepub.com/home/trr
SAGE

Micro-Simulation of Truck Platooning with Cooperative Adaptive Cruise Control: Model Development and a Case Study

Hani Ramezani[1], Steven E. Shladover[1], Xiao-Yun Lu[1], and Osman D. Altan[2]

Abstract
Cooperative adaptive cruise control (CACC) systems have the potential to improve traffic flow and fuel efficiency, but these effects are challenging to estimate. This paper reports the development of a micro-simulation model to represent these impacts for heavy trucks using CACC when they share a freeway with manually driven passenger cars. The simulation incorporates automated truck-following models that have been derived from experimental data recorded on heavy trucks driven under CACC, adaptive cruise control (ACC), and conventional cruise control (CC). The simulation includes other behavioral models for lane changing, lane change cooperation and lane use restrictions for trucks to better capture real-world traffic dynamics. The paper explains the calibration of the simulation method for a 15-mile urban freeway corridor with heavy truck traffic and significant congestion. Simulation results for different market penetration rates show that truck CACC improved traffic operations for trucks in terms of vehicle miles traveled, average speed, and flow rate. In addition, truck CACC did not adversely affect passenger car operations and in some locations it even produced considerable improvements in the general traffic conditions.

Cooperative adaptive cruise control (CACC) augments the automatic vehicle-following capabilities of adaptive cruise control (ACC) with the addition of vehicle-to-vehicle (V2V) wireless communication of data about the movements of nearby vehicles. Since it only controls the longitudinal motions of the vehicle, it is a form of Society of Automotive Engineers (SAE) Level 1 automation (*1*). Vehicles with CACC have access to speed and acceleration information about preceding vehicles that have CACC or V2V communication capabilities; as a result, CACC vehicles can react faster to speed changes and can travel safely with shorter-than-normal following distances. This could improve traffic operations and/or fuel consumption. In particular, a shorter following distance could lead to a larger throughput, and consequently it could improve traffic operations (*2–4*). In addition, a shorter following distance could reduce aerodynamic drag for a truck, and it could thereby reduce fuel consumption (*5*). Building-block studies, which will be reviewed in the following, have been conducted to estimate the benefits of truck platooning in experimental conditions; however, there are a variety of real-world conditions that have not been replicated in experimental conditions. Examples of such real-world conditions are cut-in vehicles, traffic congestion, or any other traffic phenomena that cause speed variations. To capture the effects of these phenomena and to understand conditions under which large numbers of vehicles would be using CACC, micro-simulation models should be developed to determine impacts of truck CACC on traffic flow and fuel consumption.

Past studies of truck CACC and platooning fall into two categories: (1) experimental studies and (2) micro-simulation studies. Mainstream experimental studies have focused on measuring fuel efficiency of truck platooning. Maximum fuel reduction measured in these studies varied between 6.5% and 21%, with these variations mainly caused by varying minimum gaps examined in these experiments. For instance, in the European project PROMOTE-CHAUFFEUR, a fuel saving of 21%

[1]Partners for Advanced Transportation Technology (PATH), Institute of Transportation Studies, Richmond, CA
[2]Turner-Fairbank Highway Research Center, McLean, VA

Corresponding Author:
Address correspondence to Hani Ramezani: hani.ramezani@berkeley.edu

was reported for a follower traveling with a gap of 6 m (*6*). In another study, the Energy ITS program of Japan achieved up to 18% fuel reduction using a gap of 5 m (*7*). In the SARTRE project, a follower could gain a 16% fuel reduction at a gap of 5 m (*8*). Lammert et al. reported up to 9.7% fuel reduction when a gap of 6.097 m (20 ft) was tested (*9*). Roeth reported an energy saving of 10% for a gap of 10.97 m (36 ft) (*10*). In an experiment conducted by PATH and Transport Canada, a maximum fuel consumption of 7.4% was measured for the first follower traveling with a gap of 17.4 m. This value was 11.0% for the second follower (*11*). The maximum fuel saving in a KTH study was reported to be 6.5% for a gap of 21 m (*12*).

In real traffic conditions, these fuel reductions may not be observed because truck speeds and gaps will vary because of interaction with other vehicles; as a result, characteristics such as platoon length, speed, and intra-platoon gaps may vary over the trip. This necessitates developing micro-simulation models to foresee the benefits of truck platooning, considering real-world traffic dynamics and to design management strategies that enhance truck operations.

There are limited simulation studies in this area. Müller simulated a 5 km hypothetical network with no ramps where a capacity increase of 5.5% was achieved using truck CACC (*13*). Bibeka et al. evaluated impacts of truck platooning on emission rate for a 26-mile freeway with no off-ramp and no on-ramp flows (*14*). It was concluded that the desired time gap value of 0.2 s resulted in larger emission reduction than 0.6 s. Praharaj et al. evaluated a 5 mile freeway section including ramps, and evaluated desired time gaps between 0.5 s and 1.25 s in increments of 0.25 s (*15*) . The study reported maximum gains for the shortest gap and the maximum penetration rate. Deng developed a simulation framework for truck platooning and applied the framework on a 3.5 km roadway with no ramps (*16*). Aggregated simulation results showed that truck platooning increased flow rate by at least 100 veh/lane at the expense of slight speed reductions. Van Arem et al. simulated a 6 km roadway with a lane drop and no ramps (*17*). The study concluded that a penetration rate of less than 40% would adversely affect traffic operations when there is a dedicated lane for CACC vehicles.

The objective of this study is to develop a micro-simulation model for truck CACC and platooning and to study their traffic impacts through a case study. Vehicle-following models have been developed for trucks doing CACC vehicle-following on a high-volume uncongested urban highway. The outcome of the vehicle-following study is presented in this paper. The study represents the trucks using the CACC time gap settings that were actually chosen by professional drivers using truck CACC during a human-factors experiment which was conducted over a roughly 3 h round-trip. Incorporating these developments into the simulation model, plus other features of the model, helps to more realistically simulate truck CACC operations.

The rest of this paper is organized as following: First, the mechanism of automatic vehicle-following is explained. In particular, the paper elaborates how to determine a leader and its followers as well as logic to determine an automated driving mode that could be either CACC, ACC, or conventional cruise control (CC). Thereafter, it presents driver behavior models for car following and lane changing. The simulation model is calibrated using field data and then used to study the impacts of truck CACC on traffic operations. Finally, conclusions and recommendations are drawn.

Mechanism of Automatic Vehicle-Following

The flowchart in Figure 1 demonstrates how to determine leaders and followers in a simulation based on two parameters of time gap and maximum length of CACC vehicle string. A vehicle is a follower if its time gap is shorter than 2.5 s and it is traveling in a string of $N_{max} = 5$ vehicles or fewer (including the leader); otherwise it is a leader.

This flowchart assigns a "regulation mode" to each truck. The concept of the regulation mode is then used to determine whether a vehicle travels in CACC, ACC, or CC. A vehicle may have either speed regulation mode or gap regulation mode. In speed regulation mode, a vehicle tries to reach a reference speed and its distance from the preceding vehicle is not a major factor determining acceleration rate. The reference speed could be the free-flow speed or a lower speed based on the set speed chosen by the driver. In gap regulation mode, a vehicle tries to maintain a desired gap between itself and the preceding vehicle. The major factors to determine the acceleration rate are gap and speed of the subject vehicle and speed of its preceding vehicle.

An automated truck travels in CACC when it is in the gap regulation mode and V2V communication is possible. It travels in ACC when it is in the gap regulation mode but V2V communication is not possible. Usually this happens when it follows a passenger car or a truck that lacks V2V communication capability. An automated truck travels in CC (which only has the speed regulation mode) when there is no forward vehicle within its sensor range.

The next section discusses driving behavior models including car-following models for CC, ACC, and CACC.

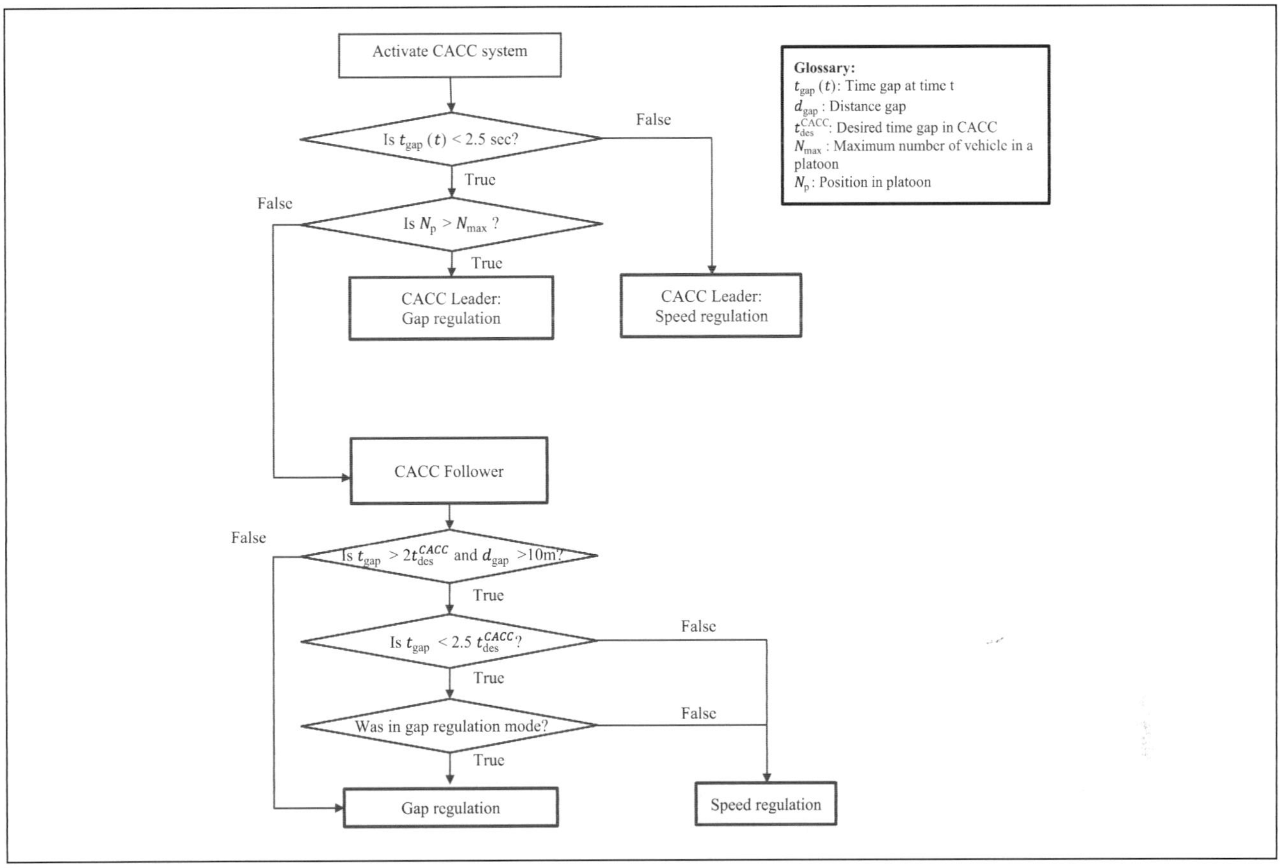

Figure 1. Simulation logic to determine leaders and followers and their regulation modes.

Car-Following Models

This section summarizes the car-following models that were used for trucks and passenger cars in the manual mode as well as for trucks in the automated modes of CACC, ACC, and CC.

Manual Mode Car-Following Models

Acceleration of a regular passenger car or a regular truck is determined as:

$$a_{\text{target}}(t) = \max(b_{\text{f}}, \text{Min}(a_{\text{F}}(t), a_{\text{N}}(t), a_{\text{G}}(t))) \qquad (1)$$

where,

b_f is maximum braking rate (m/s^2), which is used as a lower bound; $a_{\text{F}}(t)$ is acceleration rate (m/s^2) to reach the free-flow speed, which is used as an upper bound for acceleration; $a_{\text{N}}(t)$ is acceleration rate (m/s^2), which is computed based on the work of Newell (*18*);and $a_{\text{G}}(t)$ is Gipps' (*19*) deceleration component (m/s^2).

More details on these car-following models for the manual mode are reported in by Lu et al. and Liu et al. (*20*, *21*).

Automated Mode Car-Following Models

In the automated mode, target acceleration is computed as:

$$a_{\text{target}}(t) = \max(b_{\text{f}}, \text{Min}(a_F(t), a_{\text{Aut}}(t), a_{\text{G}}(t))) \qquad (2)$$

where, $a_{\text{Aut}}(t)$ is an acceleration rate for one of the automated modes of CC, ACC, or CACC. The car-following models, which will be presented in this section, have been developed using data collected from a string of three trucks using automatic vehicle-following on an uncongested urban highway with high traffic volume. More details of the model developments are given by Chou et al. (*22*).

The definitions of the other terms in Equation 2 are the same as those for the manual mode, but $a_{\text{G}}(t)$ in an automated mode is computed based on half of the human reaction time. This is because a CACC or an ACC vehicle is either capable of V2V communication or equipped with a collision avoidance warning system providing faster reaction times compared with regular vehicles.

In CC, a vehicle tries to reach or maintain a reference speed. A reference speed ($v_{\text{ref}}(t)$)is the driver's desired speed or a lower speed based on the speed of the leader.

Then, acceleration ($a_{Aut}(t)$) is determined based on the difference between the current speed and the reference speed. The following proportional control model has been proposed for CC:

$$a_{Aut}(t) = k_p(v_{ref}(t-1) - v(t-1)) \quad (3)$$

where $v(t-1)$ is the truck speed (m/s) at time $t-1$, and k_p is a coefficient equal to 0.3907 (s^{-1}), which has been determined using data collected from a truck traveling in CC mode on an urban highway.

The following linear model is proposed for ACC. The general form of the model is the same as the ACC model proposed by Milanés and Shladover for passenger cars (*23*), but its coefficient values are different, derived using truck ACC experimental data:

$$a_{Aut}(t) = k_1\left[d(t-1) - t_{des}^{ACC}v(t-1)\right] + k_2\left[v_{prec}(t-1) - v(t-1)\right] \quad (4)$$

where, $d(t-1)$ is the clearance gap in meters; t_{des}^{ACC} is the desired time gap in ACC mode; $v_{prec}(t-1)$is the speed of the preceding vehicle at time $t-1$; and k_1 and k_2 are equal to 0.0561 and 0.3393, respectively.

The CACC model has the same structure as that proposed by Milanés and Shladover for passenger cars (*23*), but the coefficient values are different, again derived from truck CACC experimental data:

$$a_{Aut}(t) = k_p e(t-1) + k_d \dot{e}(t-1) \quad (5)$$

where $e(t-1)$ is a measure to identify deviation from CACC desired time gap (t_{des}^{CACC}), computed using Equation 6:

$$e_k = d(t-1) - t_{des}^{CACC}v(t-1) \quad (6)$$

and $\dot{e}(t-1)$ is a derivative of $e(t-1)$, computed as in Equation 7:

$$\dot{e}(t-1) = v_{prec}(t-1) - v(t-1) - t_{des}^{CACC}a(t-1) \quad (7)$$

and k_p and k_d are equal to 0.0074 and 0.0805 for the first follower. For other followers (i.e., the second position or beyond), k_p and k_d values of 0.0038 and 0.0650 are used.

Switch from an Automated Mode to Manual Mode

An automated truck may switch to the manual mode if its Gipps deceleration rate is beyond a comfortable rate. A comfortable deceleration of 1.6 m/s^2 was considered as a threshold. This is 90% of 1.8 m/s^2, which is the mean value for maximum deceleration. In addition, a truck will travel in manual mode if its speed is reduced to 3 m/s^2 (10.8 km/h) or lower because the CACC system does not work at these lower speeds.

Desired Speed Computation

Mean desired speed is assumed to be $v_{SL} + v_{\delta,SL}$ mph, where speed limits v_{SL} for cars and trucks are 65 mph and 55 mph, respectively, and $v_{\delta,SL}$ is a calibration parameter. Desired speed may be adjusted for the friction effect using the following equations:

$$v_{friction}(t-1) = \min\{v_l(t-1), v_r(t-1)\} + v_{\delta,friction} \quad (8)$$

$$v_{desired}(t) = \begin{cases} \min(v_{friction}(t-1), V_{SL} + v_{\delta,SL}) & \text{if } v(t-1) > \underline{v}_{friction} \\ V_{SL} + v_{\delta,SL} & \text{Otherwise} \end{cases} \quad (9)$$

where,
$v_l(t-1), v_r(t-1)$ are the speeds of the left and right lanes; $v_{\delta,friction}$ is the speed difference between the subject lane and the adjacent lane (a calibration parameter); and $\underline{vfriction}$ is the speed threshold below which friction effect is not implemented (assumed to be 25 mph).

Lane-Changing and Cooperation Models

The lane-changing model is similar to the models presented by Lu et al. and Liu et al. (*20*, *21*). Additionally, this study considers a new type of mandatory lane change to make sure trucks are not traveling in the faster lanes. California restricts trucks to avoid using the left two lanes. In real traffic conditions, truck drivers may not restrict themselves to the lanes designated above. This may happen, for instance, when a truck passes a slower-moving vehicle. In the current version of the simulation model, lane restriction criteria have been strictly applied so trucks never use the faster lanes, even to pass a slow-moving vehicle.

Manual cars or trucks may reduce their speeds and create a longer gap to cooperate with vehicles merging from on-ramps. An automated truck may cooperate with merging vehicles if it is a CACC string leader, or it has previously disengaged the automated mode and is traveling in the manual mode. This implies that truck followers traveling in CACC mode do not deactivate the automated mode and do not disturb platoon stability to cooperate with merging vehicles.

Even with this setting, some passenger cars may cut-in between following trucks and interrupt a CACC string. A truck traveling behind a cut-in passenger car can travel in ACC mode, but if its deceleration rate is beyond the comfortable deceleration rate of 1.6 m/s^2 it will switch to the manual mode.

All the driving behavior models described above have been implemented in Aimsun using its Micro-SDK, and they were applied to conduct a case study for a freeway corridor with heavy truck traffic, as explained in the next section.

Case Study: I-710 Northbound

Effects of truck CACC on traffic operations is studied for an Interstate highway in Southern California (I-710 NB). This corridor connects the Port of Long Beach to the Interstate highway system and includes a large truck volume. The truck traffic percentage varies between 10% and 19%. In this study, only the northbound direction of this corridor is analyzed since trucks leaving the port are more likely to be heavily loaded than those traveling southbound, so this is the more severe condition.

This case study includes about 15 miles of the corridor modeled in Aimsun, starting from Milepost (MP) 1. There are 21 on-ramps and 20 off-ramps along this corridor. Between Mileposts 2 and 3, geometric characteristics are more restricted than other parts of the corridor. In particular, the mainline has three lanes and there are off-ramps and on-ramps that are closely spaced; under a high traffic volume and frequent lane-change maneuvers, congestion is likely to occur in this part of the corridor. After MP 3, the number of mainline lanes varies between four and six, and overall geometric design is less restricted; however there are major on-ramps and major off-ramps connecting I-710 to crossing Interstate highways. These major ramps can carry large traffic volumes and cause congestion on I-710.

The analysis period is 10–11 a.m. for a typical weekday, when congestion is expected to be less severe and truck volume is larger than in the peak hours. There are 11 mainline loop detectors which were healthy and did not malfunction. The loop data were downloaded from the Performance Measurment System (PeMS) and are plotted in Figure 2. Speeds at the MP2 and MP3

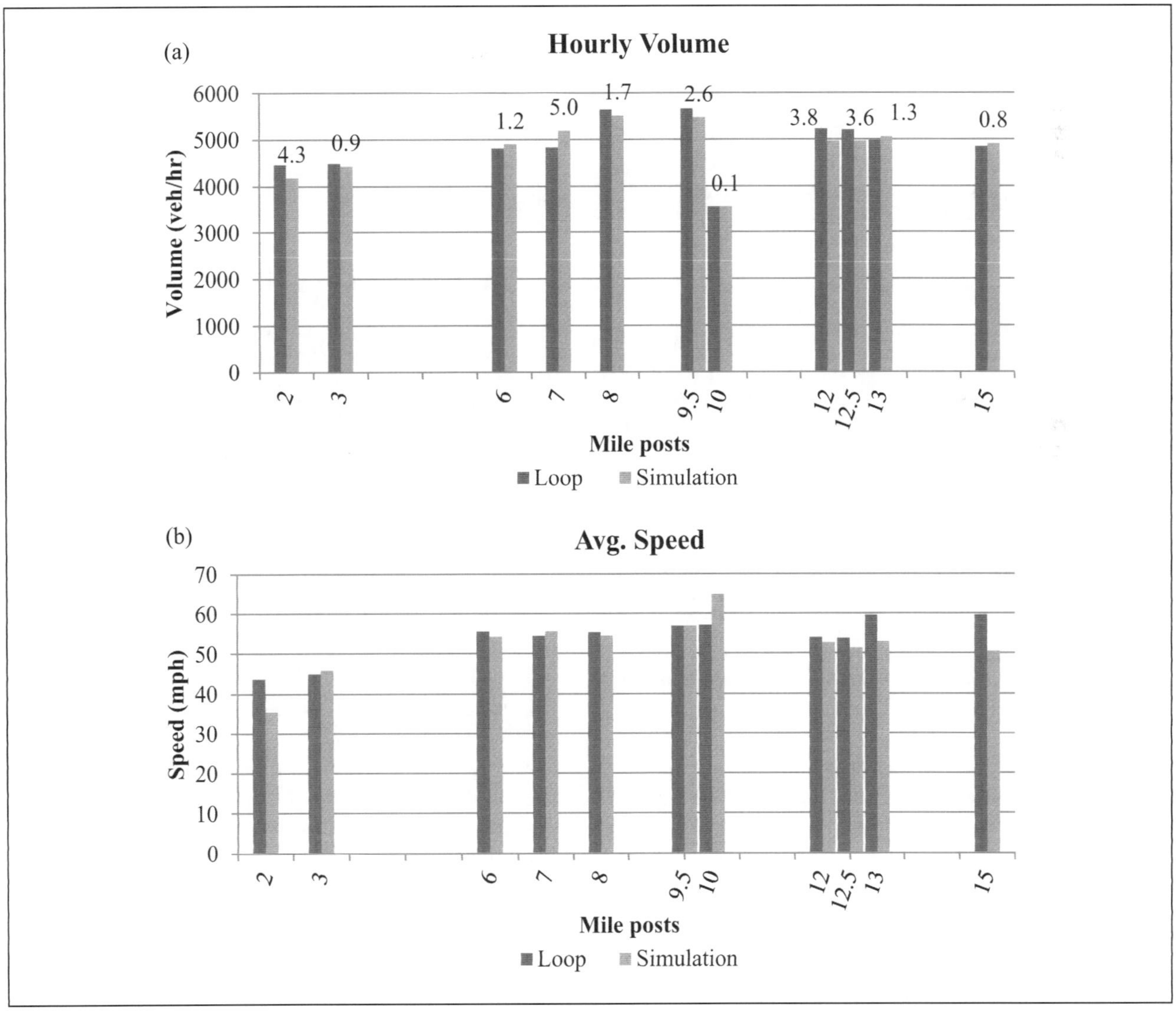

Figure 2. Comparison of simulation and field data at 11 detector stations for simulation using the calibrated parameter values.

Table 1. Maximum Acceleration Rates for Trucks (Retrieved from Pline [25])

Speed range (mph)	0–10	10–20	20–30	30–40	40–50	>50
Speed range (m/s)	0–4.4	4.4–8.9	8.9–13.3	13.3–17.78	17.78–22.22	>22.22
Max acceleration (m/s^2)	0.55	0.49	0.40	0.24	0.15	0.12

detectors are lower than those for the other detectors, which indicate the presence of congestion within MP2 and MP3, where geometric design is restricted.

Volume and speed information about this corridor is fairly limited. For the 15 mile corridor, there are only 11 healthy mainline detectors and these are not sufficient to accurately estimate ramp volumes. Thus, the initial O/D matrix was fine-tuned as part of the calibration process to replicate loop detector volumes. The parameter selection and calibration process will be explained in the next section.

Parameter Selection

Performance of trucks is governed by their limited acceleration/deceleration capabilities. The maximum acceleration rate depends on the weight-to-power ratio (w/p) and the speed of the truck. This study assumes w/p of 200 lb/hp to determine maximum acceleration rate. This w/p value was selected based on an NCHRP study that reported 200 lb/hp as the 90th percentile value of w/p distribution for California Interstate highways (*24*). Since trucks are mostly loaded in this corridor, the value of 200 lb/hp should be a good representative value. This study uses the corresponding acceleration values provided by the ITE handbook, as shown in Table 1 (*25*). The NCHRP report tabulates deceleration rates of trucks based on driver experience (*24*). Overall, deceleration rates are between 0.16 g and 0.26 g ($g = 9.81\,m/s^2$) for the worst and the best driver performance for empty tractor trailers on wet pavement. The average of these values is 0.21 g. We considered a smaller value of 0.18 g for maximum deceleration to take into account that the trucks are loaded, and as a result drivers may avoid using higher deceleration rates to have smoother speed changes.

The impacts of CACC on traffic operations depend on the time gaps that the drivers choose to use in the CACC system. Since truck CACC has not been mass produced, there are no field data to use to calibrate CACC desired time gaps. The desired time gaps assumed for this simulation study are based on the results of a very recent human-factors test of truck CACC conducted by PATH (*26*). A group of nine professional truck drivers traveled over a roughly 3 h route in a CACC string of three trucks. The driver of the leading truck was an experienced PATH driver who was familiar with the route and its traffic. The route was partially on an urban highway and partially on a rural highway. Drivers could select the gap using the driver–vehicle interface in trucks. Five time gaps between 0.9 s and 1.8 s in increments of 0.3 s were available for the drivers to choose. The test results showed that time gaps of 1.2 s or 1.5 s were the most favored time gaps for these drivers. Based on these results, it is assumed in the simulation model that 60% of drivers use the time gap of 1.2 s, and the rest use 1.5 s.

Another parameter is the desired time gap in ACC, which is expected to be longer than CACC time gaps since V2V communication is not available in ACC. An ACC desired gap of 1.8 s has been implemented for trucks in Japan (*7*). This seems to be the lower end of available ACC gaps. We decided to assume a slightly larger desired gap of 2 s for all trucks in ACC, which is in the middle range of commercially available ACC gap settings for heavy trucks.

Parameter Calibration

The Wisconsin DOT criteria, as presented by Dowling et al. (*27*), were applied to calibrate the simulation model. A goodness-of-fit measure, GEH, has been introduced to match traffic volume. GEH is defined in Equation 10, where L and S denote loop detector and simulated hourly volumes, respectively. In a good calibration, 85% of observations should have a GEH of 5 or less. The rest of the observations should not have a GEH of greater than 10:

$$\mathrm{GEH} = \sqrt{\frac{2(S-L)^2}{S+L}} \tag{10}$$

Figure 2 displays the loop detector volumes versus simulated traffic volumes with the corresponding GEH values on top of the bars. The simulated traffic volumes are computed based on 1 h of simulation and five replications with different random number seeds. These simulated volumes have been derived after fine-tuning O/D matrices and altering car-following parameters such as reaction time. All the GEH values are either 5 or less, satisfying the GEH criteria.

Wisconsin DOT criteria suggest that speed be visually matched "to analysts' satisfaction." We decided to qualitatively match speed data because these data come from

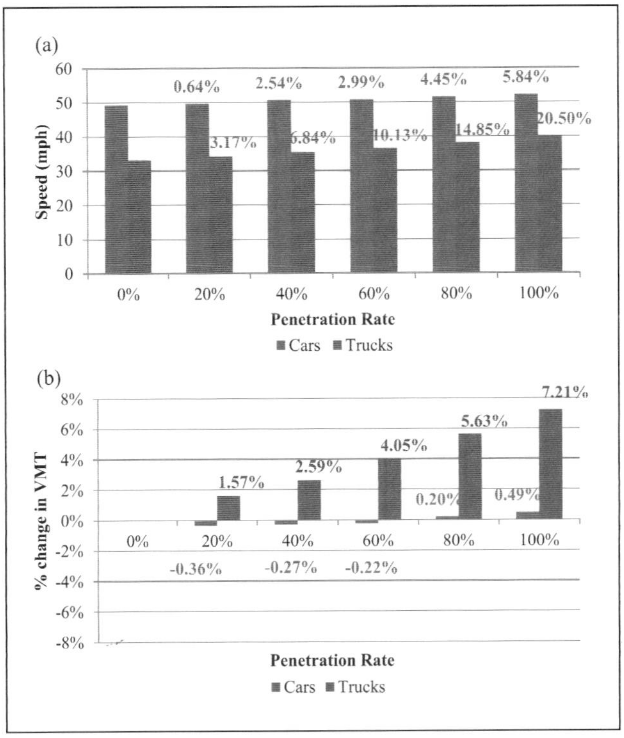

Figure 3. Effect of truck CACC market penetration rate on traffic operations for trucks and cars: (*a*) effect on speed; (*b*) effect on VMT.

single loop detectors for which the G-factor computations involve approximations because of varying length of vehicles; as a result, speed data are approximate values. Qualitative match would be sufficient for the purpose of this case study since its primary goal is comparing conditions with truck CACC versus manually driven trucks rather than designing specific alternatives for the study corridor.

Figure 2 demonstrates simulated versus observed speeds that have been averaged over 1 h. The average absolute error is equal to 4.3 mph, with a minimum of 0.3 mph and a maximum of 8.6 mph. For traffic conditions, the simulated and the loop data are in close or acceptable agreement. MP2 and MP3 detectors demonstrate congested traffic conditions as both detectors show low speed. The rest of the detectors show higher speeds and free-flow conditions.

The calibrated values of the car-following parameters are shown in Table 2.

Results

The driver behavior models have been implemented in the Aimsun framework using their Micro-SDK, and simulations have been run for five replications with different random number seeds. Effects of the CACC market penetration rate will be studied on traffic operation over the corridor. More detailed analyses at different locations will be presented for market penetration rates of 0% and 100% to highlight the impacts of automated truck platooning on traffic operation.

Table 2. Calibrated Values of Car-Following Parameters

Parameter	Calibrated value
Reaction time (τ_r) in Gipps model	1.3 s
Desired time gap (τ) for manual trucks in Newell model	2.4 s
Desired time gap (τ) for manual cars in Newell model	1.25 s
Theta (θ) in Gipps model	$0.2*\tau_r = 0.26$ s
Max acceleration for cars (a_M)	2.5 m/s^2
Max deceleration for cars (b_f)	3 m/s^2
Minimum speed difference to consider friction effect ($v_{\delta, friction}$)	10 m/s
Difference between speed limit and average desired speed ($v_{\delta, SL}$)	3 m/s

Corridor Analyses

Figure 3 shows speed and vehicle miles traveled (VMT) changes for different market penetration rates. The percentage values are computed with respect to base and are shown on top of each bar. Speeds of trucks and cars increased by up to 20.5% and 5.8%, respectively. VMT for trucks increased by up to 7.2% and changed marginally for cars. There was a very slight VMT reduction at penetration rates of 60% or lower and there was a slight VMT increase for higher penetration rates. Although this trend was weak, it can be explained. Truck CACC could influence cars in two opposite ways: first, it can increase car speeds by reducing congestion, which happened at the beginning of the corridor. Second, it could reduce car speeds or impose some waiting time on cars merging from an on-ramp to mainline. This happens because trucks mainly use the rightmost lane and they do not create gaps for merging cars if they are already in platoon. The first effect was dominant at penetration rates of 80% or larger. The second effect could be observed at lower penetration rates.

Analyses at Individual Loop Detector Locations

Figure 4 displays flow rate and speed of trucks with and without truck CACC at the loop detectors used in the model calibration. All the detector stations show speed increases ranging from 2.3 mph to 17.6 mph. To provide better understanding of the character of the speed

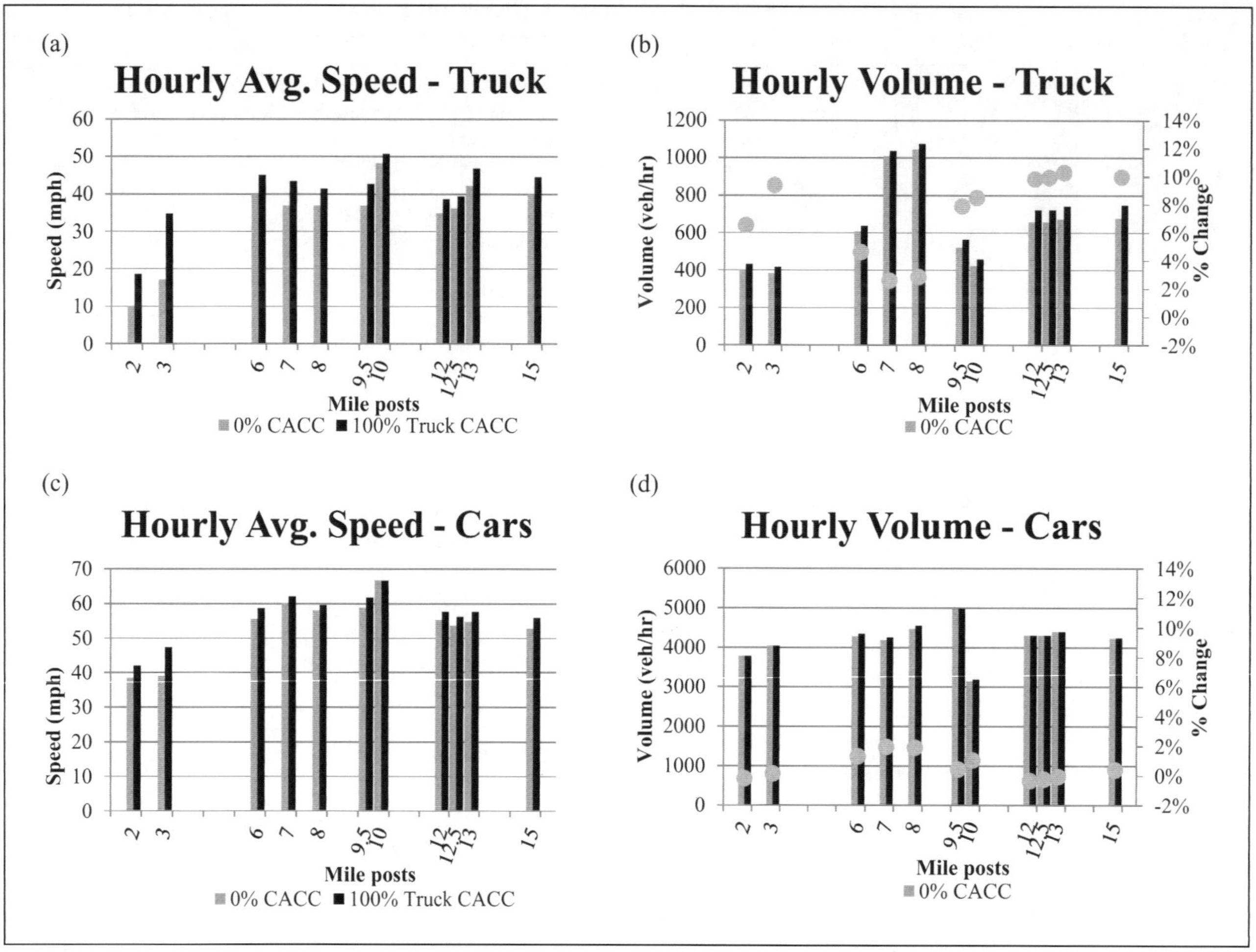

Figure 4. Effects of truck CACC at 11 detector stations for cars and trucks: (*a*) hourly average speed—trucks; (*b*) hourly volume—trucks; (*c*) hourly average speed—cars; (*d*) hourly volume—cars.

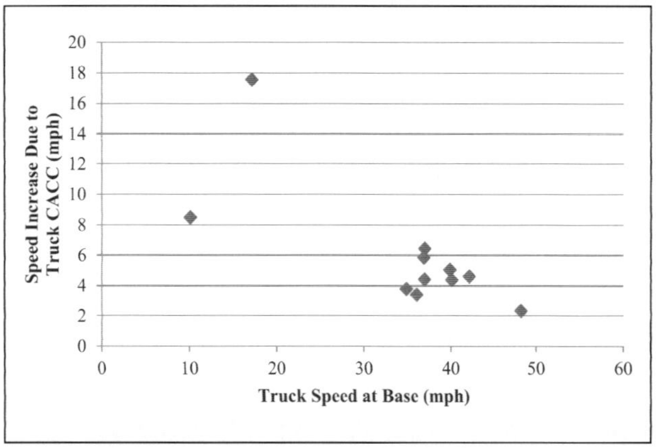

Figure 5. Speed increase caused by truck CACC versus base case speed plotted for 11 detector stations.

changes, Figure 5 plots speed-increase values versus speed values in the base case for the 11 detector stations. The plot shows that speed-increase values are noticeably larger for the two detectors that had average speeds below 30 mph and that had persistent flow breakdowns in the base condition. The speed-increase values are larger because truck CACC reduced congestion severity by postponing the onset of congestion, and this resulted in much larger speed increases.

Figure 4 illustrates that truck flow rates increased between 2.6% and 10.3%. This could be because of the speed increases discussed above as well as shorter following gaps in CACC.

Also, it could be interesting to see how truck CACC affects passenger car mobility. One of the significant effects of truck CACC is improving cars' mobility. Figure 3 suggests flow rate changes from −0.3% to 1.9% and average speed increases between 0.03 mph and 8.6 mph. Possible reasons for the slight negative flow rate changes could be the same as those presented for the corridor-level analysis.

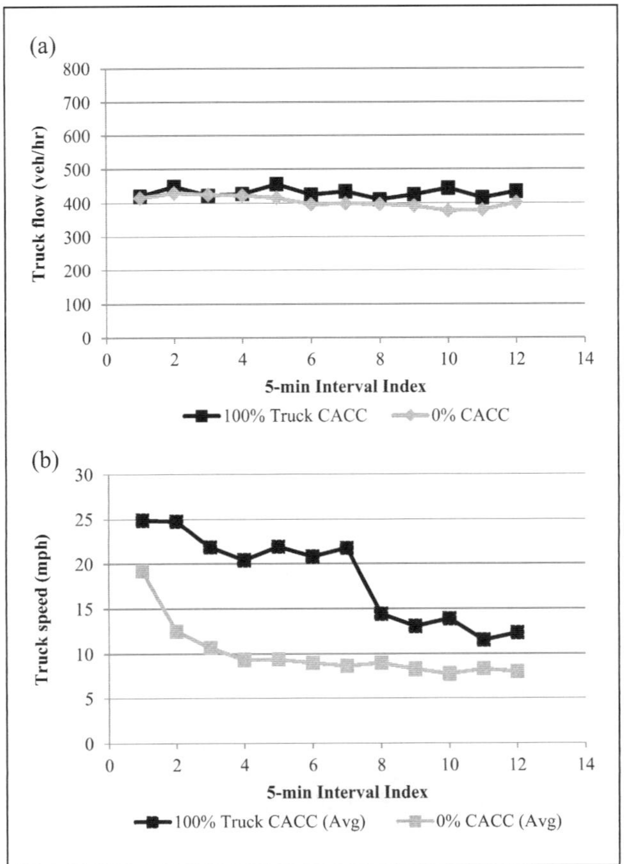

Figure 6. Speed and flow rate variations over time for the most congested loop detector: (*a*) effect on truck flow rates; (*b*) effect on truck speed.

Detailed Analyses for Most Congested Loop Detector Station

To provide more insight into how truck platooning influences congested locations, the MP2 detector was investigated because it had the lowest speed in the base condition, indicating a flow break-down condition. Two merge areas influence the traffic data collected by this loop detector. The first merge area is where the detector has been placed and the second is 270 m downstream of the detector location. The flow disturbance is mainly caused by vehicles that try to enter the rightmost lane from the on-ramps. The mainline has three lanes at these two locations, and trucks are restricted to using only the rightmost lane, but passenger cars may use this lane as well.

Figure 6 depicts simulation data, with each point representing a 5 min speed averaged over five replications. Speed data show persistent congestion at MP2 in the base condition, where average speeds varied between 7.7 mph and 19.2 mph. With truck CACC, the average speed was above 20 mph for the first 35 min. Thereafter, the queue from the downstream merging point propagated upstream and as a result the speed dropped and varied between 11.50 mph and 14.4 mph, showing higher speeds compared with base conditions. Moreover, the simulation results show that truck flow rates with CACC were consistently larger than those without CACC by an average of 26.4 truck/h.

Conclusions and Recommendations

This study has shown how traffic micro-simulation, based on well-calibrated models of human driver behavior and of automatic vehicle-following dynamics, can be used to represent the traffic impacts of widespread use of CACC on heavy trucks. The simulation method includes several innovative features derived directly from truck CACC experiments. It incorporates truck-following models that have been recently developed for the separate automated modes of CACC, ACC, and CC. The simulation logic includes switching from one automated mode to another and deactivation of automation to travel in manual mode. It also applies the results of a gap acceptance test recently conducted by PATH for truck drivers using CACC to select time gaps of 1.2 s and 1.5 s for use in the simulation. The simulation includes behavioral models such as models for lane-change maneuvers, lane-change cooperation, and lane use restriction for trucks; thus, it can realistically represent traffic dynamics such as cut-in and cut-out vehicles as well as lane-change cooperation by drivers of automated trucks.

A case study was conducted for a 15 mile urban corridor that experienced both congested and uncongested conditions during a 1 h simulation. Results of five simulation replications showed that 100% truck usage of CACC improved traffic operations for trucks and cars at different locations. In particular, truck CACC could reduce congestion propagation, and it could increase traffic speed at some uncongested locations. Corridor-level analysis for trucks demonstrated that average speed increased from 33.1 mph to 39.9 mph, and VMT increased by 7.2%. The VMT increase was due to the throughput increase in the congested part of the corridor. The results also showed that truck CACC did not adversely affect traffic operations for cars, and at some locations it even brought a significant speed increase. Corridor-level analysis for cars showed that their average speed increased from 49.3 mph to 52.2 mph. The effect on cars' VMT was marginal.

This simulation method can be used in future studies to further characterize impacts of truck CACC on traffic flow and fuel consumption and to explore management strategies that enhance the benefits of truck platooning. Thus, as future steps, the effects of market penetration rate and gap acceptance on traffic operations will be explored. The current simulation will also be extended to study impacts of truck CACC on fuel consumption.

Acknowledgments

This research was supported by the Federal Highway Administration's Exploratory Advanced Research Program under Cooperative Agreement No. DTFH61-13-R-00011, with cost sharing by the State of California Transportation Agency, Department of Transportation (Caltrans). The energy impact assessment aspects of this research were also supported by the U.S. Department of Energy's SMART Mobility Program, through the Lawrence Berkeley National Laboratory.

Author Contributions

The authors confirm contribution to the paper as follows: study conception and design: HR, SES, X-YL, ODA; data collection: HR, X-YL; analysis and interpretation of results: HR, SES, X-YL; draft manuscript preparation: HR, SES. All authors reviewed the results and approved the final version of the manuscript.

References

1. Nowakowski, C., S. E. Shladover, X.-Y. Lu, D. Thompson, and A. Kailas. *Cooperative Adaptive Cruise Control (CACC) for Truck Platooning: Operational Concept Alternatives*, PATH Project, 2015.
2. Arnaout, G. M., and J. P. Arnaout. Exploring the Effects of Cooperative Adaptive Cruise Control on Highway Traffic Flow Using Microscopic Traffic Simulation. *Transportation Planning and Technology*, Vol. 37, No. 2, 2014, pp. 186–199.
3. Kesting, A., M. Treiber, and D. Helbing. Enhanced Intelligent Driver Model to Access the Impact of Driving Strategies on Traffic Capacity. *Philosophical Transactions of the Royal Society of London A: Mathematical, Physical and Engineering Sciences*, Vol. 368, No. 1928, 2010, pp. 4585–4605.
4. VanderWerf, J., S. Shladover, M. Miller, and N. Kourjanskaia. Effects of Adaptive Cruise Control Systems on Highway Traffic Flow Capacity. *Transportation Research Record: Journal of the Transportation Research Board*, 2002. 1800: 78–84.
5. Tsugawa, S., S. Jeschke, and S. E. Shladover. A Review of Truck Platooning Projects for Energy Savings. *IEEE Transactions on Intelligent Vehicles*, Vol. 1, No. 1, 2016, pp. 68–77.
6. Bonnet, C., and H. Fritz. *Fuel Consumption Reduction in a Platoon: Experimental Results with Two Electronically Coupled Trucks at Close Spacing*. SAE Technical Paper No. 2000-01-3056. SAE International, Warrendale, PA, 2000.
7. Tsugawa, S.Energy ITS; What We Learned and What We Should Learn. In: *Transportation Research Board Workshop on Road Vehicle Automation*, Irvine, CA, July 24–26, 2012.
8. Chan, E. SARTRE Automated Platooning Vehicles. In *Towards Innovative Freight and Logistics* (Blanquart, C., U. Clausen, and B. Jacob, ed.), Wiley, New York, NY, 2014, pp. 137–150.
9. Lammert, M. P., A. Duran, J. Diez, K. Burton, and A. Nicholson. Effect of Platooning on Fuel Consumption of Class 8 Vehicles Over a Range of Speeds, Following Distances, and Mass. *SAE International Journal of Commercial Vehicles*, Vol. 7, 2014, pp. 626–639.
10. Roeth, M. *CR England Peloton Technology Platooning Test Nov. 2013*. North American Council for Freight Efficiency, 2013. http://nacfe.org/wp-content/uploads/2013/12/CREngland.pdf.
11. McAuliffe, B. R., C. Mark, A.-B. Mojtaba, and R. Arash, *Fuel-Economy Testing of a Three-Vehicle Truck Platooning System*. Report No. LTR-AL-2017-0008. http://nparc.cisti-icist.nrc-cnrc.gc.ca/eng/view/fulltext/?id=d21e1097-5d30-4a0f-b742-35ffad931c2f.
12. Alam, A., B. Besselink, V. Turri, J. Martensson, and K. H. Johansson. Heavy-Duty Vehicle Platooning for Sustainable Freight Transportation: A Cooperative Method to Enhance Safety and Efficiency. *IEEE Control Systems*, Vol. 35, No. 6, 2015, pp. 34–56.
13. Müller, S. The Impact of Electronic Coupled Heavy Trucks on Traffic Flow. *Proc., 40th European Transport Conference (ETC)*, Glasgow, Scotland, 2012.
14. Bibeka, A., P. Songchitruksa, and Y. Zhang. Assessing Environmental Impacts of Truck Platooning on Multilane Freeways. Presented at 96th Annual Meeting of the Transportation Research Board, Washington, D.C., 2017.
15. Praharaj, A., R. E. Turochy, and D. M. Bevly, *Modeling and Analysis of Freeway Traffic Operations with Deployment of Driver Assistive Truck Platooning Technology*, Presented at 96th Annual Meeting of the Transportation Research Board, Washington, D.C., 2017.
16. Deng, Q. A General Simulation Framework for Modeling and Analysis of Heavy-Duty Vehicle Platooning. *IEEE Transactions on Intelligent Transportation Systems*, Vol. 17, No. 11, 2016, pp. 3252–3262.
17. Van Arem, B., C. J. Van Driel, and R. Visser. The Impact of Cooperative Adaptive Cruise Control on Traffic-Flow Characteristics. *IEEE Transactions on Intelligent Transportation Systems*, Vol. 7, No. 4, 2006, pp. 429–436.
18. Newell, G. F. A Simplified Car-Following Theory: A Lower Order Model. *Transportation Research Part B: Methodological*, Vol. 36, No. 3, 2002, pp. 195–205.
19. Gipps, P. G. A Behavioural Car-Following Model for Computer Simulation. *Transportation Research Part B: Methodological*, Vol. 15, No. 2, 1981, pp. 105–111.
20. Lu, X. Y., X. D. Kan, S. E. Shladover, D. Wei, and R. A. Ferlis. An Enhanced Microscopic Traffic Simulation Model for Application to Connected Automated Vehicles. Presented at 96th Annual Meeting of the Transportation Research Board, Washington, D.C., 2017.
21. Liu, H., X. Kan, S. E. Shladover, X. Lu, and R. Ferlis. Impact of Cooperative Adaptive Cruise Control (CACC) on 2 Multilane Freeway Merge Capacity. *Transportation Research Record: Journal of the Transportation Research Board*, 2018.
22. Chou, F., H. Ramezani, X. Lu, and S. E. Shladover. Modeling Vehicle-Following Dynamics of Heavy Trucks Under

Automatic Speed Control Based on Experimental Data. *Transportation Research Record: Journal of the Transportation Research Board*, 2018.

23. Milanés, V., and S. E. Shladover. Modeling Cooperative and Autonomous Adaptive Cruise Control Dynamic Responses Using Experimental Data. *Transportation Research Part C: Emerging Technologies*, Vol. 48, 2014, pp. 285–300.
24. Harwood, D. W. *NCHRP Report 505: Review of Truck Characteristics as Factors in Roadway Design*. Transportation Research Board of the National Academies, Washington, D.C., 2003.
25. Pline, J. L. (ed.). *Traffic Engineering Handbook*, 5th ed. Institute of Transportation Engineers, Washington, D.C., 1999.
26. Yang, S., S. Shladover, X. Lu, H. Ramezani, and K. Aravind. Investigating Drivers' On-the-Road Experience of Using Cooperative Adaptive Cruise Control for Truck Platooning. *Transportation Research Record: Journal of the Transportation Research Board*, 2018.
27. Dowling, R., A. Skabardonis, and V. Alexiadis. *Traffic Analysis Toolbox Volume III: Guidelines for Applying Traffic Microsimulation Modeling Software*, No. FHWA-HRT-04-040. Washington, D.C., 2004.

The Standing Committee on Intelligent Transportation Systems (AHB15) peer-reviewed this paper (18-05698).

The contents of this paper reflect the views of the authors, who are responsible for the facts and accuracy of the data presented here. The contents do not necessarily reflect the official views or policies of the Federal Highway Administration or the State of California.

Article

TRR
JOURNAL OF THE TRANSPORTATION RESEARCH BOARD

Transportation Research Record
2018, Vol. 2672(19) 66–78

DOI: 10.1177/0361198118799012
journals.sagepub.com/home/trr
SAGE

Cybersecurity Attacks in Vehicle-to-Infrastructure Applications and Their Prevention

Mhafuzul Islam[1], Mashrur Chowdhury[1], Hongda Li[2], and Hongxin Hu[2]

Abstract

A connected vehicle (CV) environment is comprised of diverse computing infrastructure, data communication and dissemination, and data collection systems that are vulnerable to the same cyberattacks as all traditional computing environments. Cyberattacks can jeopardize the expected safety, mobility, energy, and environmental benefits from CV applications. As cyberattacks can lead to severe consequences such as traffic incidents, it has become one of the primary concerns in CV applications. In this paper, we evaluate the impact of cyberattacks on the vehicle-to-infrastructure (V2I) network from a V2I application point of view. Then, we develop a novel V2I cybersecurity architecture, named CVGuard, which can detect and prevent cyberattacks on the V2I applications. In designing CVGuard, key challenges, such as scalability, resiliency and future usability were considered. A case study using a distributed denial of service (DDoS) attack on a V2I application, "Stop Sign Gap Assist (SSGA)" application, shows that CVGuard was effective in mitigating the adverse safety effects created by a DDoS attack. In our case study, because of the DDoS attack, conflicts between the minor and major road vehicles occurred at an unsignalized intersection, which could have caused crashes. A reduction of conflicts between vehicles occurred because CVGuard was in operation. The reduction of conflicts was compared based on the number of conflicts before and after the implementation and operation of the CVGuard security platform. Analysis revealed that the strategies adopted by CVGuard were successful in reducing the conflicts by 60% where a DDoS attack compromised the SSGA application at an unsignalized intersection.

The driving force behind a country's economy is a surface transportation system that enables reliable and efficient transportation of passengers and goods. Despite remarkable improvements in vehicle design and performance, which has improved vehicle safety, the number of fatalities is still very high. Annually more than 30,000 fatalities occur on US highways alone (*1*), and in the European Union almost 25,700 road fatalities were reported (*2*). To reduce many roadway crashes and the associated societal costs, different countries have been promoting connectivity between vehicles, known as vehicle-to-vehicle communication or V2V, and between vehicles and transportation infrastructure components, which is vehicle-to-infrastructure communication or V2I. This type of Intelligent Transportation System has the potential to reduce roadway crashes significantly.

V2V and V2I enabled vehicles are emerging as next-generation vehicles for surface transportation systems, leveraging the rapidly growing information and communication technology. Connected vehicles (CVs) share information with other vehicles and with transportation infrastructure using wireless communication that increases traffic safety, provides efficient mobility services and reduces environmental impacts (*3*). However, the risk of cyberattacks increases as vehicles become more connected through the Internet, and wireless networks. One of the cyberattack gateways to CVs is V2I. Cyberattacks on V2I communication can have devastating consequences if V2I systems are not properly secured. V2I applications present a variety of vulnerabilities that create an attractive target for hackers. For example, hackers could take control of traffic signals, create hazards, and even cause a breakdown of the traffic

[1]Glenn Department of Civil Engineering, Clemson University, Clemson, SC
[2]Division of Computer Science, School of Computing, Clemson University, Clemson, SC

Corresponding Author:
Address correspondence to Mhafuzul Islam: mdmhafi@g.clemson.edu

system. Thus, it is critical to develop new security solutions to protect the V2I environment.

The Connected Vehicle Reference Implementation Architecture (CVRIA) (*4*), as developed by the US Department of Transportation (USDOT), has defined hundreds of CV applications to date. Among them, more than 20 applications are related to V2I. Many of these applications are related to V2I safety (e.g., curve speed warning, in-vehicle signage, and Stop Sign Gap Assist [SSGA]). According to CVRIA, all these applications share some common sub-applications (e.g., speed warnings, intersection safety), and use common processes (e.g., collecting roadside safety data and processing collected vehicle safety data) to support this application layer. For this reason, data flows can pose security risks when shared between these applications. Thus, if a cybersecurity risk is present in any of the shared sub-applications or processes, other applications might be at risk.

Much research has been conducted on enabling privacy, maintaining authentication, and providing integrity by targeting in-vehicle and V2V communication. However, very few studies focus on V2I-level security as a part of the in-vehicle and V2V security solutions (*5–7*). Therefore, research that is solely focused on V2I security solutions needs more attention. Primarily, in this paper, we evaluate the significance of V2I application security. Then, we present an innovative V2I security architecture, CVGuard, which can identify, prevent, and provide countermeasures against security threats and protect V2I applications from being compromised by cyberattacks.

The remainder of the paper is organized as follows: the next section provides a literature review followed by the CVGuard development method. The next section provides an overview of the CVGuard operation and benefits. This is followed by case study conducted by selecting one of the cyberattacks, a distributed denial of service (DDoS) attack on a V2I application, SSGA application. We then articulate the detection and prevention techniques adopted in the CVGuard for a DDoS attack. Moreover, based on the case study on the SSGA application, we also depict the capability and effectiveness of the CVGuard based on its ability to reduce conflicts caused by a DDoS attack. We conclude with a review of future work and limitations.

V2I Security-Related Studies

In this section, the primary goal is to summarize the cybersecurity aspects of the CV environment. Before developing a cybersecurity solution, it is necessary to determine cybersecurity requirements, different types of potential V2I attacks, and existing V2I security solutions. Also, a review of the emerging technologies is imperative while developing a cybersecurity solution for V2I applications.

V2I Security Requirements

In a CV environment, vehicles can communicate with internal (i.e., V2S [vehicle-to-sensor]) and external environments, such as, V2V and V2I, including roadside units (RSUs) that use dedicated short-range communication (DSRC) (*8*). An on-board unit (OBU) placed inside a vehicle transmits information to the surrounding environment. RSUs collect data from vehicles, and applications installed in an RSU deliver the requested service (*9*). According to CVRIA, all security solutions related to V2I and V2V must focus on three core elements: confidentiality, integrity, and availability (*10*).

Confidentiality. The content of the messages exchanged in a V2I environment must be kept confidential, that is, the content cannot be accessed by unauthorized and unintended users. However, most messages in a CV environment are public, particularly the exchanged messages between vehicles and infrastructure. Data confidentiality and message confidentiality need to be taken into consideration when designing a secure V2I environment (*11*).

Integrity. Messages exchanged between a vehicle and infrastructure must be protected from any unauthorized alteration or modification. This ensures the accuracy, reliability, and trustworthiness of the messages. Every security solution must provide protection from unauthorized intentional or unintentional modifications. Loss of integrity can result in the degradation of services provided by a V2I environment (*12*).

Availability. Availability ensures that systems and information are accessible and usable to authorized individuals. In a V2I environment, all RSUs must be available at all times. For example, the critical latency for an intersection collision warning must be less than 100 milliseconds (*13*). Also, all security solutions must provide an operational system even in the presence of faults or risky conditions.

V2I Cyberattacks

Classifications of malicious activities and security threats as well as attackers in the vehicular environment have been discussed in the existing literature. In (*14*) and (*15*), security attackers are classified as internal or external based on membership functionality. Activity levels determine whether the attackers are active or passive. Assessment of intention or reason for an attack has classified attackers as rational or malicious. However,

Table 1. Different Types of Cyberattacks in V2I Environment (*10, 11, 13–15*)

Attack type	Compromised security element			Effect of attack on vehicle-to-infrastructure environment	Likelihood
	Availability	Integrity	Confidentiality		
Distributed denial of service (DDoS)	HIGH	MODERATE	LOW	Unavailability of service, network collapse, defeating service integrity.	HIGH
Impersonation	LOW	MODERATE	HIGH	Disturbing the network, hiding identity, and gaining privileges are the primary motives of the attacker.	HIGH
Message alternation	MODERATE	HIGH	MODERATE	Affects integrity and confidentiality of the V2I environment, thereby impacting the safety service provided by RSUs. Responding to this type of attack depends on anti-theft solutions provided by vehicles or RSUs.	MODERATE
Malware and spam	MODERATE	LOW	HIGH	Malware can cause potential serious disruption in service for an RSU. The impact is considered high due to its long-lasting outages. Malware can be injected into the system during a software update of RSUs.	LOW
Eavesdropping	LOW	LOW	HIGH	Both active and passive attackers can steal sensitive and private information, which violates the confidentiality of drivers, and vehicles.	MODERATE

existing work focuses solely on unraveling the problems behind the V2V security attacks. This research is aimed at V2I cyberattacks in the context of emerging CV applications. Table 1 shows the cyberattacks that are likely to happen in a V2I interface based on the vulnerability of confidentiality, integrity, and availability with the projected likelihood of an attack ranked as HIGH, MODERATE, or LOW. This table provides an overall understanding of the effect of potential cyberattacks on a V2I interface (*15*).

Emerging Technologies

Currently, several new technologies have emerged to make communication networks more secure, more scalable, and capable of achieving more fine-grained control. In the following sections, we briefly discuss the use of these emerging technologies in terms of V2I security solutions.

Edge Computing. Edge computing is a method where the data is processed close to the data source. Edge computing offers dynamic content management, proper resource allocation, and low latency. It ensures high bandwidth by distributing the computational tasks to different edges (*16, 17*). In the context of V2I, security modules need to be near the source of data (e.g., RSU) for faster processing, which can be achieved by edge computing.

Software-Defined Networking (SDN). Traditional network management is complex, brittle, and error-prone (*18*). Those issues are caused by the strong coupling of the data plane (used for data forwarding) and control plane (used for routing logic, packet management, and access control). By separating the control plane and data plane, and by introducing programmability, SDN simplifies the network management tasks (e.g., reconfiguring network topology, changing routing logic) and provides more flexibility with respect to data distribution. In a CV environment, SDN can be used to dynamically reroute the data to different destinations as necessary. (*19*) shows an example of leveraging SDN to provide more flexible protection strategies and better resource management in the defense of cyberattacks while minimizing network congestion and user-perceived latency. We believe that SDN can aid in the construction of V2I threat protection systems.

Network Functions Virtualization (NFV). Network functions such as firewalls and intrusion detection systems (IDSs) are currently implemented in specialized hardware such as field-programmable gate arrays (FPGA) (*20*). Although specialized hardware gives better performance, it also leads to high cost, maintenance complexity, and inflexibility. NFV has emerged to overcome these problems (*21*). NFV reduces cost and provides new opportunities in scalability and implementation of network functions in commodity hardware having fine-grained control (*22*). NFV and SDN could play a major role in securing the CV environment in the near future, as leading industry players (e.g., Verizon, AT&T) are embracing these modern technologies (*19*, *23*).

This paper focuses on V2I network-level security and presents a newly developed distributed computational platform, CVGuard, which ensures confidentiality, integrity, and availability by leveraging emerging edge computing, SDN, and NFV technologies.

Development of CVGuard

To achieve our research objective, the first step was to define the components and functionalities of the CVGuard system and to create a formulation of the CV environment to detect and prevent networked-vehicle cyberattacks.

System Overview

The CVGuard is a new software-based security architecture designed to protect V2I applications, leveraging three emerging technologies: edge computing, SDN, and NFV. The primary goal of the CVGuard is to detect and isolate any cyberattacks in a V2I environment before they can negatively affect vehicles or transportation networks, which could lead to crashes and impede the adoption of CV technologies. Figure 1 shows a high-level vision of the CVGuard, which contains two major components. The first primary component consists of customized microBoxes (small network-security functions). These small software modules are situated in each fixed edge (e.g., RSU) and act as security gateways for each vehicle (mobile edge). They also provide dynamic attack capturing and analysis abilities for V2I interfaces. The second major component is the CVGuard controller, which is logically centralized and resides in the cloud as a regional edge where it monitors the contexts of different vehicles, RSUs as well as V2I communications. The CVGuard controller, being inside a cloud space as a regional edge, can virtualize all classes of network nodes (e.g., routers, switches), network functions and services (e.g., firewalls, IDS and intrusion prevention systems [IPS]), and network address translation (NAT) gateways. The CVGuard controller also identifies security threats, analyzes the threats, and controls microBoxes to eliminate the threats. RSUs represent the entry point for potential cyberattacks, and the microBoxes in each RSU do edge computing on incoming data to detect cyberattacks in the first layer of the CVGuard. Because of resource constraints, and to provide better service, security functions need to consume as few resources as possible (e.g., memory, processing time). The small software security functions in microBoxes ensure an efficient use of resources in an RSU by being created only when needed and erased when no longer necessary. Also, communication networking between RSUs is necessary to disseminate the attack information among neighboring RSUs. This communication is facilitated by an SDN to propagate the attack

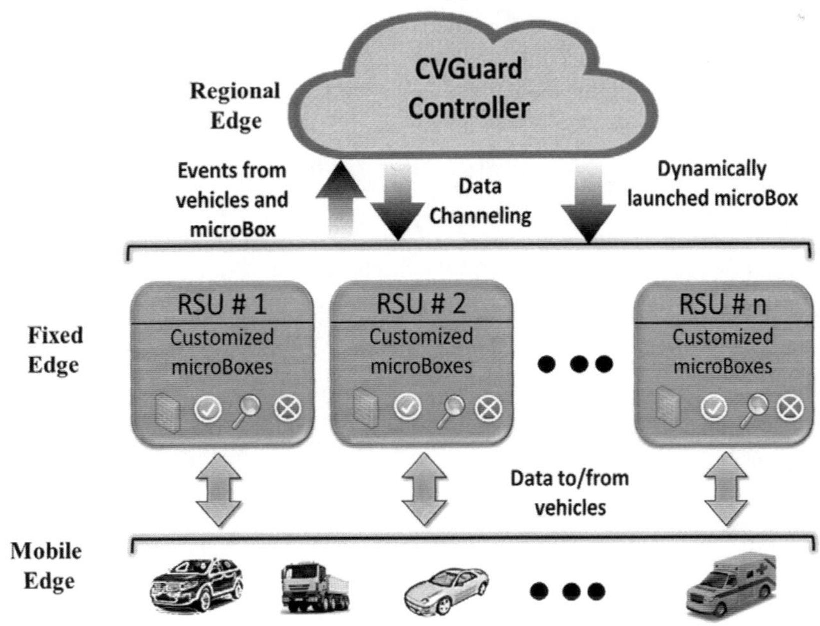

Figure 1. CVGuard: connected vehicle V2I security architecture.

information and prevention policies to and from the CVGuard controller and nearby RSUs. By leveraging NFV and SDN technologies, the CVGuard controller can create or erase security functions flexibly and quickly. The CVGuard controller can also dynamically route attack information and prevention policies to the relevant security functions. Furthermore, the CVGuard controller acts as a centralized security function controller to enable dynamic interoperation and automatic reconfiguration of software security functions, which are distributed in different RSUs.

The CVGuard supports two types of network-security functions: i) attack detection as provided by IDS (*24*) and DDoS detectors (*25*), as well as ii) attack prevention as provided by firewalls and the IPS (*24*). Once the attack detection functions discover a potential threat, the CVGuard identifies the problem and dynamically launches attack prevention functions to adopt the necessary resolution strategies. Both attack detection and prevention mechanisms in the CVGuard are discussed in the following subsections.

Attack Detection. The CVGuard attack detection is based on dynamic security policies and rules, which are developed based on a realistic CV environment where all vehicles are in motion and continuously broadcast the basic safety messages (BSM)—Part 1 at a rate of 10 Hz (10 packets/second) (*26*). Using wireless communication, each vehicle can communicate within a certain communication range. From the RSU range of influence, being a static component in the CV environment, vehicles come into the range of communication with the RSU and leave the range after a certain period. To quantify the CV environment, representation of the properties and functionalities of CV components in a proper and realistic manner is essential. For example, each vehicle can be represented as a moving vehicular node having properties, such as speed (S_i), and location ($L_i[x,y]$). Table 2 denotes the common properties of the components found in the CV environment.

Based on the properties of a vehicular node as listed in Table 2, we can formulate the rules needed in a CV environment. The following rules hold true representing the normal behavior of a vehicle without cyberattack:

a) Each RSU has a communication range based on the medium of communication (e.g., the recommended DSRC range is up to 300 m or 984 ft [*27*]). Thus, an RSU will only receive data from those vehicles, which are within the DSRC range. An anomalous location can be detected using the following rule:

$$|L_i(x,y) - L_{rsu}(x,y)| < R_{rsu} \quad (1)$$

Table 2. Properties of a Vehicular Node

Symbol	Description
$L_i(x,y)$	Location of the ith vehicle with a longitude of x and a latitude of y
$L_{rsu}(x,y)$	Location of the RSU with a longitude of x and a latitude of y
$S_i(x,y)$	Two-dimensional speed of the ith vehicle
S_i	Resultant speed of the ith vehicle
R_i	Communication range of the ith vehicle
R_{rsu}	Communication range of an RSU
$T_{i,start}$	Time when the ith vehicle comes into an RSU's communication range
$T_{i,end}$	Time when the ith vehicle leaves an RSU's communication range
$D_{i,j}$	Distance between the ith vehicle and the jth vehicle
$e_{i,j}$	Neighbor relationship between the ith vehicle and the jth vehicle
$\tilde{N}$	Possible maximum capacity of a neighbor's vehicle
N_{rsu}	Number of vehicles within the communication range of an RSU
DRR_i	Data receiving rate of the ith vehicle
S_{max}	Maximum speed of a roadway
S_{min}	Minimum speed of a roadway

b) The RSU contains the roadway geometry information (e.g., road location, map information) of its surrounding area. Also, the location of each vehicle will be located inside the roadway geometry:

$$L_i(x,y) \in Road\ geometry \quad (2)$$

c) A predictive approach to the location of vehicles can be computed. If $L_{current}(x,y)$denotes the current location of a vehicle, and $L_{prev}(x,y)$ denotes the previous location of the same vehicle. Then, the relationship between them can be represented as follows, where δ is a constant dependent on the traffic situation (e.g., traffic volume, queue):

$$|L_{current}(x,y)| \leq |L_{prev}(x,y) + \delta \quad (3)$$

d) Usually, each vehicle will maintain a minimum driving distance (ϵ) from other vehicles (Equation 4), and headway (h_i) from the vehicle immediately in front (Equation 5):

$$|L_i(x,y) - L_j(x,y)| > \epsilon| \quad (4)$$

$$|L_i(x,y) - L_{i+1}(x,y)| \approx h_i \quad (5)$$

e) Each vehicle will broadcast BSM at a certain rate, and RSUs will receive that data at a rate no higher than its dissemination rate, C1.

$$DRR_i \leq C1 \qquad (6)$$

f) RSUs will receive data from each vehicle. To provide a proper service by RSU, the data receiving rate (DRR) must be higher than the threshold value C2:

$$DRR_i \geq C2 \qquad (7)$$

g) Having a defined communication range, an RSU can accommodate a certain number of vehicles within its communication range:

$$N_{rsu} \leq \textit{Capacity of vehicles within the RSU range} \qquad (8)$$

h) Vehicles can communicate with a certain number of vehicles inside its communication range:

$$\tilde{N} < \textit{Capacity of neighbors of each vehicle} \qquad (9)$$

i) The network structure or topology can be determined by defining the neighborhood where two vehicles (V_i *and* V_j) are defined as neighbors if their physical distance is less than a defined range (μ):

$$V_i \textit{ and } V_j \textit{ are neighbors}(e_{i,j} = 1), \quad if |L_i(x, y) - L_j(x, y)| < \mu \qquad (10)$$

j) For a certain road section or corridor, the speed of each vehicle will be within a certain boundary:

$$S_{min} \leq S_i \leq S_{max} \qquad (11)$$

k) A vehicle (i^{th} *vehicle*) will come into the range of communication with an RSU at time $T_{i,start}$ and leave the communication range at $T_{i,end}$. Because of a defined communication range(R_{rsu}) for the RSU, the difference between $T_{i,end}$ and $T_{i,start}$ will be less than constant, τ:

$$T_{i,end} - T_{i,start} \leq \tau \qquad (12)$$

In the CVGuard, these rules reside inside the microBoxes that monitor the context of each vehicle. The incoming vehicles' data and a policy set consisting of behavior rules and violation rules are given as input into a microBox. Both the vehicles' data and policy set act as a small software security module. The set of policies are dynamically created by the CVGuard controller and fed to the microBoxes. Based on the set of rules and types of violations, a microBox can detect the attack types performed by different attackers. Figure 2 depicts the abstract idea of the detection method. In Figure 2, a cyberattack X type is identified if the vehicular data violates policy P, which consists of rules:{a, b}. Similarly, violation of a policy set Q that consists of a rule set {b, c, d} results in a Y type cyberattack.

Attack Prevention. Depending on the type of attack, the CVGuard can take countermeasures aimed at prevention. On the detection of an attack, a microBox at an RSU will steer the data from the attackers and legitimate vehicles into the prevention module as shown in Figure 2. Then the attacker's data is quarantined, and legitimate vehicles' data is passed to a CV application. For example, if a vehicle performs a fake location attack, the RSU can perform an estimation of the correct location of the attackers based on the data collected from other vehicles. Then this localized value can be fed to CV applications. Prevention policies can be changed dynamically based on the type of attack using the CVGuard controller, which makes the system more flexible, robust, and scalable. For example, for a DDoS attack, the violation set includes {e, f}, which creates a higher DRR for one vehicle, causing the DRR of the other vehicles to lower. In the prevention of DDoS, the CVGuard can drop or filter the data based on the application requirements to ensure a proper output from a CV application.

Case Study: SSGA Application

In this section, we conducted a case study from the vantage point of a V2I application (i.e., an SSGA application). We are going to define and develop a DDoS attack model and later use that model to evaluate the adverse effects on the SSGA application.

SSGA Application Overview

In our case study, we have selected a V2I safety application, an SSGA. The goal of the application is to improve the safety of an unsignalized intersection (having a posted stop sign) by providing safety-related warning and advisory messages to the incoming vehicles at the intersection. In an SSGA application, minor road vehicles are approaching the stop sign, and an RSU is providing safety messages to minor road vehicles based on the state of the intersection (e.g., available safe gap of the major road vehicles). At the intersection, an SSGA application alerts and warning messages indicate unsafe gaps based on the approaching major road vehicle's speed and distance from the intersection. As a result of real-time warning and alert messaging, the number of right-angle crashes will be reduced in the intersection (*27*, *28*). The performance of the SSGA application depends on the data availability from the vehicles traveling on the major and minor roads. Being a safety-related application, if compromised, the related misinformation can cause crashes at an unsignalized intersection. Because of the

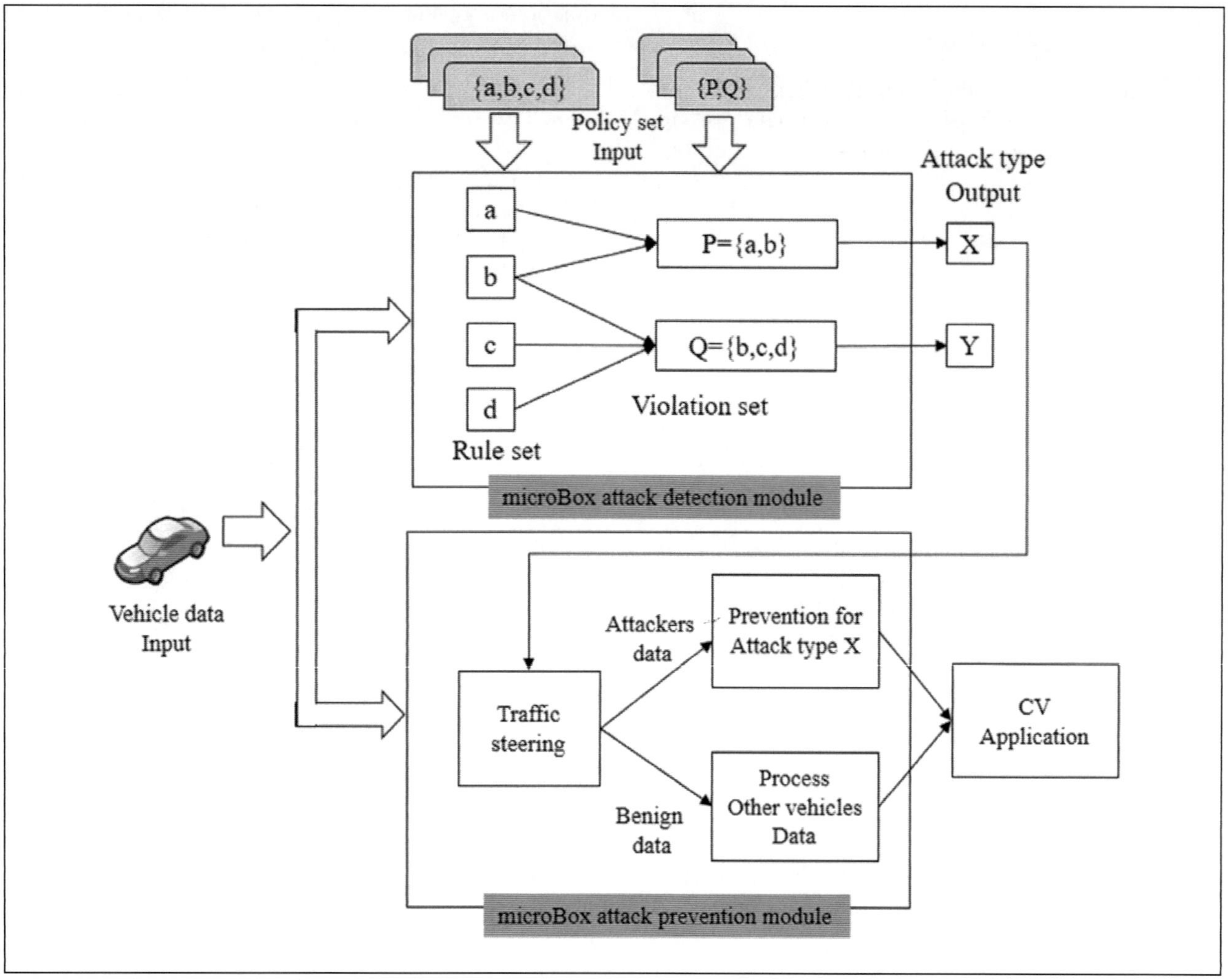

Figure 2. MicroBox workflow for both attack detection and prevention.

safety significance of the SSGA application, this application is suitable for our case study in determining the effectiveness of the CVGuard system.

Development of DDoS Attack Model

We focused on a general DDoS (e.g., SYN flood) attack against a CV application running in the roadside infrastructure. The attackers' aim is to exhaust the network and compromise the CV applications. The SYN flood attack is one kind of DDoS attack where a huge number of unnecessary data is transmitted by the attacker (*29*). By creating the SYN flood attack, the services from the CV application can be made unavailable, as this will exhaust the computation resource of the applications. We assume, the advisory model controls its own single or multiple OBUs and can control other vehicles' OBUs to create this attack. The attackers' goal is to compromise safety applications to cause a catastrophic situation. Because a safety application, for example, an SSGA application only takes BSM to compute the safety operations, we assume that the attacker will create SYN flood with BSM only. For simplicity, we restricted this study to focus on only one element of the roadside infrastructure, RSU.

In the development process of the DDoS attack model, a formulation of a DDoS attack was first developed to evaluate the feasibility of the attack. Second, from existing DSRC devices, the value of parameters (e.g., DSRC channel availability, packet size, transmission latency, and capacity) needed for creating the DDoS attack were extracted. Third, using these extracted parameters, in a simulation environment, the attack was launched against the SSGA application. Details of these steps are as follows.

Formulation of the DDoS Attack Model. Attacker vehicles launch the DDoS attack by flooding the communication channel. Typically, an attacker uses its maximum transmission capacity to flood the network. To create a breakdown of V2I application, attackers need to transmit more data than the receiver's (e.g., RSU) maximum receiving capacity that can be calculated as:

$$Attackers' \ transmission \ capacity \geq Receivers' \ receiving \ capacity \quad (13)$$

$$N_{attackers} * (Tx_a) \geq CRx_{max} \quad (14)$$

where $N_{attackers}$ is the number of attackers, Tx_a is the attacker's transmission rate, and CRx_{max} is the maximum message receiving capacity of an RSU. Once the attackers flood the network, the RSU will not be able to receive data from legitimate vehicles, and will be unable to provide proper services (e.g., gap alert and warnings) to legitimate vehicles. The ability to perform a successful attack depends on the attackers' transmission rate ($\tilde{x}$), the receivers' receiving capacity (e.g., RSU) (x), average packet size (y), transmission overhead ($\tilde{o}$), and percentage of service channel (SCH) availability (ρ). By using these dependent parameters as shown in Equation 14, we can then use Equation 15 to estimate the number of attackers ($N_{attackers}$) needed to create a DDoS attack on a V2I application:

$$N_{attackers} \geq \rho \left(\frac{1}{x} \tilde{x} + \tilde{o} \times \frac{10^6}{8y} \times x \right) \quad (15)$$

Parameter Extraction from the Real Devices. The next step is to verify the feasibility of creating a DDoS attack when the attackers are using real devices based on Equation 15. This feasibility analysis was conducted on commercially available RSUs. The goal was to find the attacker's capacity (e.g., minimum overhead and latency of transmission) with different configurations; then, the extracted values were used to create DDoS attacks in the simulation. Our experiment shows that the minimum transmission overhead at approximately 2.456 milliseconds (ms) with 1 millisecond from the inter-packet gap (IPG) when the device is configured at 3 Mbps. Also, on the receiver side, for a single radio, a 46 ms window is available for an SCH in a 100 ms time window; thus, the available SCH is 46% of the time of an RSU (16). Using this latency and other standard configurations as shown below, Equation 15 gives the minimum number of OBUs ($min(N_{attackers}) = 3$) that are considered as attackers to create a DDoS attack in our experiment, which consists of the following parameters:

- Average packet size, $y = 220$ bytes
- Attackers' transmission rate, $\tilde{x} = 3$ Mbps
- Receiver's receiving capacity, $x = 12$ Mbps
- Total transmission overhead, $\tilde{o} =$ IPG + Transmission overhead = (1 + 2.456) ms, So, $\tilde{o} = 3.45\,6$ ms = 0.003465 sec.
- SCH availability, $\rho = 46\%$
- Minimum number of attackers required, $N_{attackers} \geq 2.83$ or $min(N_{attackers}) = 3$

Development of a V2I Simulation Environment. Using the Network Simulator Version 3 (ns-3) (*30*), the simulation environment has been developed to create the CV environment. Vehicles are represented as wireless mobile nodes that use Wave Short Message Protocol (WSMP) to communicate with each other. A stationary wireless node acts as an RSU. The mobility of nodes was created from the "Simulation of Urban Mobility (SUMO)" (*31*) software to mimic real vehicular movement. In the simulation, Nakagami-m distribution was used as the propagation model, and we set the transmission power as 23 dBm, packet size as 250 bytes, and transmission range as approximately 300 m.

CV-Application Level Impact under DDoS Attack

Using the latency parameters from real devices as described above, different attack scenarios were created to evaluate the impact of DDoS attacks on an SSGA application in the simulation. With a different number of attackers, the performance of an RSU was evaluated based on the DRR by the application. As the attackers were flooding the RSU, a significant amount of data from legitimate vehicles was not received by the RSU. As shown in Figure 3, with the increasing number of attackers, DRR by the application decreased due to the DDoS attack. With five attackers, the DRR dropped to 10%, which is the lowest. Because of the data loss from other legitimate vehicles, the CV safety application will be unable to determine the location of legitimate vehicles. This will cause safety application failure to produce accurate output. For example, if the SSGA application produces incorrect output, the conflict between vehicles might have a negative consequence and conflict can potentially lead to a crash. In our experiments, a conflict was recorded as occurring when two vehicles are in close proximity with a time headway of less than 1.5 s (*32, 33*).

Attack Detection by CVGuard

As discussed above, the CVGuard monitors the context of vehicles. When rules defined for those vehicles are violated, a malicious activity occurs. The CVGuard can detect DDoS attacks and identify the attacker in the CV environment by using the detection policies (e.g., rule set {e, f}) on missing data, and the attacker's DRR. In the case of a DDoS attack, the DRR of the attacker will be

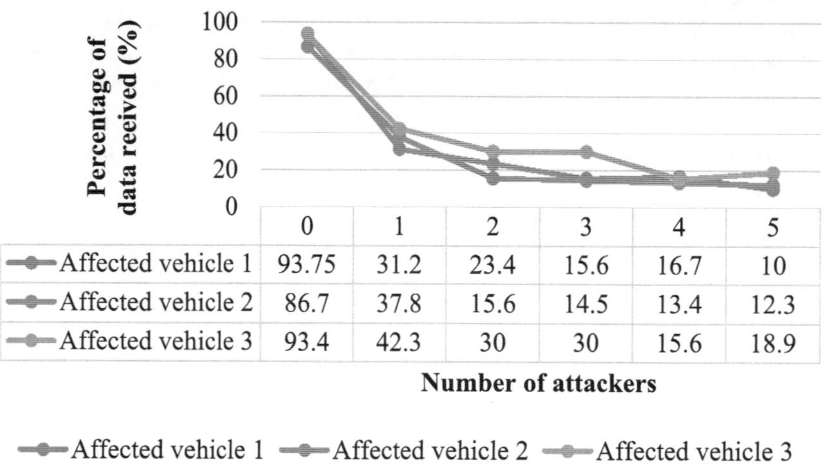

	0	1	2	3	4	5
Affected vehicle 1	93.75	31.2	23.4	15.6	16.7	10
Affected vehicle 2	86.7	37.8	15.6	14.5	13.4	12.3
Affected vehicle 3	93.4	42.3	30	30	15.6	18.9

Figure 3. Effect on data receiving rate with different numbers of DDoS attackers.

high, and the DRR of the other vehicles will decrease as shown. This will cause a violation of policy set {e, f}. According to the rule {e}, the RSU will get more data than expected from an attacker. In addition, because of flooding the network, the RSU will not receive the data from other vehicles that will cause the violation of rule {f}. Thus, it will get a lesser amount of data from other vehicles to run the application properly thereby making the application service unavailable. However, by monitoring the violation of the policy set:{e, f}, microBox will be able to detect the DDoS attack.

Attack Prevention by CVGuard

The CVGuard not only identifies the attackers but can also take countermeasures to mitigate the effect of attacks. In the CV environment, affected components cannot just block the attackers as typically happens in traditional DDoS prevention systems, since to perform safety operations, the location information of attackers is also necessary. Upon the detection of the DDoS attack, the CVGuard creates a microBox to prevent the DDoS attack. The purpose of this DDoS prevention microBox is to limit the data flow to the safety application. If we can limit the data flow to the application, the application will not be overloaded by unnecessary BSM messages from attackers. We can limit the boundary of message consumption of the application as follows:

$$\alpha \leq DRR_{application} \leq \frac{C}{Number\ of\ vehicles} \times \beta \tag{16}$$

where, $DRR_{application}$ is the DRR for operating the application properly, α is the minimum DRR required for operating the application and giving a correct output, C is the maximum capacity of the receivers and β represents the limiting factor for the upper bound data rate. The values of α and β depend on the requirements of the CV applications and the attackers' sending rate. For example, an SSGA application operating at 10 Hz. (*28*) means the receiver should receive data in every 100 ms which can also be referred to as sampling time. For the SSGA application, the requirement consists of the safe distance and the average velocity of the corridor. The determination of sampling time $T_{interval}$ is crucial in providing a better service to the other vehicles. The DDoS prevention module for the SSGA application will decide the value of sampling time. The selection of $T_{interval}$ can be mathematically expressed as:

$$T_{interval} \geq \frac{D_{safe}}{V_{avg}} \tag{17}$$

where $T_{interval}$ is the sampling time for collecting data from legitimate vehicles, D_{safe} is the minimum distance for safe operations that depends on the roadway conditions, and V_{avg} is the average vehicle speed on the subject roadway that is extracted from the data of other vehicles. We chose the minimum of 100 ms and $1/T_{interval}$ as α in Equation 16 to ensure that the SSGA application can operate properly. As shown in Figure 2, the prevention module within a microBox will first classify the data into to two categories, the benign data and the attackers' data. The benign data will be processed normally and delivered to the app. The attackers' data will be processed by the prevention component. In the case of a DDoS attack, the data will be sampled in $T_{interval}$ according to Equation 17. The primary actions taken by the CVGuard upon the detection of a malicious DDoS attack are described as follows.

(a) Action 1: According to the detection result, microBox launches the prevention module for a DDoS attack. The prevention module consists of three components: traffic classification, attack mitigation, and normal processing.
(b) Action 2: The attack mitigation component calculates $T_{interval}$ based on the attackers' condition and other vehicles conditions by considering α and β.
(c) Action 3: The mitigation component collects legitimate vehicle data for time $T_{interval}$ and attackers' data from $1/T_{interval}$ time interval; it then feeds these data to the CV application.

Evaluation of CVGuard under DDoS Attack

Performance evaluations of the CVGuard system under DDoS attacks were conducted using an SSGA application with and without the CVGuard to evaluate the effectiveness of the CVGuard system. In our experiment, three DDoS attackers (see above) compromised the SSGA application, which caused conflicts at the intersection because the SSGA application was unable to determine the location of both the major and minor road vehicles. In real-world traffic, the roadway traffic condition changes. Our simulations consisted of 50 major road vehicles and a variable number of minor road vehicles (varying from 5 to 45). Using the minor road vs. major road vehicle ratio, we were able to determine the effect of the roadway traffic condition (e.g., number of conflicts) during a DDoS attack. We set $T_{interval}$ at 100 ms as the minimum sampling time with $\beta = 1$ in our simulation. As shown in Figure 4*a*, using three attackers and each of them transmitting at 500 packets/sec, the average percentage of conflict was about 28.35% before the CVGuard intervention, and the average percentage of conflict was approximately 12.65% after the CVGuard deployment. In addition, the rate of conflicts depends on the attackers' message transmission capability (e.g., transmission rate). Figure 4*b* shows that if the attackers' transmission rate increases the percentage of conflicts in the intersection increases proportionally. However, after the CVGuard integration, the percentage of conflicts decreased providing more safety under the DDoS attack scenario with three attackers. For instance, we observed a 46.87% to 18.75% drop in the rate of conflicts in our experiment when each of the attacker's data transmission rate was 1000 packets/sec. Figure 4*c* shows the percentage-reduction of conflicts when the CVGuard was deployed under the DDoS attack. After the CVGuard integration, the percentage of conflict was reduced by 60% on average.

Conclusion

Because the CV environment is evolving rapidly, the need for cybersecurity protection is necessary for the near future. Because of cost and maintenance issues, hardware-based security solutions (e.g., hardware security modules) are not feasible. Software-based solutions are much cheaper and dynamically configurable. However, the challenge lies in selecting appropriate methods to protect CVs against cybersecurity threats. Emerging technologies, such as edge computing, SDN, and NFV, have great potential to make innovations for effective cybersecurity solutions in a CV environment. This paper presents a novel secure architecture called CVGuard to mitigate V2I cybersecurity threats by leveraging new emerging technologies. First, a case study was conducted encompassing a simulated DDoS attack to determine associated adverse impacts on an SSGA application due to the attack. The simulation analyses show that because of the DDoS attack, the DRR by V2I applications dropped to 10% during a DDoS attack compared with DDR of 93% during the normal operation (i.e., without the DDoS attack). Second, from a V2I application point of view, the drop in the receiving data rate caused the SSGA application to malfunction and created conflicts in an unsignalized intersection. Third, following the case study, the CVGuard proved its capability by detecting and preventing the DDoS attack and by mitigating its adverse effects. Our analysis shows that the strategies adopted by the CVGuard were successful in reducing the conflicts by 60% where a DDoS attack compromised the SSGA application at an unsignalized intersection. Currently, our study is limited to a DDoS attack. However, it is possible to develop detection and prevention mechanisms based on CVGuard in a flexible and programmable way to deal with other types of attacks listed in Table 1. As discussed, rule and policy sets can be used in different combinations corresponding to the detection and prevention of different cyberattacks. These combinations are defined by the CVGuard controller and can be adjusted dynamically depending on the types of attacks. Note that we have considered 100% CVs in our simulation, but in real-world traffic, a mixed traffic situation may exist with connected and non-connected vehicles in the traffic stream. Future study will include mitigation of all the most common types of cyberattacks mentioned in this paper, and the creation of different mixed traffic scenarios to evaluate the solutions by the CVGuard. We will develop more modules for the CVGuard system to make the system more robust, scalable, flexible, and elastic. Because the CVGuard is a software-based solution, it will be

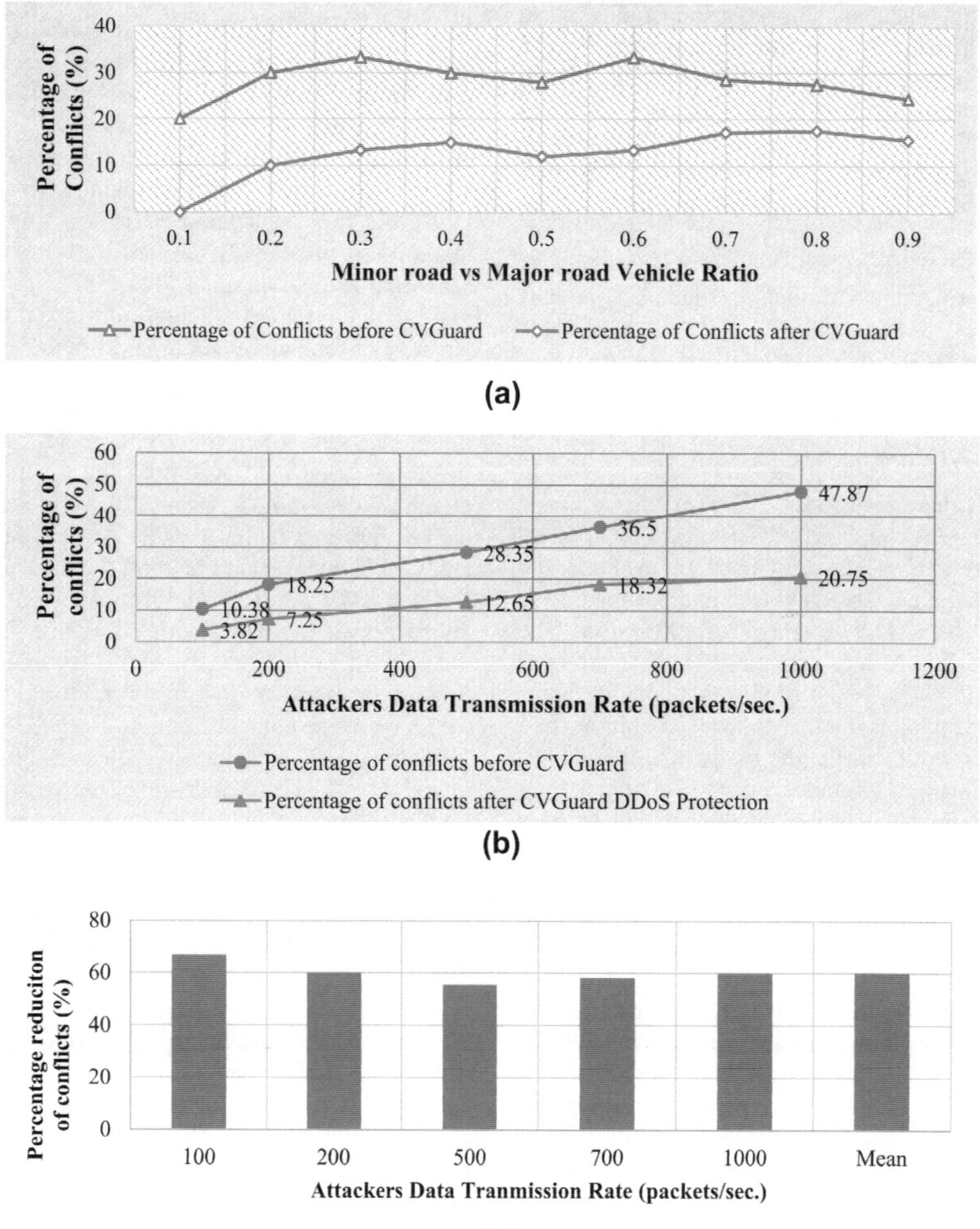

Figure 4. Improvement of conditions before and after the deployment of the CVGuard: (*a*) comparison of percentage of conflicts before and after the deployment of the CVGuard having $N_{attackers} = 3$ and attackers' data transmission rate= 500 packets/sec; (*b*) comparison of percentage of conflicts before and after the deployment of the CVGuard having $N_{attackers} = 3$; (*c*) intersection safety improvement by the CVGuard.

possible to develop different modules by other researchers and developers to contribute to the continuous development of the CVGuard platform. Also, more extensive experiments will be conducted in the near future in the real-world CV environment to evaluate the effectiveness and efficiency of the CVGuard system.

Acknowledgments

This paper is based upon research supported by the Southeastern Transportation Center (STC) under O/E Grant 2016–1017, and the Center for Connected Multimodal Mobility (C^2M^2, a USDOT Tier 1 University Transportation Center) grant.

Author Contributions

The authors confirm contribution to the paper as follows: Study conception and design: MC, MI; data collection: MI, HL; analysis and interpretation of results: MI, MC, HL, HH; draft manuscript preparation: MI, MC, HL, HH. All authors reviewed the results and approved the final version of the manuscript.

References

1. NHTSA. *Fatality Analysis Reporting System (FARS) Encyclopedia.* https://www-fars.nhtsa.dot.gov/Main/index.aspx. Accessed May 8, 2017.
2. European Commission. Road Safety in the European Union - Trends, Statistics and Main Challenges. Technical Report. European Commission, Brussels, Belgium, 2015.
3. Delgrossi, L., and T. Zhang. *Connected Vehicles and Cybersecurity.* U.S. Department of Transportation, Washington, D.C., 2012, pp. 32–43.
4. Connected Vehicle Reference Implementation Architecture. *Applications.* U.S. Department of Transportation, Washington, D.C. www.iteris.com/cvria/html/applications/applications.html. Accessed May 8, 2017.
5. Engoulou, R. G., M. Bellaïche, S. Pierre, and A. Quintero. VANET Security Surveys. *Computer Communications*, Vol. 44, 2014, pp. 1–13.
6. Xiao, B., and B. Yu. Detection and Localization of Sybil Nodes in VANETs. *Proc., DIWANS'06, Dependability Issues in Wireless Ad Hoc Networks and Sensor Networks*, Los Angeles, CA, 2006, pp. 1–8.
7. He, L., and W. T. Zhu. Mitigating DoS Attacks against Signature-Based Authentication in VANETs. *Proc., 2012 IEEE International Conference on Computer Science and Automation Engineering*, Zhangjiajie, China, 2012, pp. 261–265.
8. Li, Y. (Jeff). An Overview of the DSRC/WAVE Technology. Springer, Berlin, Heidelberg, Germany, 2012, pp. 9–21.
9. Rawat, D. B., C. Bajracharya, Y. Gongjun, G. Yan, Y. Gongjun, and G. Yan. Toward Intelligent Transportation Cyber-Physical Systems: Real-Time Computing and Communications Perspectives. *Proc., IEEE SoutheastCon 2015*, Fort Lauderdale, FL, IEEE, New York, NY, 2015, pp. 1–6.
10. Raw, R. S., M. Kumar, and N. Singh. Security Challenges, Issues and Their Solutions for VANET. *International Journal of Network Security & Its Applications*, Vol. 5, No. 5, 2013, pp. 95–105.
11. Mejri, M. N., J. Ben-Othman, and M. Hamdi. Survey on VANET Security Challenges and Possible Cryptographic Solutions. *Vehicular Communications*, Vol. 1, No. 2, 2014, pp. 53–66.
12. U.S. Government Accountability Office. *Intelligent Transportation Systems: Vehicle-to-Infrastructure Technologies Expected to Offer Benefits, but Deployment Challenges Exist.* Washington, D.C., 2015, p. 57.
13. Karagiannis, G., O. Altintas, E. Ekici, G. Heijenk, B. Jarupan, K. Lin, and T. Weil. Vehicular Networking: A Survey and Tutorial on Requirements, Architectures, Challenges, Standards and Solutions. *IEEE Communications Surveys and Tutorials*, Vol. 13, No. 4, 2011, pp. 584–616.
14. Mishra, R., A. Singh, and R. Kumar. VANET Security: Issues, Challenges and Solutions. *Proc., 2016 International Conference on Electrical, Electronics, and Optimization Techniques (ICEEOT)*, Chennai, India, IEEE, 2016.
15. Sumra, I. A., I. Ahmad, H. Hasbullah, and J. L. Bin Ab Manan. Classes of Attacks in VANET. *Proc., 2011 Saudi International Electronics, Communications and Photonics Conference (SIECPC)*, Riyadh, Saudi Arabia, IEEE, New York, NY, 2011.
16. Hu, Y. C., M. Patel, D. Sabella, N. Sprecher, and V. Young. *Mobile Edge Computing a Key Technology toward 5G.* ETSI, Sophia Antipolis, France, 2015, pp. 1–16.
17. Ahmed, A., and E. Ahmed. A Survey on Mobile Edge Computing. *Proc., 10th International Conference on Intelligent Systems and Control (ISCO)*, Coimbatore, India, 2016, pp. 1–6. https://ieeexplore.ieee.org/document/7727082.
18. Greenberg, A., G. Hjalmtysson, D. A. Maltz, A. Myers, J. Rexford, G. Xie, H. Yan, J. Zhan, and H. Zhang. A Clean Slate 4D Approach to Network Control and Management. *ACM SIGCOMM Computer Communication Review*, Vol. 35, No. 3, 2005, p. 41.
19. AT&T. *ECOMP (Enhanced Control, Orchestration, Management & Policy) Architecture White Paper.* 2016, p. 31.
20. Clark, C., W. Lee, D. Schimmel, D. Contis, M. Koné, and A. Thomas. A Hardware Platform for Network Intrusion Detection and Prevention. In P. Crowley, M. A. Franklin, H. Hadimioglu, and P. Z. Onufryk (eds.), *Network Processor Design*, 2005, pp. 99–118. https://doi.org/10.1016/B978-012088476-6/50007-1.
21. Han, B., V. Gopalakrishnan, L. Ji, and S. Lee. Network Function Virtualization: Challenges and Opportunities for Innovations. *IEEE Communications Magazine*, Vol. 53, No. 2, 2014, pp. 90–97. https://ieeexplore.ieee.org/document/7045396.
22. Peter, S., J. Li, I. Zhang, D. R. K. Ports, D. Woos, A. Krishnamurthy, T. Anderson, and T. Roscoe. Arrakis: The Operating System Is the Control Plane. *ACM Transaction on Computer Systems*, Vol. 33, No. 4, 2016, pp. 11:1–11:30. https://doi.org/10.1145/2812806.
23. AT&T. *AT&T Vision Alignment Challenge Technology Survey.* AT&T Domain 2.0 Vision Whitepaper, 2014.
24. Leinmuller, T., E. Schoch, and C. Maihofer. Security Requirements and Solution Concepts in Vehicular Ad Hoc Networks. *Proceedings of 4th Annual Conference on Wireless On Demand Network Systems and Services (WONS 2007)*, Kennesaw, GA, 2007, pp. 84–91.
25. Hasbullah, H., I. A. Soomro, and J. A. Manan. Denial of Service (DOS) Attack and Its Possible Solutions in VANET. *International Journal of Electrical, Computer, Energetic, Electronic and Communication Engineering*, Vol. 4, No. 5, 2010, pp. 411–415.
26. Ahmed-Zaid, F., F. Bai, S. Bai, C. Basnayake, B. Bellur, S. Brovold, G. Brown, L. Caminiti, D. Cunningham, H. Elzein, K. Hong, J. Ivan, D. Jiang, J. Kenney, H. Krishnan, J. Lovell, M. Maile, D. Masselink, E. McGlohon, P.

Mudalige, Z. Popovic, V. Rai, J. Stinnett, L. Tellis, K. Tirey, and S. VanSickle. *Vehicle Safety Communications – Applications (VSC-A): Final Report*. National Highway Traffic Safety Administration, Washington, D.C., 2011. https://www.nhtsa.gov/sites/nhtsa.dot.gov/files/811492a.pdf.

27. IEEE Vehicular Technology Society. *Trial-Use Standard for Wireless Access in Vehicular Environments (WAVE) - Resource Manager*. 2007.
28. U.S. Department of Transporation, and FHWA Office of Safety Research and Development. *Safety Applications Performance Requirements, Vol . 7, Stop Sign Gap Assist (SSGA)*. Final Report, Vol. 7, 2015.
29. Fayaz, S. K., Y. Tobioka, V. Sekar, M. Bailey, and M. Bailey. Bohatei: Flexible and Elastic DDoS Defense. *Proc., 24th USENIX Conference on Security Symposium*, Washington, D.C., 2015, pp. 817–832.
30. "ns-3." [Online]. https://www.nsnam.org/. Accessed March 16, 2018.
31. DLR, and Institute of Transportation Systems. *SUMO – Simulation of Urban MObility*. http://www.dlr.de/ts/en/desktopdefault.aspx/tabid-9883/16931_read-41000/. Accessed June 5, 2017.
32. Lu, X. Y., and J. Wang. Multiple-Vehicle Longitudinal Collision Avoidance and Impact Mitigation by Active Brake Control. *Proc., 2012 IEEE Intelligent Vehicles Symposium*, Alcala de Henares, Spain, IEEE, New York, NY, 2012, pp. 680–685.
33. Javed, M. A., and J. Y. Khan. Performance Analysis of a Time Headway Based Rate Control Algorithm for VANET Safety Applications. *Proc., 7th International Conference on Signal Processing and Communication Systems (ICSPCS)*, IEEE, New York, NY, 2013, pp. 1–6.

The Standing Committee on Intelligent Transportation Systems (AHB15) peer-reviewed this paper (18-06492).

Article

TRR
JOURNAL OF THE TRANSPORTATION RESEARCH BOARD

Transportation Research Record
2018, Vol. 2672(19) 79–90

DOI: 10.1177/0361198118787636
journals.sagepub.com/home/trr
SAGE

Investigating Cybersecurity Issues in Active Traffic Management Systems

Zulqarnain H. Khattak[1], Hyungjun Park[1], Seongah Hong[1], Richard Atta Boateng[1], and Brian L. Smith[1]

Abstract

Active traffic management (ATM) systems have been introduced by transportation agencies to manage recurrent and non-recurrent congestion. ATM systems rely on the interconnectivity of components made possible by wired and/or wireless networks. Unfortunately, this connectivity that supports ATM systems also provides potential system access points that results in vulnerability to cyberattacks. This is becoming more pronounced as ATM systems begin to integrate Internet of Things (IoT) devices. Hence, there is a need to rigorously evaluate ATM systems for cyberattack vulnerabilities, and explore design concepts that provide stability and graceful degradation in the face of cyberattacks. In this research, a prototype ATM system along with a real-time cyberattack monitoring system were developed for a 1.5-mile section of I-66 in Northern Virginia. The monitoring system detects deviation from expected operation of an ATM system by comparing lane control states generated by the ATM system with lane control states deemed most likely by the monitoring system. This comparison provides the functionality to continuously monitor the system for abnormalities that would result from a cyberattack. In the case of any deviation between two sets of states, the monitoring system displays the lane control states generated by the back-up data source. In a simulation experiment, the prototype ATM system and cyberattack monitoring system were subject to emulated cyberattacks. The evaluation results showed that the ATM system, when operating properly in the absence of attacks, improved average vehicle speed in the system to 60 mph (a 13% increase compared to the baseline case without ATM). However, when subject to cyberattack, the mean speed reduced by 15% compared to the case with the ATM system and was similar to the baseline case. This illustrates that the effectiveness of the ATM system was negated by cyberattacks. The monitoring system allowed the ATM system to revert to an expected state with a mean speed of 59 mph and reduced the negative impact of cyberattacks. These results illustrate the need to revisit ATM system design concepts as a means to protect against cyberattacks in addition to traditional system intrusion prevention approaches.

Cybersecurity is a broad term referring to processes and practices designed to protect networks, computers, programs, and data from attack, damage, or unauthorized access (*1*). In today's connected world, cybersecurity is an emerging threat in every field that relies on communications. Therefore, transportation operation and management systems utilizing wired and wireless communications for managing roadways are also at significant risk of such cyberattacks. While the use of smart devices and communications provides new features and functionality, they also introduce new risks.

Arabo stated that with the emergence of Internet of Things (IoT) devices, many of which are beginning to be used in traffic management systems, the traditional strategy of isolating control systems from other networks will no longer be feasible (*2*). Similarly, Jang-Jaccard and Nepal also stated that the exponential growth of internet connections has led to a significant growth of cyberattacks (*3*). Traditionally malware attacks (which are secretly acting against the will of users) used to happen at a single point among vulnerable components. Therefore, rather than protecting each asset, the perimeter defense strategy (using firewall or antivirus) was utilized, but such efforts are now ineffective. The major drawbacks of greater use of information and communications technology (ICT) are increased cybersecurity risks (*4*).

Transportation operations and management systems in the past were closed proprietary systems (isolated systems), having very limited cyber vulnerabilities. With the emergence of cloud or network computing and continued reliance on emerging technology such as IoT, those proprietary systems have now transformed into more open systems with increased accessibility (*5*). Furthermore, the National Transportation Communication for ITS Protocol (NTCIP) uses center-to-center communications that rely on request-based

[1]Center for Transportation Studies, Department of Civil and Environmental Engineering, University of Virginia, Charlottesville, VA

Corresponding Author:
Address correspondence to Hyungjun Park: hpark@email.virginia.edu

protocols through XML messages (*6*). These protocols have no built-in security and rely on the assumptions that most attacks are from the inside, and that hackers make up only a small portion of total intrusions, thus further increasing the risk over proprietary systems (*7*). In this regard, the U.S. DOT has taken a huge initiative to develop a security credential management system (SCMS) (*8*)—a message security solution for vehicle-to-vehicle (V2V) and vehicle-to-infrastructure (V2I) communication.

Active traffic management (ATM) involves the ability to manage recurrent and non-recurrent congestion using real-time and predictive operational strategies. An ATM system can be deployed either in isolation to address a specific need such as the use of adaptive ramp metering to control the flow of traffic entering the freeway, or can be used in conjunction with other systems to meet network-wide needs of congestion management, incident management, traveler information, and safety resulting in synergetic gains. Motorists rely on information received via ATM systems for making travel decisions. Thus, the validity of such information is critical, but these modern open protocol systems rely on continuous communication between roadway and traffic management centers (TMCs). This dependency on communication opens up a wide array of entry points, which makes TMCs vulnerable to cyberattacks and the least understood in terms of cybersecurity.

The research presented in this paper is based on the premise that perfect protection from cyberattacks is not realistic. Furthermore, a system may also be attacked internally by a disgruntled employee. Therefore, this research chose to develop a real-time threat-monitoring system to continuously check to see if the system is behaving as expected. This paper is further structured as follows. First, a literature review is presented to analyze relevant studies and how different fields perceive cybersecurity. Then, the next section describes the efforts made to develop a prototype ATM system. This is followed by a section that identifies various attack points in the prototype ATM system and presents examples of architecture for the monitoring system. The next section explains the development of a monitoring system and finally an evaluation is conducted for identifying the impact of a possible attack on the prototype ATM system, and benefits of reverting back to a safe lane control state using a monitoring system.

Literature Review

Different fields have realized the importance of cybersecurity and various studies have been conducted on cyber issues of control systems, but the transportation industry is lagging in this field. Cherdantseva et al. reviewed state-of-the-art practice in cybersecurity risk assessment of supervisory control and data acquisition (SCADA) systems and proposed that security was not the primary concern in standalone SCADA systems in the past, which was achieved by controlling physical access to system components (*9*). However, the situation has now changed and several standards and directives relating to the cybersecurity of SCADA systems have emerged. These authors examined the advantages and drawbacks of the analyzed risk-assessment methods and concluded by stating that despite many methods for risk assessment of SCADA systems, there is still room for future research and improvement towards overcoming failure and developing monitoring systems.

Woo and Kim provided quantitative assessment of cybersecurity risk of SCADA system based on optimal power flow in a smart grid system, and quantified vulnerabilities by first defining relevance of threat to each component and then assigning a vulnerability index to each component based on historical data (*10*). The study finally calculated risk based on monetary terms as a product of probabilities of a threat and vulnerabilities and the cost of an asset, and concluded that this was a starting stage for quantifying cyber risks.

C'ardenas et al. used a chemical reactor for developing an anomaly-detection monitoring system based on cyberattacks that change the behavior of the targeted control system (*11*). The paper used simulation to test attacks and their impact on the Tennessee Eastman process control system model using an irreversible reaction inside a reactor of fixed volume, and proposed an automatic response mechanism that monitors the system. Another study evaluated the disruption of electric power systems caused by cyberattacks while stating that cyber systems form the backbone of the nation's critical infrastructure. The paper stated that interdependence among computers, communication, and power infrastructure have increased the risk of cyberattacks, and proposed a real-time monitoring and anomaly-detection system along with impact analysis. The impact analysis involved analysis of intrusion behaviors and evaluation of consequences of cyberattacks based on potential loss of power (*12*).

Wu et al. also used an integrated adaptive cybersecurity monitoring system to provide cyberattack awareness for scientific and military applications (*13*). The paper further used sensor data from networks and many simulations to evaluate the performance of the proposed monitoring system. Similarly, Ashok et al. proposed applications based on a wide-area monitoring system for power grid security (*14*). The paper provides architectural concepts pertaining to monitoring systems for anomaly detection and mitigation.

In recent years, several attacks have been reported on transportation operation and management systems, compromising the safety and security of these systems (*15*). During summer 2011, a cyberattack carried out by a group known as "Anonymous" against a metropolitan transit system compromised personal identifiable information (PII) such as customer names, addresses, emails, ID numbers, and passwords for more than 200 customers (*16*). During July 2009, several hardware experts revealed the vulnerabilities of a parking meter by hacking such a parking meter in San Francisco, obtaining an unlimited credit parking card using

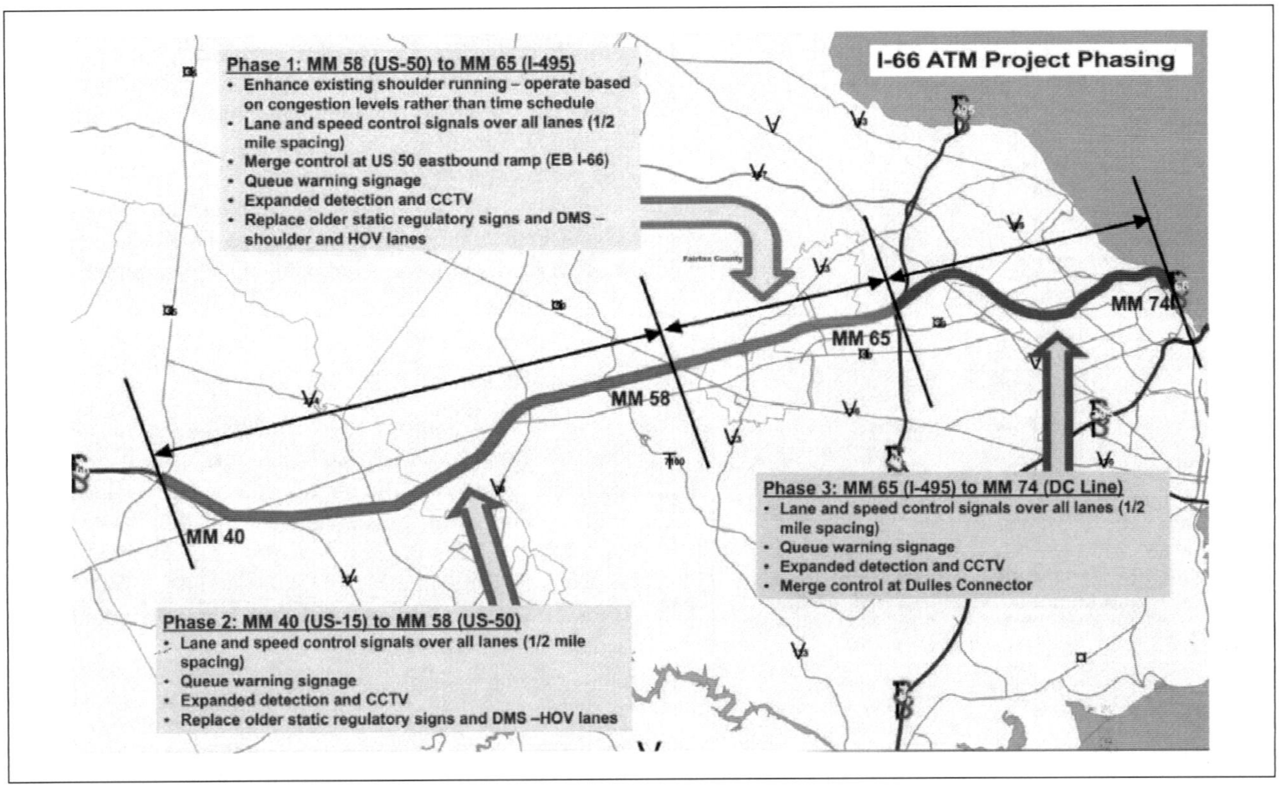

Figure 1. I-66 ATM system location (7).

reverse-engineering techniques (*17*). More recently, a group of experts demonstrated their ability to hack a Jeep Cherokee connected car equipped with the Uconnect system. Using the IP address, the two hackers were able to turn the engine of the car off, activate and deactivate the brakes, take remote control of the vehicle's information display and entertainment system, and activate the windshield wipers (*18*). Hacking of dynamic message signs (DMS) are frequently in the news, and on the recent Memorial Day Weekend, again DMS signs were hacked, with signs ranging from political slogans to "Party Yardy Hall!" and a week later displaying a statement about a recent Cincinnati Zoo gorilla incident (*19*).

With these attacks on transportation infrastructure systems over the past several years, the conventional wisdom is that cybersecurity is a critical and growing issue. Yet, there has been little research in the area. Folk first provided a brief overview of cybersecurity issues in transportation systems along with a history of some past attacks (*20*), and later provided fundamentals of common cyberattack processes and offered ideas on how to prepare and respond to an incident (*21*). Similarly, Amoozadeh et al. evaluated the effects of cyberattacks on communication channel as well as sensor tampering of a connected vehicle (CV) stream equipped to achieve cooperative adaptive cruise control (CACC) using a simulation model (*22*). Their results revealed that an insider attack can cause significant instability in the CACC vehicle stream, and they also proposed various countermeasures, such as downgrading to adaptive cruise control (ACC), that could be used to improve security and safety of CVs (*23*). Since there is no way to eliminate cyberattacks completely and no rigorous study exists on cyber vulnerabilities of transportation operations and management systems, this paper tries to fill this gap in the literature by providing a first attempt at evaluating the impact of cyber vulnerabilities of transportation operation and management systems and proposing a real-time monitoring system for detecting cyberattacks to avoid the consequences of a compromised system. It should be noted that deploying a monitoring system is a well-accepted concept in cybersecurity risk assessment for detecting anomalies (*12–14*).

Study Location

The ATM system deployed on I-66 from the Washington, D.C. line to Route 29 in Gainesville, VA was used as a case study in this research and is presented in Figure 1. The ATM system consists of speed harmonization, queue warning, hard shoulder running, and lane control states. These systems were deployed for improving safety and incident management at heavy bottleneck locations that averaged 60,000 vehicles per day between US-15 and the Fairfax line. The number increased to 90,000 vehicles per day between the

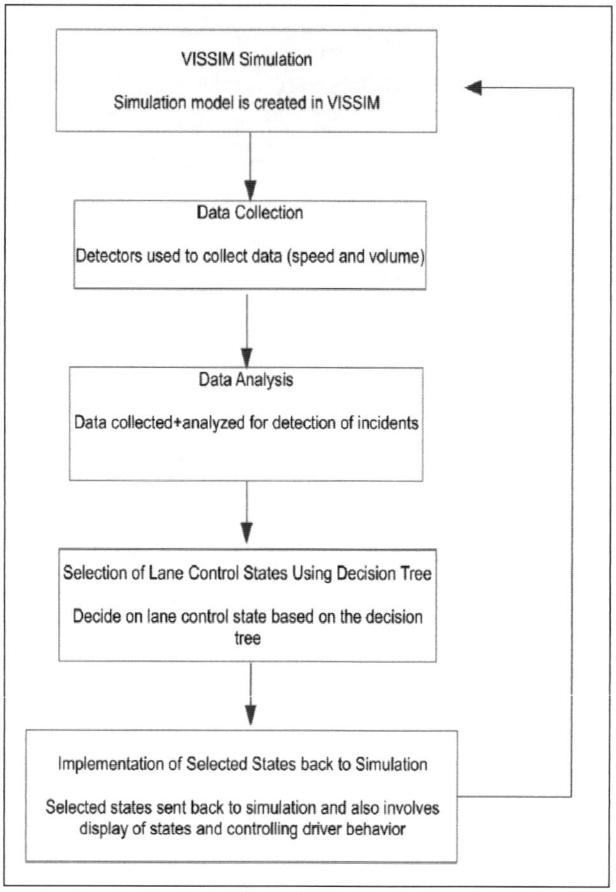

Figure 2. Prototype ATM system flow chart.

Fairfax line and I-495 and reduced to 64,500 vehicles per day between I-495 and the Arlington County line. The highest crash frequency existed at SR 7100 and I-495. This study modeled a 1.5-mile section of the I-66 ATM system in VISSIM for cybersecurity evaluation.

Development of a Prototype ATM System

A prototype ATM system was developed first based on the concept of operations of the actual ATM system on I-66. The various steps involved are presented in Figure 2, and a detailed description for each step is provided below.

VISSIM Simulation

A microscopic traffic simulation of the corridor was developed to serve as the "real world" in our prototype system. This allowed the research team to see the impact of the ATM system, both without attacks and under attack, on the transportation system. This step involved the modeling of a testbed network in microsimulation software VISSIM, shown in Figure 3. The testbed network consists of a 1.5-mile segment of a freeway corridor with three lanes in one direction. For the variable lane control state system, two gantries (each with a set of three states displaying either open, closed, or merge) were placed at the mileposts 0.5 and 1 mile along the corridor. A traffic incident was modeled within the most downstream section between the mileposts at 1 and 1.5 miles. The network was loaded with a vehicular volume of 3000, 4500, and 6000 vehicles per hour to represent low, medium, and high congestion conditions of traffic.

Data Collection

During this step, various detector stations were embedded in the VISSIM network at each lane near each gantry. The detectors recorded traffic measurements such as speed and traffic volume, which were aggregated within the predetermined data collection interval (every 10 s). Using the component object model (COM) interface, the traffic measurement data was retrieved at every update instance and sent to the ATM algorithm.

Data Analysis

In this step, the data collected from the simulation were processed and analyzed to identify whether there was a disruption of flow on the network. The analysis was based on speed of vehicles dropping below a threshold or speed at one instant of time being lower than the speed at the preceding instant of time. The algorithm used for incident detection is given by:

$$U_{i,t} == 0,$$
$$Q_{i,t} == 0 \; or,$$
$$U_{i,t} < C_1 * U_{i,t-1} \text{ and } U_{j,t} < \pm(1+C_2) * U_{j,t-1}$$

where U denotes speed, Q denotes volume, i represents the lane under study, j denotes the other two lanes, t represents a particular instant of time, and $t-1$ represents the immediately preceding instant of time. The algorithm detects incidents in which speeds or volume on a lane under consideration drop to zero or the speed at one particular instant of time is lower than another instant of time and the speed in the opposing two lanes at that instant of time is greater than the speed at the preceding instant of time.

Selection of Lane Control States using Decision Tree

During this step, the lane control states were selected based on the set of decision trees developed based on the concept of operations of the actual ATM system on I-66 – see Figure 4. The lane control policy was decided based on the traffic speed collected through detector stations embedded in the

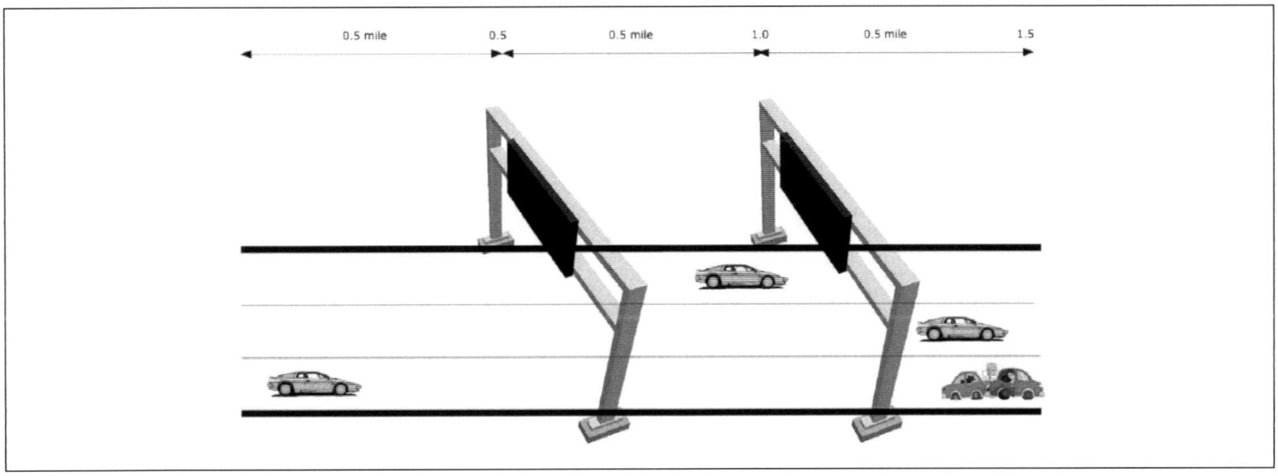

Figure 3. Testbed network.

VISSIM model, while the variable lane control state application was modeled using the VISSIM application programming interface (API) and C# language. This policy is consistent with ATM control used for operation on I-66 (*24*). The decision tree used for selecting lane control states consists of four critical decision variables: speed, event, distance to the event, and the lane on which the event is occurring. The first critical variable to be screened for deciding the appropriate lane control state for display is speed; if, for instance, a speed greater than 20 mph is selected, the next critical variable of the event is screened. It has three decision variables: no event, paving work on the road, and an incident on the road. After selecting the type of event, the third critical decision variable of distance to the event is screened by the algorithm. If, for instance, the event is less than 0.2 miles away from the gantry location, the last decision variable is screened to find the lane on which the event is occurring. Once these critical variables are screened and appropriate decision variables are selected, the applicable lane control states for display are selected for each of the three lanes.

Implementation of Selected States Back to the Simulation

Once the states for lane control were selected using the data from the simulation model and its screening for appropriate traffic conditions based on the decision tree, they were transferred back to the VISSIM network for developing a continual feedback loop. The driving behavior of the vehicles was also controlled according to the selected lane control states using the VISSIM API. The vehicles at the upstream location, which were required to comply with the variable lane control states, were controlled to conduct lane merging to the adjacent lane when it was recommended by the decision of the lane control states application, and were not allowed to travel in specific lanes when the lane control state algorithm indicated a lane closure.

Development of a Monitoring System

The objective of this step was to develop a monitoring system, starting with the identification of entry points for cyberattacks within the ATM system, and then ending with proposing examples for the development of a monitoring system architecture based on the identified attack points. Based on the results of this step, a prototype monitoring system was finally developed.

Identification of Entry Points

The identification of entry points requires clear description of the ATM architecture and its data flow. The data flow diagram in Figure 5 shows the functional aspects of lane control states along with likely entry points for cyberattacks. The lane control states are connected to the network switches in the cabinets provided at each of the gantry locations, from where they are again connected into the existing fiber-optic cable that runs to the TMC where the ATM software is installed. The ATM system is built around data collection from various roadway traffic detectors, environmental condition detectors, and monitoring through CCTV. The data collected are used to detect any unusual behavior such as congestion and incidents, which are processed by the ATM software using predefined logic to come up with appropriate response plans for which lane control states to display on the road network. Additional connections to private-sector probe data providers, such as Inrix, and IoT devices also exist. The testbed network developed in VISSIM mimics the data flow architecture of the I-66 ATM system, with vehicle detectors as the main data source. The attack points that are likely to be compromised by an adversary in the prototype system includes point A (the ATM

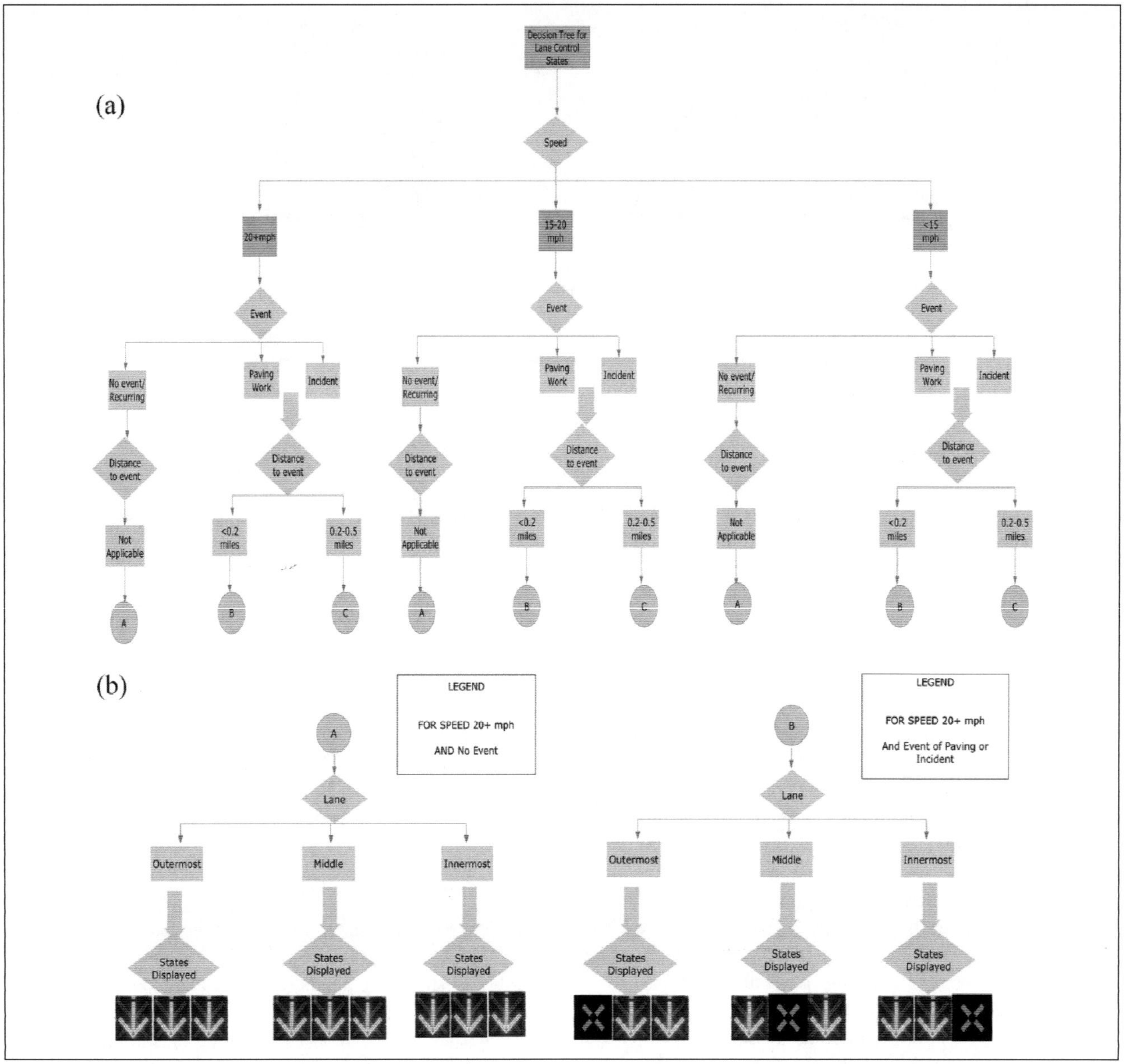

Figure 4. Decision tree for lane control states: (*a*) First three decision variables for lane control state; (*b*) an example of fourth decision variable for speed of 20+ mph.

software itself), point B (data source such as detectors), and point C (connection between field device, lane control states, and the cabinet at the gantry location). The adversary can affect the normal operation of the lane control states by gaining access to any of these three attack points—for instance, gaining access to the ATM software through point A can allow the adversary to change the final decision generated by the ATM software algorithm and send any desired set of states for display on the gantries. Similarly, gaining access to the detectors through point B can allow the adversary to send manipulated traffic data to the ATM software that may lead to generation of incorrect states for display. Finally, gaining access to the gantry display through point C may allow the adversary to manipulate the decision for display of lane control states received from the ATM software algorithm and display any desired states on the gantries.

Development of Monitoring System Architecture

It is important to make it clear that the monitoring system developed in this research is tasked with monitoring the operation of the ATM system itself. We traditionally think of monitoring in traffic management as the use of detectors to measure speed and volume of vehicles. This provides an

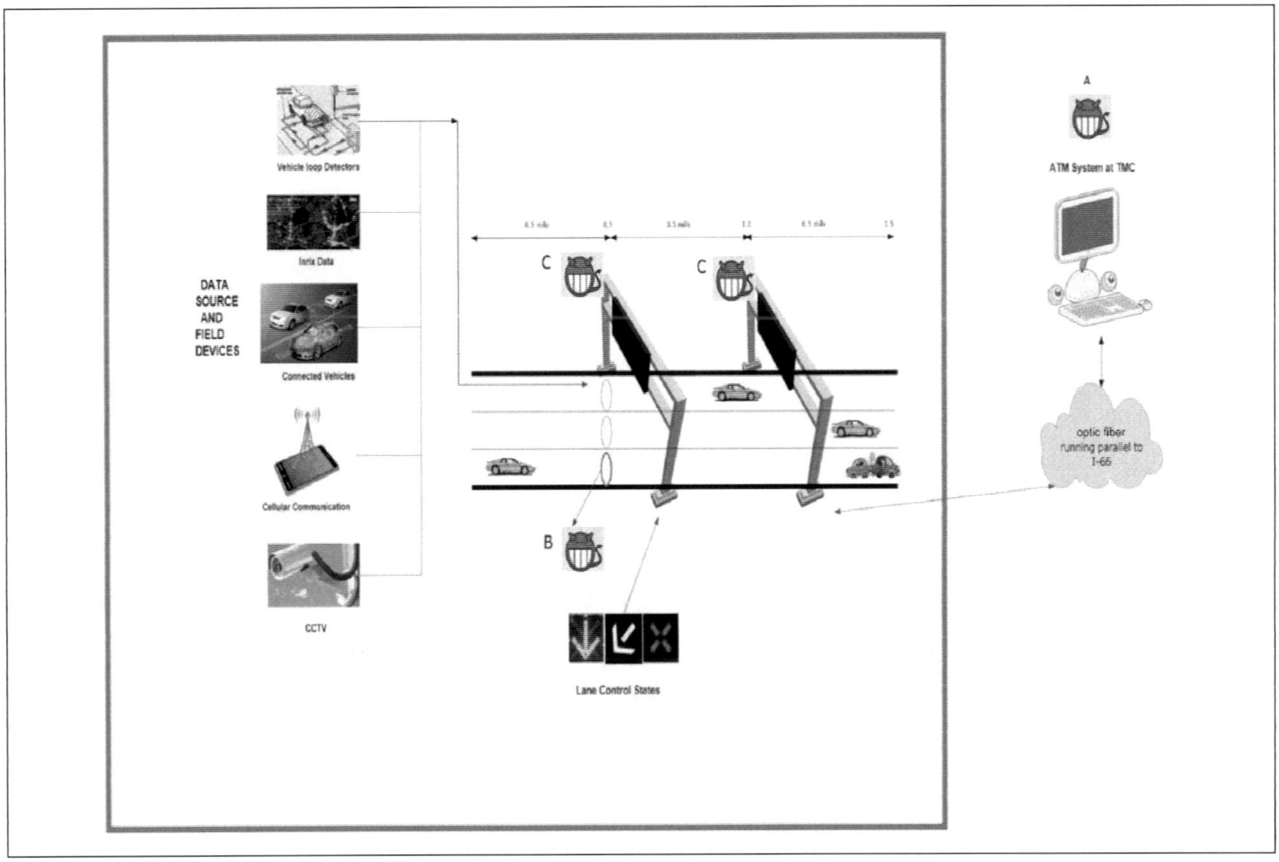

Figure 5. Data flow diagram of ATM system with attack points.

understanding of the system state. In a cybersecurity monitoring system, we are monitoring the functioning of the ATM system itself to be sure it is behaving as expected. Furthermore, the main purpose of developing a monitoring system is to add further redundancy to the existing system so as to make it difficult for an attacker to achieve his or her motives. The architecture of the proposed monitoring system for the identified attack points is shown in Figure 6. Figure 6*a* and *b* provide examples of separate monitoring systems for attack points A and B. The proposed monitoring systems consists of two channels that process the traffic data independently: (a) a channel associated with the operations of the ATM application; and (b) a channel associated with the monitoring system, which decides the control decision and display states independently and matches the decision with that of the other channel associated with the ATM application. The two channels are communication paths, receiving data from two separate or the same sources that are further processed for deciding lane control states. The difference between the two-individual monitoring systems shown in Figure 6*a* and *b* is in the data source processed by the monitoring channel—that is, the monitoring system for attack point A has only one data source of detectors, which is fed into both monitoring and ATM channels, while the monitoring system for attack point B uses a different data source—such as historical data, CV data, or data from CCTV—for the monitoring channel.

The example for the architecture of the monitoring system proposed for attack point C is provided in Figure 6*c*. The monitoring system will again operate on the same principle, with two different channels processing data independently: (a) a channel associated with the operation of the ATM application; and (b) a channel associated with the monitoring system, which decides the control decision and display lane control states independently and matches the decision with that of the other channel associated with the ATM application. Here, the states being displayed on the gantry are viewed through CCTV and matched with the states generated by the channel associated with the monitoring system. Based on the matching step, the possibility of a cyberattack will be decided as in the previous monitoring system; in case of a cyberattack, lane control states selected by the monitoring channel will be displayed on the network and the operator will also be alerted to a possible cyberattack.

Development of a Prototype Monitoring System

Based on the example architectures proposed in the previous section, a prototype monitoring system was developed for

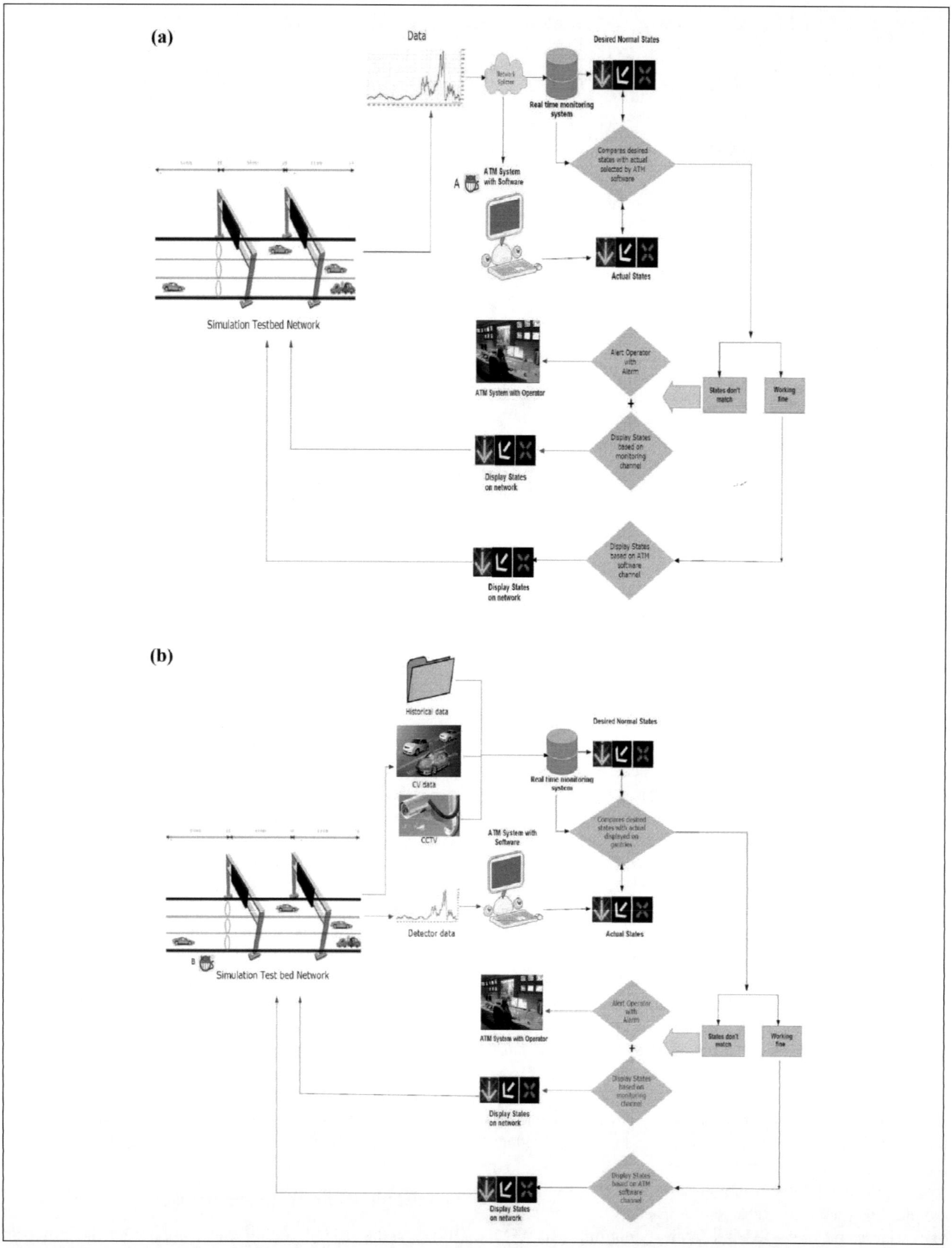

Figure 6. Example architecture of monitoring systems: (*a*) architecture of monitoring system for attack point A (ATM software at TMC); (*b*) architecture of monitoring system for attack point B (detectors); (*c*) architecture of monitoring system for attack point C (gantry locations).

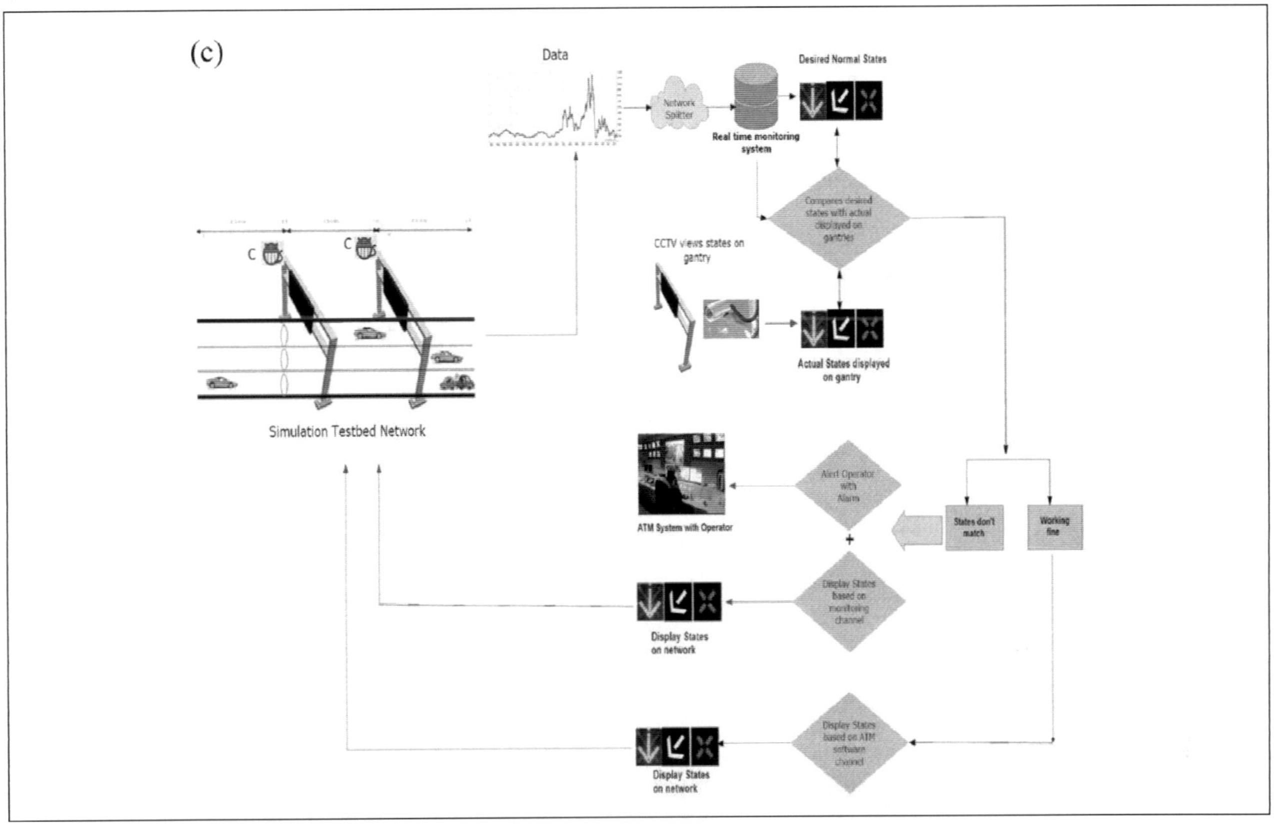

Figure 6. (continued)

the ATM system. Figure 7 represents the architecture of the prototype monitoring system for attack points A and B developed in this project. The traffic data are processed by two different channels: (a) a channel associated with the operations of the ATM application that receives data from detectors embedded in the network; and (b) a channel associated with the monitoring system, which receives data from CVs. Both channels decide the control decision and display lane control states independently. The lane control states generated by the two independent channels are matched by the monitoring system for detection of anomalies. It should be noted that both the monitoring channel and the ATM channel operate on the same logic but utilize the data from different sources. A 100% CV penetration was assumed in the simulation environment and data from individual vehicles were sent to the monitoring channel for generating lane control states. If any type of deviation is observed during the matching step, the monitoring system reports it as an anomaly, or a cyberattack is possibly occurring. In the case of anomaly detection, the monitoring system cedes the states generated by the ATM channel and displays the lane control states generated by the monitoring channel over the network. The monitoring system also provides an alert to the operator, but if no deviation is observed during the matching step, the lane control states generated by the ATM algorithm are displayed on the network.

Evaluation Results

The impact of a possible attack on the prototype ATM system, reverting back to a safe lane control state using a monitoring system and running the ATM system without a monitoring system, was evaluated using VISSIM microscopic simulations. In this evaluation, four cases were prepared and used, depending on the existence of an ATM system, cyberattacks, and a monitoring system. These cases were:

- Baseline: without an ATM system, cyberattacks, and a monitoring system
- Case 1: with an ATM system, but without cyberattacks, and a monitoring system
- Case 2: with an ATM system and cyberattacks, but without a monitoring system
- Case 3: with an ATM system, cyberattacks, and a monitoring system

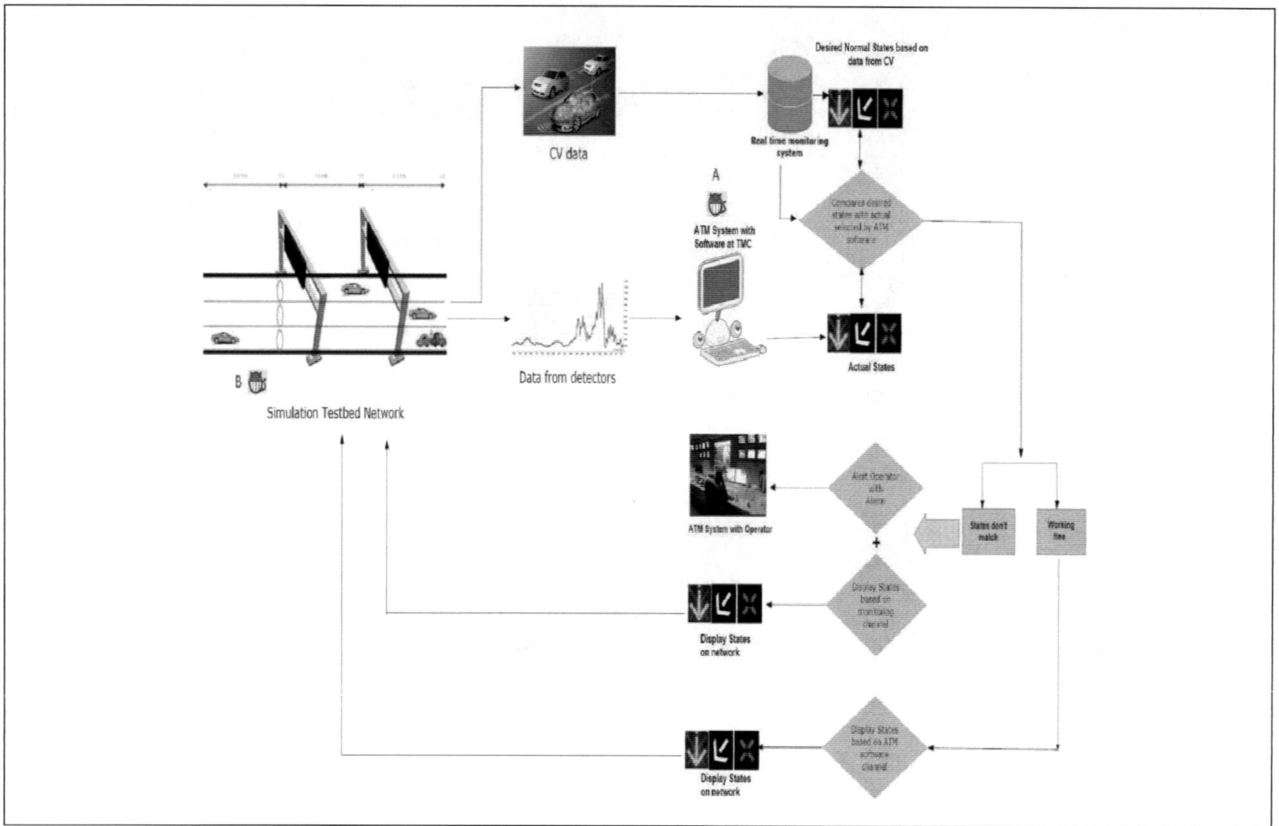

Figure 7. Architecture of monitoring system covering attack points A and B.

Table 1. Evaluation Results for the ATM Monitoring System

Scenario	Presence of: ATM	Presence of: Cyberattack	Presence of: Monitoring system	Mean speed (mph)	Standard deviation	Percentage changes *p*-value	Percentage changes *p*-value	Percentage changes *p*-value
Baseline	No	No	No	53	20	Base value compared to	N/A	N/A
Case 1	Yes	No	No	60	14	13% <0.01	Base value compared to	N/A
Case 2	Yes	Yes	No	51	24	–4% >0.11	–15% <0.01	Base value compared to
Case 3	Yes	Yes	Yes	59	15	11% <0.01	–2% >0.11	16% <0.01

For all the cases, an incident was created at the far downstream location of the 1.5 mile test freeway segment, in an attempt to create a situation in which an ATM system is needed. The ATM system described in the previous section was modeled in VISSIM and cyberattacks were randomly generated during the simulation using the COM interface and C# in Visual Studio. For the attacks at point A, the lane control states were randomly changed while, for the attacks at point B, the detector data were randomly scrambled. Lastly, if any anomalies in lane control states were detected, the monitoring system reverted the compromised lane control states back to the ones created by the monitoring channel using a different source of data (i.e., CV data). We completed 55 replications for each of the four cases, with each replication running for 10 min with a warm-up period of 2 min. The results are provided below.

This evaluation basically involved the comparison of mean speeds from all four cases using paired *t*-tests, assuming a null hypothesis that the two means were equal. The results are presented in Table 1 and the interpretations are provided below:

- Baseline: without an ATM system, cyberattacks, and a monitoring system:
  - This is the base case without an ATM system. Therefore, no attacks can happen and thus a monitoring system is not needed.
  - A mean speed of 53 mph was obtained in this baseline case.
- Case 1: with an ATM system, but without cyberattacks, and a monitoring system:
  - Case 1 added an ATM system on top of the baseline case. The ATM system provided a set of lane control states to better manage and operate traffic according to the prevailing traffic situations.
  - It was found out that, with an ATM system in operation, the mean speed improved to 60 mph (13% increase from 53 mph of baseline). This is statically significant ($p < 0.01$).
- Case 2: with an ATM system and cyberattacks, but without a monitoring system:
  - In Case 2, random cyberattacks were generated that compromised the original ATM lane control states and resulted in a set of "random" lane control states that may have a negative impact on traffic.
  - The mean speed in Case 2 under attacks was degraded to 51 mph, a reduction of 15% from the 60 mph of Case 1. This degradation was found to be statistically significant ($p < 0.01$).
  - The Case 2 mean speed (51 mph) was similar to the baseline mean speed (53 mph), implying that the effectiveness of the ATM was negated by attacks.
- Case 3: with an ATM system, cyberattacks, and a monitoring system:
  - Case 3 added a monitoring system to Case 2. The monitoring system detected anomalies in lane control states created by attacks, and rectified the compromised states as needed.
  - With the monitoring system, the mean speed was improved back to 59 mph, a 16% increase compared to Case 2, with statistical significance ($p < 0.01$).
  - The Case 3 mean speed of 59 mph was also found to be similar to the Case 1 speed of 60 mph. The implication is that the monitoring system effectively overcame the negative impact created by cyberattacks and allowed the ATM system to function as intended.

In summary, this research has demonstrated that a cyberattack monitoring system has the potential to limit the impact of attacks on an ATM system and improve freeway operations.

Conclusions

The research effort presented in this paper is a first step of a huge undertaking—investigating, analyzing, and addressing cybersecurity issues expected in various transportation-related systems in the very near future, from a holistic point of view. The goal here was to demonstrate the issues and a possible solution to highlight the significance of the cybersecurity analysis, rather than solving all the problems. For this, we took the ATM system as an example and the simulation results were presented for this purpose.

Furthermore, this study is one of the first attempts to investigate cybersecurity issues in ATM systems. A 1.5-mile section of the ATM system deployed on I-66 in Northern Virginia was selected as a case study. First, a prototype ATM system was developed based on the concept of operations of the actual ATM system in Northern Virginia. Then a prototype monitoring system was developed to detect anomalies in ATM operations and to avoid the consequences of a compromised system by cyberattacks. The monitoring system leverages real-time data by comparing the lane control states generated by two different channels: the ATM channel and the monitoring channel. In the case of any deviation between two sets of lane control states, the monitoring system starts displaying the lane control states generated by the CV data and alerts the operator.

The evaluation results showed that the ATM system could increase the mean vehicle speed in the system by 13% compared to the baseline case. However, when subject to cyberattack through manipulation of traffic data or the lane control states through the ATM software, mean speed was similar to the baseline case and reduced by 15% compared to the case with the ATM system. This illustrates that the effectiveness of the ATM system was negated by cyberattacks. The monitoring system, however, allowed the ATM system to revert to an expected state with a mean speed of 59 mph and reduced the negative impact of cyberattacks. Therefore, it is reasonable to say that the proposed monitoring system was successful in avoiding the severe consequences of cyberattacks and has significant potential in improving freeway operations. Therefore, it is necessary to further continue the effort for developing and evaluating more rigorous monitoring systems, along with revisiting ATM system design concepts as a means to protect against cyberattacks in addition to traditional system intrusion prevention approaches.

Author Contributions

ZHK contributed by providing conceptual design and methodological framework, helping with simulation coding and data analysis, writing the whole draft manuscript, and carrying out subsequent revisions. SH helped with the simulation coding. RAB helped with analyzing the data that came out of the simulation. BLS and HP supervised the project and provided research directions. All authors reviewed the results and approved the final version of the manuscript.

References

1. Rouse, M. *Cybersecurity*. http://whatis.techtarget.com/definition/cybersecurity. Accessed on April 7, 2017.

2. Arabo, A. Cyber Security Challenges within the Connected Home Ecosystem Futures. *Procedia Computer Science*, Vol. 61, 2015, pp. 227–232.
3. Jang-Jaccard, J., and S. Nepal. A Survey of Emerging Threats in Cybersecurity. *Journal of Computer and System Sciences*, Vol. 80, No. 5, 2014, pp. 973–993.
4. Langer, L., F. Skopik, P. Smith, and M. Kammerstetter. From Old to New: Assessing Cybersecurity Risks for an Evolving Smart Grid. *Computers and Security*, Vol. 62, 2016, pp. 165–176.
5. Khattak, Z. H., M. J. Magalotti, and M. D. Fontaine. Estimating Safety Effects of Adaptive Signal Control Technology Using the Empirical Bayes Method. *Journal of Safety Research*, Vol. 64, 2018, pp. 121–128. https://doi.org/10.1016/j.jsr.2017.12.016.
6. NTCIP9010. *National Transportation Communications for ITS Protocol- XML in ITS Center-to-Center Communications*. AASHTO, ITE, NEMA, Washington, D.C., 2003.
7. NTCIP1105. *National Transportation Communications for ITS Protocols-CORBA Security Service Specification*. AASHTO, ITE, NEMA, Washington, D.C., 2011.
8. Kreeb, B., and K. Gay. *Security Credential Management System (SCMS) Proof of Concept (POC)*. U.S. Department of Transportation, Washington, D.C., 2014.
9. Cherdantseva, Y., P. Burnap, A. Blyth, P. Eden, K. Jones, H. Soulsby, and K. Stoddart. A Review of Cyber Security Risk Assessment Methods for SCADA Systems. *Computers & Security*, Vol. 56, 2015, pp. 1–27.
10. Woo, P. S., and B. H. Kim. A Study on Quantitative Methodology to Assess Cyber Security Risk of SCADA Systems. *Advanced Materials Research*, Vols. 960–961, No. 1, 2014, pp. 1602–1611.
11. C'ardenas, A. A., S. Amin, Z. S. Lin, Y. L. Huang, C. Y. Huang, and S. Sastry. Attacks Against Process Control Systems: Risk Assessment, Detection, and Response. *Proc., 6th ACM Symposium on Information, Computer and Communications Security (ASIACCS'11)*, Hong Kong, China, 2011, pp. 355–366.
12. Ten, C. W., G. Manimaran, and C. C. Liu. Cybersecurity for Critical Infrastructures: Attack and Defense Modeling. *IEEE Transactions on Systems, Man, and Cybernetics Part A: Systems and Humans*, Vol. 40, No. 4, 2010, pp. 853–865.
13. Wu, Q., D. Ferebee, Y. Lin, D. Dasgupta, W. Qishi, and L. Yunyue. An Integrated Cyber Security Monitoring System Using Correlation-Based Techniques. *Proc., 2009 IEEE International Conference on System of Systems Engineering (SoSE)*, Albuquerque, NM, IEEE, New York, 2009, pp. 1–6.
14. Ashok, A., M. Govindarasu, and J. Wang. Cyber-Physical Attack-Resilient Wide-Area Monitoring, Protection, and Control for the Power Grid. *Proceedings of the IEEE*, Vol. 105, No. 7, 2017, pp. 1389–1407.
15. Khattak, Z. H., and M. D. Fontaine. Evaluating the Safety Effects of Span Wire to Mast Arm Signal Conversion. *Transportation Research Record: Journal of the Transportation Research Board*, 2018. Advance online publication. https://doi.org/10.1177/0361198118768525.
16. Hartman, K. Hacker Group Anonymous Attacks San Francisco's Transit System, 2011. http://www.digitaljournal.com/article/310370. Accessed January 1, 2017.
17. McMillan, R. Meter Hackers Find Free Parking in San Francisco, 2009. http://pcworld.com/article/169376/meter_hackers.html.
18. Paganini, P. Hackers Take Over Remotely a Fiat Chrysler Connected Car, 2015. http://securityaffairs.co/wordpress/38739/hacking/connected-car-hacking.html. Accessed January 1, 2017.
19. Mckay, M. Hacked Road Signs: What's Really the Message Here?, 2016. http://web.sesamerica.com/hacked-road-signs-whats-really-the-message-here. Accessed January 1, 2017.
20. Folk, E. *An Introduction to Modern Cybersecurity Issues in Transportation Systems*. Institute of Transportation Engineers Journal, Washington, D.C., 2013.
21. Folk, E. *Protecting Your Transportation Management Center*. Institute of Transportation Engineers Journal, Washington, D.C., 2015.
22. Amoozadeh, M., A. Raghuramu, C. N. Chuah, D. Ghosal, H. Michael Zhang, J. Rowe, and K. Levitt. Security Vulnerabilities of Connected Vehicle Streams and Their Impact on Cooperative Driving. *IEEE Communications Magazine*, Vol. 53, No. 6, 2015, pp. 126–132.
23. Khattak, Z. H., M. J. Magalotti, J. S. Miller, and M. D. Fontaine. Using New Mode Choice Model Nesting Structures to Address Emerging Policy Questions: A Case Study of the Pittsburgh Central Business District. *Sustainability*, Vol. 9, No. 11, 2017, pp. 1–16. https://doi.org/10.3390/su9112120.
24. Iteris. *I-66 Active Traffic Management System ConOps*. Virginia Department of Transportation, Richmond, VA, 2011.

The Standing Committee on Intelligent Transportation Systems (AHB15) peer-reviewed this paper (18-03501).

Article

TRR
JOURNAL OF THE TRANSPORTATION RESEARCH BOARD

Transportation Research Record
2018, Vol. 2672(19) 91–101

DOI: 10.1177/0361198118776113
journals.sagepub.com/home/trr
SAGE

Understanding the Highway Safety Benefits of Different Approaches of Connected Vehicles in Reduced Visibility Conditions

Md Sharikur Rahman[1], Mohamed Abdel-Aty[1], Ling Wang[2], and Jaeyoung Lee[1]

Abstract

This study evaluated the effectiveness of connected vehicle (CV) technologies in adverse visibility conditions using microscopic traffic simulation. Traffic flow characteristics deteriorate significantly in reduced visibility conditions resulting in high crash risks. This study applied CV technologies on a segment of Interstate I-4 in Florida to improve traffic safety under fog conditions. Two types of CV approaches (i.e., connected vehicles without platooning (CVWPL) and connected vehicles with platooning (CVPL) were applied to reduce the crash risk in terms of three surrogate measures of safety: the standard deviation of speed, the standard deviation of headway, and rear-end crash risk index (RCRI). This study implemented vehicle-to-vehicle (V2V) communication technologies of CVs to acquire real-time traffic data using the microsimulation software VISSIM. A car-following model for both CV approaches was used with an assumption that the CVs would follow this car-following behavior in fog conditions. The model performances were evaluated under different CV market penetration rates (MPRs). The results showed that both CV approaches improved safety significantly in fog conditions as MPRs increase. To be more specific, the minimum MPR should be 30% to provide significant safety benefits in terms of surrogate measures of safety for both CV approaches over the base scenario (non-CV scenario). In terms of surrogate safety measures, CVPL significantly outperformed CVWPL when MPRs were equal to or higher than 50%. The results also indicated a significant improvement in the traffic operation characteristics in terms of average speed.

It is known that reduced visibility due to fog has caused serious traffic safety and flow issues. Florida had experienced a total of 4,954 fog-related crashes between the year of 2008 and 2012, of which 132 crashes were fatal, and about 30% of the total fog-related crashes were fatal and injury crashes (*1*). It is worth mentioning that fog-related crashes tend to result in more severe injuries and involve more vehicles compared with clear conditions crashes (*2*, *3*). Fog affects roadway safety by increasing crash risk. Therefore, it is necessary to evaluate the appropriate countermeasures to enhance traffic safety under fog conditions.

There has already been a lot of research conducted on traffic safety under normal weather conditions. On the other hand, traffic safety under fog conditions has attracted much less attention. However, some researchers have already proposed the traditional approach of variable speed limits (VSL) or variable message signs (VMS) to decrease the crash risk in reduced visibility conditions (*1*, *3*, *4*). This approach can possibly improve traffic safety and mitigate traffic crashes by adjusting vehicle speeds and decreasing speed variation among vehicles in reduced visibility conditions. Nevertheless, the success of VSL or VMS is dependent on the level of compliance (*5*–*7*). Therefore, the VSL would not guarantee the improvement of traffic safety if drivers do not follow the new speed limit.

An innovative feature of this study was to apply the connected vehicle (CV) technologies in adverse visibility conditions under a microsimulation environment. To be more specific, this research aims to contribute to the implementation of two CV approaches, that is,

[1]Department of Civil, Environmental and Construction Engineering, University of Central Florida, Orlando, FL
[2]College of Transportation Engineering, Tongji University, Shanghai Shi, China

Corresponding Author:
Address correspondence to Md Sharikur Rahman:
sharikur@knights.ucf.edu

connected vehicle without platooning (CVWPL) and connected vehicle with platooning (CVPL), to improve traffic safety in reduced visibility conditions. CVPL concept is an extension of CVWPL approach in which several CVs form a "platoon" that behaves as a single unit. A car-following model for both CV approaches was also used in fog conditions with an assumption that applied CVs would follow this car-following behavior in the simulation. The most significant difference of CVs driving behavior between the two approaches was joining vehicles to maintain a platoon. In the near future, the market penetration rate (MPR) will not achieve 100%. Meanwhile, the penetration will increase gradually. Hence, it is worthwhile to study the safety benefits of CV technologies under different MPRs.

Background

Florida is among the highest ranked states in the United States with regards to traffic safety problems resulting from adverse weather conditions, especially in fog. As an example, a fog-related severe crash caused 5 fatalities, several injuries, and left a pileup of 70 vehicles on I-4, Polk County, Florida (*4*). The injury and death rates (per 100 crashes) for fog-related crashes and were found to be 3.75 and 2.25 for the corresponding type of crashes occurring in normal weather conditions, respectively (*8*). Previous studies have been examined to evaluate the traffic characteristics in fog conditions. Abdel-Aty et al. (*9*) conducted a comprehensive study with an effort to examine the traffic characteristics in fog conditions. The study concluded that speed and headway decreased significantly under reduced visibility conditions. Furthermore, the standard deviation of speed and headway increased in fog conditions compared to clear conditions. A more recent study by Peng et al. (*1*) identified that reduced visibility would significantly increase the standard deviation of speed and headway which intensifies traffic crash risk. It was also observed that time to collision decreased significantly in reduced visibility conditions, which means that the crash risk would be higher under reduced visibility conditions. They also found that the impact of low visibility on crash risk was different for different vehicle types and for different lanes. The crash risk is higher for passenger vehicles compared with heavy vehicles, and the inner lane (close to the median) has higher crash risk compared with the middle and outer lanes. Other studies also pointed out that headway distance was reduced in fog conditions and sometimes reduced headway would have a perceptual control benefit to the driver in terms of reduction in response time under fog conditions (*10*, *11*). Brooks et al. (*12*) examined the effect of fog conditions on lane-keeping ability using driving simulator. It was shown that lane keeping performances were significantly degraded by the existence of fog.

There is relatively little work in the literature describing the countermeasures in reduced visibility conditions. The findings of the previous studies provided several recommendations as guidelines to improve safety in reduced visibility conditions. Based on a questionnaire survey, Hassan et al. (*4*) suggested that changeable message signs can be a good countermeasure to reduce driving speed. Pang et al. (*13*) used a simulation-based study to examine traffic safety and operation in fog conditions. The study showed that fog-related crashes were reduced by controlling upstream traffic flow (decreasing upstream traffic volume) and implementing VSL. Peng et al. (*1*) suggested that implementing the algorithms in real-time with intelligent transport system (ITS) measures, such as VSL and VMS, can reduce the crash risk in reduced visibility conditions. Speed variance would be lower with the implementation of VSL, which in turn decrease crash risk (*14*–*17*). In terms of safety, VSL has been used during inclement weather to decrease both the mean and the standard deviation of speed (*18*, *19*). However, the success of the VSL application is more dependent on the compliance level. In the low level of compliance, the VSL might fail to improve traffic safety (*5*–*7*). The research by Abdel-Aty et al. (*15*) also evaluated that the implementation of VSL might reduce the rear-end and lane-change crash risks in uncongested traffic conditions but not successfully reduce the crash risk in congested situations. Therefore, the success of the VSL is also dependent on the level of congestion.

The new ITS technology, CV, has been recently recognized as an auspicious approach which proved its potential to improve traffic safety, including mitigating crash severity and declining the possibility of crashes by offering vehicle to vehicle (V2V) and vehicle to infrastructure (V2I) communication. Most of the previous research was concerned with the mobility and the traffic operations under CV environment but did not focus on traffic safety. Fyfe and Sayed (*20*) combined VISSIM and the surrogate safety assessment model (SSAM) with the application of the cumulative travel time (CTT) algorithm which evaluates the safety under CV environment. The study showed a 40% reduction of rear-end conflict frequency at a signalized intersection with the application of CV. Olia et al. (*21*) experimented with CV technology in PARAMICS and estimated that the safety index improved up to 45% under CV environment. Paikari et al. (*22*) also used PARAMICS which combined the V2V and V2I technologies and obtained higher safety and mobility enhancement on freeways under the CV environment. Vehicle platooning with CV technology is another key element of future transportation systems which helps to enhance traffic operations and safety simultaneously. Tian et al. (*23*) proposed a stochastic model to evaluate the collision probability for the heterogeneous vehicle platoon which can deal with the intervehicle distance distribution. The results have great potential to decrease chain collisions and alleviate the severity of chain collisions in the platoon at the same time. However, until this point, no researcher has potentially analyzed CV technologies which are expected to decrease the crash risk in reduced visibility conditions.

When compared with the previous studies, this study is unique in a sense that it reflects the fog conditions in microscopic simulation and applies CV technologies which are expected to improve traffic safety in reduced visibility conditions.

Data Preparation

A section of Interstate, a main arterial for the Orlando metropolitan area, was selected for this study. The studied section had experienced severe fog-related crashes (*3*). Data from two different sources were collected for this study. Weather data were collected from Fog Monitoring System (FMS), a new visibility detection system, installed in the segment of I-4. And, real-time traffic data were collected from Regional Integrated Transportation Information System (RITIS) augmented with a device installed close to the FMS. RITIS indicates the basic traffic characteristics of the selected road segment, while the added device captures both regular traffic parameters and the headway between each vehicle on each lane.

The collected weather data contain 21 variables, including visibility distance, air temperature, surface moisture, dew point, wind speed, barometric pressure, rainfall, and so forth. Among these parameters, visibility distance is significant for fog conditions. The traffic data were collected from RITIS detectors. The traffic dataset comprises eight important variables related to traffic flow characteristics, including vehicle speed, vehicle length, duration of detection, and lane assignment. In this study, vehicles were classified into two categories: (1) passenger car (PC), and (2) heavy goods vehicle (HGV). A vehicle was considered as a PC if its length was equal to or less than 7.32 m (24 ft). According to the weather data, the visibility distance from 6:45 a.m. to 7:45 a.m. on February 2, 2016 (Tuesday) was the lowest among all days of field data collection between the observed months of January to May in 2016. And this hour's maximum and minimum visibility distance were recorded as 88 m and 45 m, respectively. Referring to the traffic flow data, the data of traffic volume and traffic speed in the same period, 6:45 a.m. to 7:45 a.m. on February 2, were chosen for basic simulation model development.

VISSIM Simulation Model

A well calibrated and validated VISSIM network replicating the fog conditions was one of the most important parts of this study. Simulations were conducted in PTV VISSIM, version 8.0. The testbed was a 10-mile section of I-4 which had experienced a severe fog-related crash. The traffic information on the simulation network, including traffic volume (aggregated into 15 min), PC and HGV percentages, and desired speed distribution were obtained from the RITIS detectors. In addition to that, the "Look Ahead Distance" was changed in VISSIM driving behavior to replicate reduced visibility conditions based on field visibility distance. The simulation time was set from 6:15 a.m. to 8:15 a.m. in VISSIM. After excluding first 30 min of VISSIM warmup time and last 30 min of cooldown time (no statistics were collected during this time), 60 min of VISSIM data was used for calibration and validation. Geoffrey E. Heavers (GEH) statistic was used to compare the field volumes with simulation volumes. The GEH statistic is a modified Chi-square statistic that incorporates both relative and absolute differences. The definition of GEH is

$$\text{GEH} = \sqrt{\frac{(M_{\text{obs}}(n) - M_{\text{sim}}(n))^2}{0.5 \times (M_{\text{obs}}(n) + M_{\text{sim}}(n))}} \quad (1)$$

where $M_{\text{obs}}(n)$ is the observed volume of field detectors and $M_{\text{sim}}(n)$ is the simulated volume obtained from the simulation network. The simulated volume would precisely reflect the field volume if more than 85% of the measurement locations' GEH values are less than 5 (*5*, *24*). As for speed, the absolute speed difference between simulated speeds and field speeds should be within 5 mph for more than 85% of the checkpoints (*25*). The simulated traffic volumes and speeds were aggregated to 15-min intervals and then compared with the corresponding field traffic data. Ten simulation runs with different random seeds worth of results showed that 91.25% of observed GEHs were less than 5, and 92.50% of the aggregated speeds in the simulation were within 5 mph of field speeds. The results above proved that the traffic calibration and validation satisfy the requirements and indicate that the network was consistent with that of the field traffic conditions.

Further Calibration to Reflect Fog Conditions

To reflect the fog conditions, there was a need to revalidate the VISSIM network with respect to both traffic and safety. For further validation, headway was used to validate the VISSIM network using two-sample t-test and the result showed that the mean simulated headway was significantly different from the mean field headway when all the driver behavior parameters in VISSIM were set as default. Previous studies considered only "Look Ahead Distance" as one of the most essential simulation parameters in VISSIM to replicate the fog conditions (*26*, *27*). Hence, changing only the "Look Ahead Distance" in VISSIM driving behavior may not reflect the fog conditions. To simplify the further calibration process, a sensitivity analysis was conducted on VISSIM driver behavior parameters in simulation models to reflect the fog conditions. The 10 sets of the car-following parameters (CC_0 to CC_9) were tried and each set was run 10 times with different random seeds. For each parameter, a range of values (9 values), which includes the default, was determined based on previous studies and engineering judgment (*28, 29*). A total of 730 simulation runs ([1 base-models + 9 × 8 car-following parameters] times 10 random seeds) were conducted. Toward this end, the standard deviation of speed (significant traffic characteristic in fog condition) was selected to compare the field and simulated value with two-sample t-test at 5% significance level. For each value of parameters, the results of t-test with 10 different random seeds proved whether the distribution of the field and simulated standard deviation of speed were identical or not. The sensitivity analysis results showed that three most important parameters were vital to reflect the fog conditions. These were CC_0 (standstill distance), CC_1 (headway time), and CC_2 (following variation). From the results of sensitivity analysis, the safety distance parameters (i.e. CC_0, CC_1, CC_2) decreased compared to the default values in fog conditions. The default value of CC_0, CC_1, CC_2 in VISSIM were 1.5 m, 0.9 sec, and 4 m whereas the calibrated values were found to be 1 m, 0.7 sec, and 3 m, respectively. Thus, the safety distance of the calibrated network has lower value compared to the uncalibrated network. Therefore, the safety distance between two vehicles has been reduced in fog conditions. For further validation, headway was again used to validate the new calibrated VISSIM network using two-sample t-test. After replicating the fog conditions, there was no significant difference between the simulated mean headway and the field mean headway. Therefore, the simulation network was well calibrated and validated with respect to both traffic and safety.

Methodologies

To assess the safety performance in fog conditions, this paper tested two distinct CV approaches including CVWPL and CVPL on the segment of I-4. Therefore, the understanding of the car-following behavior of CV technologies is essential for studying the impact on traffic safety in fog conditions under microsimulation. A car-following model for both CV approaches was used in fog conditions with an assumption that applied CVs would follow this car-following behavior in the simulation.

Car-Following Model in Fog Conditions

A car-following model is a prerequisite to regulate the driving behavior of CVs in microsimulation. The desired model should be able to simulate user defined driving behavior significantly differing from the traditional ones (i.e. Wiedemann model). The basic Intelligent Driver Model (IDM) which was proposed by Treiber et al. (*30*) has been used as machine driving by many researchers (*31, 32*) Many researchers have already used IDM or modified IDM in order to simulate their own machine driving platform named Adaptive Cruise Control (ACC) and Cooperative Adaptive Cruise Control (CACC) (*31, 32*). The basic IDM model is a nonlinear car-following model in which the acceleration ($\dot{v}_{IDM}$) is the function of desired gap distance s^* and the speed difference between leading and following vehicles Δv, expressed by

$$\dot{v}_{IDM}(t + t_a) = \max\left\{ b_m, a_m\left[1 - \left(\frac{v}{v_o}\right)^{\delta} - \left(\frac{s^*}{s}\right)^2\right]\right\}. \quad (2)$$

where

$s^* = s_0 + max\left[0, vT + \frac{v\Delta v}{2\sqrt{a_m b}}\right]$

and

t_a = perception-reaction time,
b_m = maximum deceleration,
a_m = maximum acceleration,
v = speed of the following vehicle,
v_0 = desired speed,
δ = acceleration exponent,
s = gap distance between two vehicles,
s_0 = minimum gap distance at standstill,
T= the safe time headway, and
b = the desired deceleration

In this study, this IDM model was used for CVs car-following behavior in fog conditions. The parameter settings for this model were potentially determined according to previous studies (*31–33*). The desired speed (v_0), acceleration exponent (δ), maximum acceleration (a_m), desired deceleration (b), minimum gap distance at standstill (s_0), safe time headway (T), maximum deceleration (b_m), and time delay (t_a) were selected 120 km/h, 4, 1 m/sec^2, 2 m/sec^2, 2 m, 0.6 sec, 2.8 m/sec^2, and 1.5 sec, respectively.

Additionally, CVs were implemented as a platooning concept (CVPL), in which several vehicles form a "platoon" that behaves as a single unit. However, the IDM model was followed by CVs in both approaches (i.e., CVWPL and CVPL) under fog conditions. The most significant difference of CVs driving behavior between the two approaches was joining vehicles to maintain a platoon. A minimum five CVs were considered to maintain a platoon in this study. Three grouping schemes for CVs, such as rear, front, and cut-in joins, as depicted in Figure 1, were implemented to maintain the platoon. The rear join leads a new CV following the last vehicle of a CV group driving along the most adjacent lane of the joining vehicle. The front join performs the same

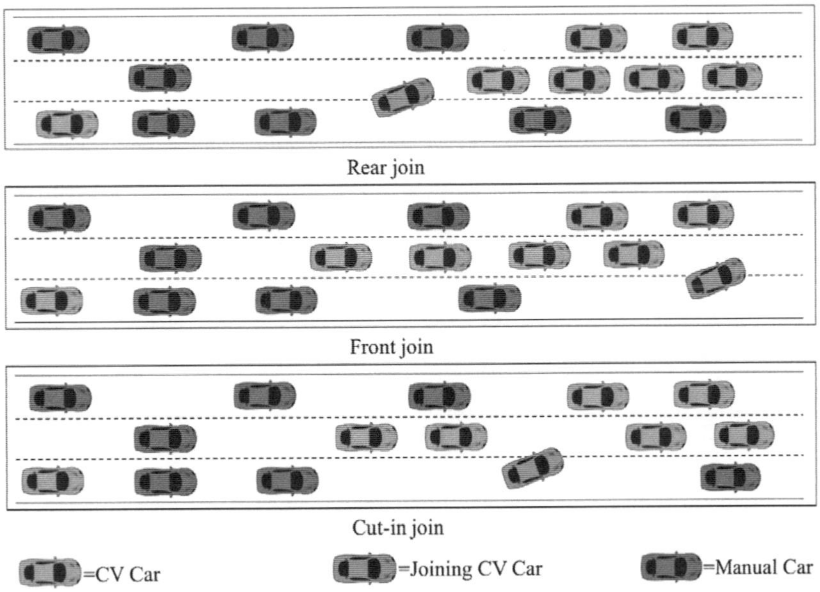

Figure 1. Joining of CVs to maintain a platoon.

process to allow a new CV to join into an existing CV group except that it leads the joining vehicle to the front of the first vehicle in the CV group. The cut-in join method is implemented by cooperatively adjusting the maneuvers of the joining vehicle and a CV in the group. As shown in Figure 1, once the joining vehicle identifies a target CV group, it approaches the group and determines a proper position to be inserted based on its current driving information such as speed, position, and so forth. Then the deceleration rate of a CV in the target group is adjusted to create a safe gap for the joining vehicle while the leading vehicle maintains its current speed. If the safe gap is satisfied for the lane change behavior of the joining vehicle, which is governed by VISSIM's lane changing model, the joining vehicle begins to change lane.

A high-level control algorithm architecture was developed for CVPL approach as shown in Figure 2. It is worth mentioning that the algorithm continuously adjusted the acceleration or deceleration rates using the above-mentioned Equation 2 between the leading and the subject vehicles using two-way communications under CV environment which offers a dedicated short-range communication (DSRC) of 300 m (1000 ft).

The aforementioned two driving behavior models were implemented as dynamic link library (DLL) plug-in for both approaches, which overrides the VISSIM default driving behavior. These two DLL were written in C++ which offers VISSIM an option to replace the internal driving behavior. During the simulation, the DLL file is called up in each time step and then controls the behavior of the vehicle for all or part of the vehicles depending on the MPRs. Note that the car-following and the lane-changing behavior of non-CVs were determined by VISSIM's default driving behavior model.

Surrogate Measures of Safety

Traffic crashes are rare events which involve numerous human factors along with the road environment and vehicle factors. A surrogate safety assessment technique should be adopted to measure safety as microsimulation software cannot be directly used to measure crashes or traffic safety. A number of previous studies used surrogate measures including speed variance, headway variance, time to collision, post-encroachment time, and RCRI (*1*, *15*, *34*). From the above-mentioned literature review the crash risk increased in fog conditions compared to normal weather conditions as the standard deviation of speed and headway increased significantly. Additionally, the rear-end crash is the significant type of crash in reduced visibility conditions (*2*, *8*, *35*). A rear-end crash may occur if the leading vehicle stops suddenly, and the following vehicle does not decelerate in time because of the low visibility. Maintaining insufficient safety distance between the leading and the following vehicle is the primary cause of rear-end crashes. To avoid the rear-end crash the stopping distance of the following vehicle should be smaller than the leading vehicle. An RCRI is proposed by Oh et al. (*36*) in which the dangerous condition can be mathematically expressed as

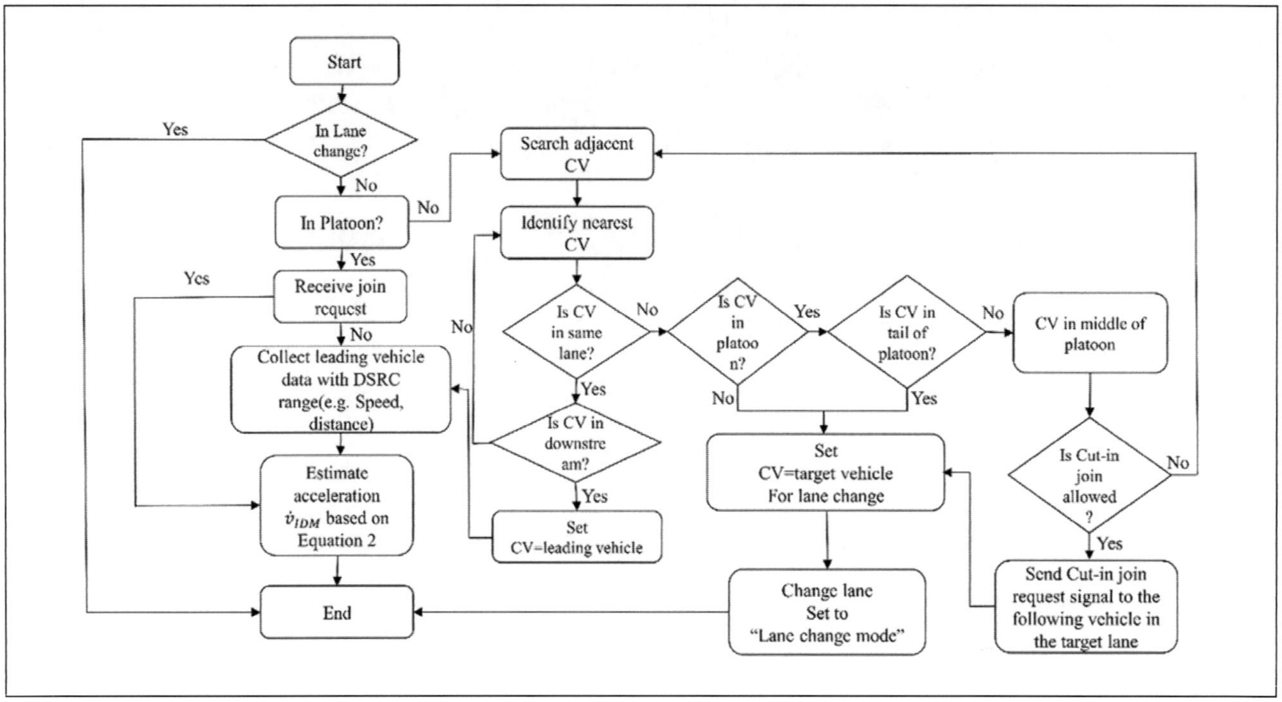

Figure 2. Control algorithm of CVs to maintain a platoon.

$$\mathrm{SD}_F > \mathrm{SD}_L \tag{3}$$

$$\mathrm{SD}_L = v_L \times h + \frac{v_L^2}{2 \times a_L} + l_L \tag{4}$$

$$\mathrm{SD}_F = v_F \times \mathrm{PRT} + \frac{v_F^2}{2 \times a_F} \tag{5}$$

where
SD_L = stopping distance of the leading vehicle,
SD_F = stopping distance of the following vehicle,
l_L = length of the leading vehicle,
v_L = speed of the leading vehicle,
v_F = speed of the following vehicle,
PRT = perception–reaction time,
h = time headway,
a_L = deceleration rate of the leading vehicle, and
a_F = deceleration rate of the following vehicle.

As mentioned earlier, for the VISSIM model, two types of vehicles were used: PC and HGV. Therefore, different deceleration rates were employed to estimate the reliable safe distance for the leading and following vehicles. The deceleration rates of PC and HGV were selected as 3.42 m/s^2 and 2.42 m/s^2 respectively, while the PRT was used as 1.5 s. These values are generally accepted by AASHTO (*37*). So, the RCRI is defined by the formula

$$\mathrm{RCRI} = \begin{cases} 1\ (\mathit{Dangerous})\ \mathit{If}\ \mathrm{SDF{>}SDL} \\ 0\ (\mathit{safe})\ \mathit{Otherwise} \end{cases} \tag{6}$$

In a nutshell, the standard deviation of speed, the standard deviation of headway, and RCRI were considered as surrogate measures of safety to evaluate the safety performances in fog conditions in a microsimulation platform.

Results and Discussions

Three surrogate measures of safety were considered to evaluate the safety performances in fog conditions under two CV approaches. Two-sample t-test was applied to compare the differences between two group means and their average variations in 2 min. This technique provides a method for comparing the mean standard deviation of speed and headway between base scenario and CV scenarios. Base scenario comprised of 100% regular vehicles, and CV scenarios were comprised of two types of CV approaches (i.e., CVWPL and CVPL) with different MPRs. The Chi-square test was also applied to analyze the significance in the difference of RCRI between base scenario and CV scenarios. Ten simulation runs, each with a different random number seed were conducted to eliminate random effects for each scenario.

Table 1 illustrates the summary of two surrogate measures of safety, (i.e., standard deviation of speed, standard deviation of headway) with the implementation of CVWPL and CVPL techniques. Compared to the base scenario, the standard deviation of speed and headway decreased significantly in both CV approaches. Model

Table 1. Summary of Measure of Effectiveness

		Speed (km/h)	Standard deviation of speed in 2 mins (km/h)		Standard deviation of headway in 2 mins (s)	
MPR	Comparisons	Mean difference (*P*-value)	Mean difference (*P*-value)	% Reduction	Mean difference (*P*-value)	% Reduction
20%	Base vs CVWPL	−0.288 (0.0322)	0.264# (0.1915)	2.78	0.139# (0.2645)	3.31
	Base vs CVPL	−0.398 (0.0030)	0.375# (0.0642)	3.96	0.184# (0.1371)	4.39
	CVWPL vs CVPL	−0.108# (0.4062)	0.111# (0.4997)	1.20	0.045# (0.6519)	1.11
30%	Base vs CVWPL	−0.570 (<0.0001)	0.597 (0.0042)	6.29	0.344 (0.0060)	8.19
	Base vs CVPL	−1.149 (<0.0001)	0.769 (0.0002)	8.12	0.453 (0.0003)	10.79
	CVWPL vs CVPL	−0.579 (<0.0001)	0.174# (0.2503)	1.96	0.109# (0.2055)	2.83
50%	Base vs CVWPL	−1.334 (<0.0001)	0.848 (<0.0001)	8.95	0.456 (0.0002)	10.87
	Base vs CVPL	−2.457 (<0.0001)	1.476 (<0.0001)	15.57	0.764 (<0.0001)	18.21
	CVWPL vs CVPL	−1.125 (<0.0001)	0.626 (0.0005)	7.25	0.308 (<0.0001)	8.24
70%	Base vs CVWPL	−2.395 (<0.0001)	1.745 (<0.0001)	18.41	0.584 (<0.0001)	13.92
	Base vs CVPL	−3.275 (<0.0001)	2.536 (<0.0001)	26.76	1.005 (<0.0001)	23.95
	CVWPL vs CVPL	−0.880 (<0.0001)	0.793 (<0.0001)	10.24	0.421 (<0.0001)	11.66
100%	Base vs CVWPL	−4.897 (<0.0001)	2.700 (<0.0001)	28.49	0.784 (<0.0001)	18.68
	Base vs CVPL	−5.535 (<0.0001)	3.687 (<0.0001)	38.90	1.394 (<0.0001)	33.22
	CVWPL vs CVPL	−0.637 (<0.0001)	0.988 (<0.0001)	14.58	0.610 (<0.0001)	17.88

Note: #Difference is insignificant at the 5% level.

performances were evaluated for three different condition sets (Base, CVWPL and CVPL) each under five different MPRs (20%, 30%, 50%, 70%, and 100%). To find out the safety impact of the applied technologies the mean values of the surrogate safety measures were compared with the base condition. In 100% MPR, the standard deviation of speed and the standard deviation of headway were found to be reduced by 28.49% and 18.68%, respectively, in CVWPL compared to base condition. On the other hand, in CVPL, the reductions were found to be 38.90% and 33.22%, respectively. The results revealed that the applied CV technologies enhanced traffic safety by decreasing the surrogate measures of safety in fog conditions. From Table 1 it was found that the maximum significant improvement resulted at 100 % MPR, while the improvement below 30% MPRs was insignificant at 5% level of significance.

For each of the 15 scenarios listed in Table 1, the mean differences of standard deviation of speed and standard deviation of headway were higher for CVPL than CVWPL. It was also found that the CVPL achieved significant reductions in the standard deviation of speed and headway compared to CVWPL when the MPRs were equal or greater than 50%. For instance, standard deviation of speed and standard deviation of headway for CVPL were 14.58% and 17.88% lower respectively than CVWPL at 100% MPR. Thereby, CVPL approach clearly outperformed CVWPL approach in terms of safety improvement. In terms of traffic operation, simulation results demonstrated higher speed for CVWPL and CVPL compared to the base condition. Additionally, compared to CVWPL, the average speed was higher in CVPL. Hence, for both traffic safety and operation the CVPL approach outperformed CVWPL approach.

Figure 3 shows the decreasing trend of standard deviation of speed and headway for CVWPL and CVPL approaches with increasing MPRs. As seen from the figure, the higher the percentage of the CVs implemented, the lower were the standard deviations of speed and headway, and therefore the higher were the safety benefits achieved.

Apart from statistical significance, Figure 4*a* and *b* compares the profile of both the surrogate measures of safety under base, CVWPL and CVPL scenario in 100 % MPR. For every 2-min time interval which is denoted in the *x* axis, the standard deviation of speed and standard deviation of headway (denoted in *y* axis) were calculated. Figure 4*a* and *b* illustrates that both CV approaches not only reduced the standard deviation of speed and headway but were able also to stabilize the profile. With lower variances in standard deviation of speed and headway these CV technologies are expected to decrease the crash risks.

The RCRI was considered as another surrogate measure for rear-end crashes. The Chi-square test was applied to test the significance in differences of RCRI between base scenario and CV scenarios. The percentages of vehicles under potential rear-end crash risk for different scenarios are listed in Table 2 with the Chi-square significance test. It can be seen from Table 2 that the percentages of potential rear-end crash observations were

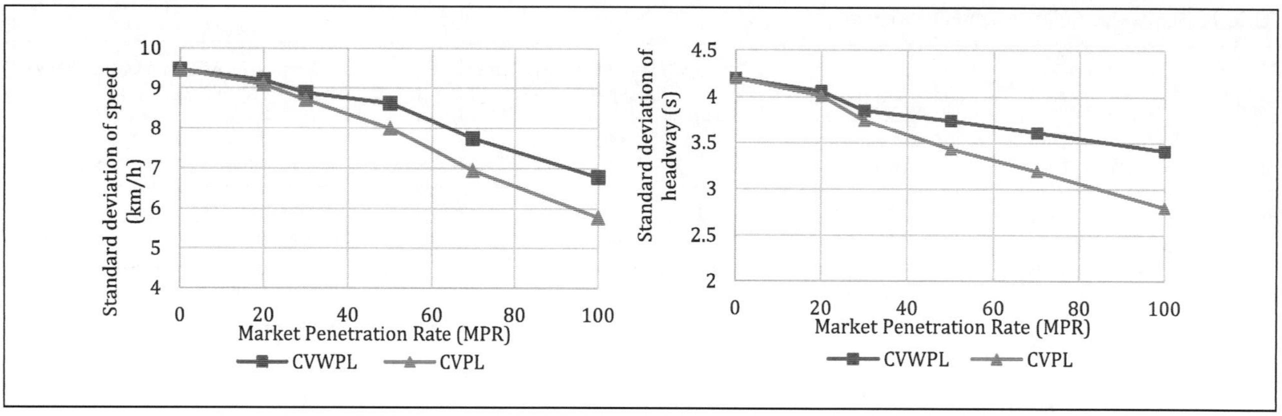

Figure 3. Reduction of surrogate measures of safety with different MPRs.

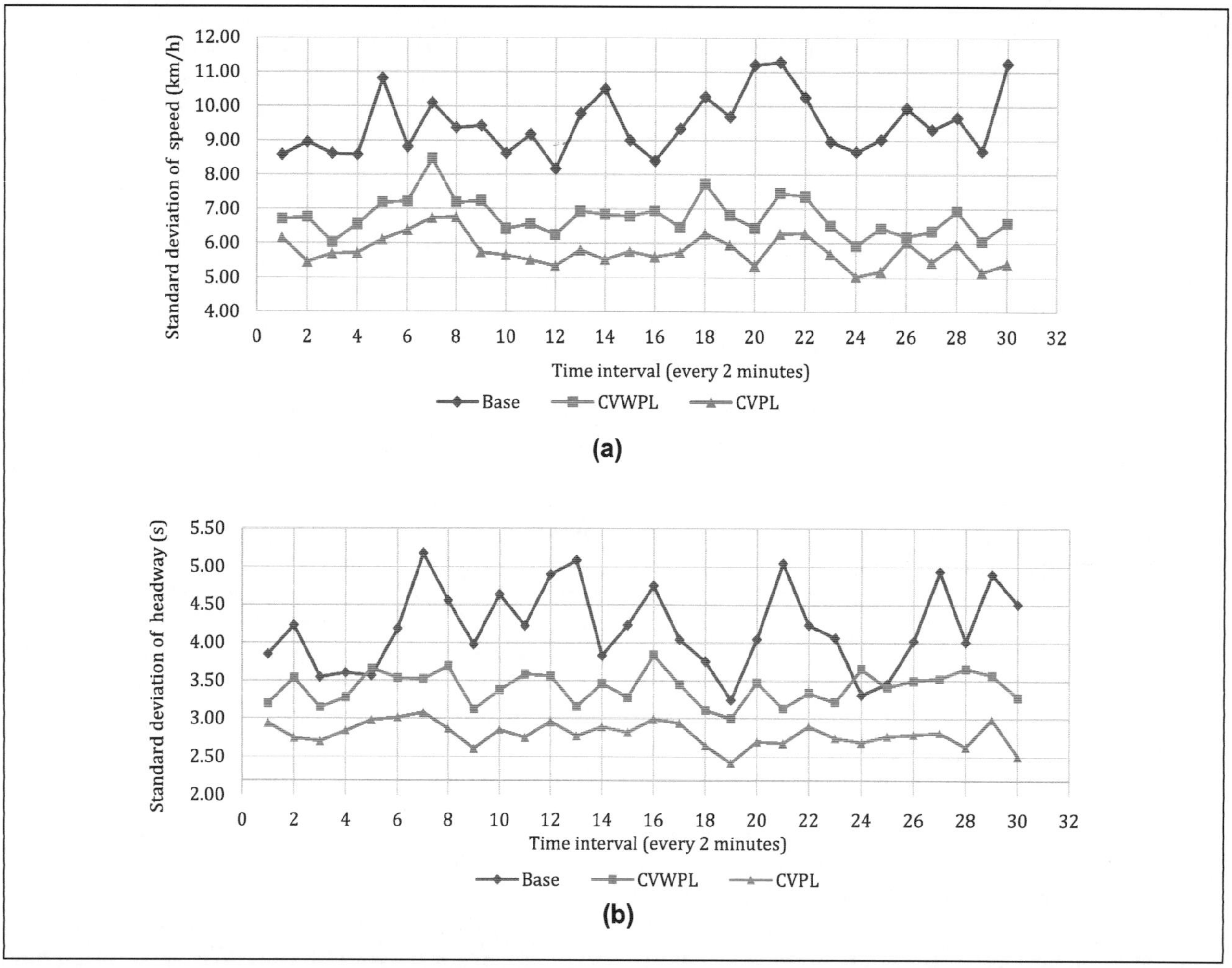

Figure 4. Stabilize profile of surrogate measures of safety at 100% MPR.

lower for CVWPL and CVPL than the base condition. At 100% MPR, the percentage of vehicles with potential rear-end crash risks were 11.55% lower in CVWPL and 14.67% lower in CVPL compared to the base condition. Hence, the rear-end crash risk decreased with the application of CV technologies. Also, the CVPL approach performed better than the CVWPL approach for each MPR in terms of RCRI. It was also found that at least 30%

Table 2. Rear-end Crash Risk with Different MPRs

MPR	Classification	Total observation	Number of potential rear-end crash observation	Comparison	Chi-square	P-value
20%	Base	10035	4161 (41.46%)	Base vs CVWPL	0.780#	0.3770
	CVWPL	10034	4099 (40.85%)	Base vs CVPL	3.274#	0.0704
	CVPL	10035	4035 (40.21%)	CVWPL vs CVPL	0.858#	0.3544
30%	Base	10035	4161 (41.46%)	Base vs CVWPL	23.487	<0.0001
	CVWPL	10037	3823 (38.12%)	Base vs CVPL	39.848	<0.0001
	CVPL	10030	3725 (37.11%)	CVWPL vs CVPL	2.151#	0.1425
50%	Base	10035	4161 (41.46%)	Base vs CVWPL	75.775	<0.0001
	CVWPL	10035	3561 (35.49%)	Base vs CVPL	118.091	<0.0001
	CVPL	10035	3414 (34.03%)	CVWPL vs CVPL	4.704	0.0301
70%	Base	10035	4161 (41.46%)	Base vs CVWPL	169.646	<0.0001
	CVWPL	10035	3270(32.59%)	Base vs CVPL	264.023	<0.0001
	CVPL	10031	3055 (30.46%)	CVWPL vs CVPL	10.548	0.0012
100%	Base	10035	4161 (41.46%)	Base vs CVWPL	291.941	<0.0001
	CVWPL	10040	3003 (29.91%)	Base vs CVPL	480.641	<0.0001
	CVPL	10037	2689 (26.79%)	CVWPL vs CVPL	24.045	<0.0001

Note: #Difference is insignificant at the 5% level.

MPR was needed to have significant reduction in rear-end crash risk. Additionally, CVPL achieved higher reductions of RCRI compared to CVWPL when the MPRs were equal or greater than 50%. It is worth mentioning that, the higher the MPRs implemented, the lower were the potential rear-end crash observations, and therefore the higher were the safety benefits achieved.

Overall, the deployment of CVs in reduced visibility conditions would significantly decrease the standard deviation of speed, standard deviation of headway, and RCRI; thereby might decrease the probability of crashes.

Conclusion

Traffic flow characteristics deteriorate significantly in fog conditions compared to normal weather conditions which might result in high crash risk. In order to improve traffic safety in fog conditions, two CV strategies were applied in microsimulation. The strategies include CVWPL and CVPL. A car-following model for both approaches was used with an assumption that the CVs would follow this car-following behavior in fog conditions. Three surrogate measures of safety including the standard deviation of speed, the standard deviation of headway, and RCRI were considered as safety indicators in this study. The safety benefits were observed under different MPRs for both approaches. In general, both CV approaches improved safety in fog conditions by providing significant reductions in standard deviation of speed, standard deviation of headway, and RCRI. It was found that the higher the MPRs of CV implemented, the higher were the safety benefits achieved. Maximum improvement was found to be at 100% MPR. A minimum of 30% MPR was needed to observe significant safety benefits of the applied CV approaches compared with the base scenario. The results showed that the CVPL significantly outperformed CVWPL in terms of three surrogate measures of safety. It was also found that at least 50% MPR was needed to achieve the safety benefits for the CVPL compared with CVWPL. To be more specific, both approaches of CV technology achieved significant safety benefits over the base scenario with at least 30% MPR. Additionally, CVPL achieved higher safety benefits compared with CVWPL when the MPRs were equal to or greater than 50%. From the profiles of standard deviation of speed and headway, it was found that the variances of these values decreased thereby providing a stabilized flow with fewer crash risks. On the other hand, simulation results asserted that speed was higher in both CV approaches compared with the base scenario. Therefore, the CV technologies not only improved traffic safety but also enhanced traffic operation. However, the average speed was higher in CVPL compared with CVWPL. Hence, taking both traffic safety and operation into consideration, the CVPL approach performed better than CVWPL approach.

For the car-following model, this study considered several parameters implemented in previous studies. However, the optimization of these parameters was out of the scope for this study. This study can be a good platform for further analysis with a combination of VSL and CV technologies. In this regard, V2I protocol might be useful with a combination of V2V communication under CV environment.

As a follow-up study, a full-scale field experiment may be considered. Nevertheless, the experiment will be limited for several reasons. First of all, the effects of CV by market penetration rate (MPR) were tested in this study. Nevertheless, the full market penetration of CVs will not be accomplished in the near future. Thus, it is difficult to incorporate the effective full-scale field experiment with V2V communication. A full-scale field experiment with a small group of experimental cars with V2V communication might be needed to substantiate and extend the results of this simulation study. That would be very important to policy makers or researchers working toward the implementation of CV technologies.

Acknowledgments

All opinions expressed in this study are those of the authors and not the sponsor. The authors acknowledge the support of THE TIER 1 UTC SAFER-SIM. SAFER-SIM is a RITA funded UTC Center lead by the University of Iowa.

Author Contributions

The authors confirm contribution to the paper as follows: study conception and design: Md Sharikur Rahman, Mohamed Abdel-Aty, Ling Wang; data collection: Md Sharikur Rahman; analysis and interpretation of results: Md Sharikur Rahman, Mohamed Abdel-Aty, Ling Wang, Jaeyoung Lee; draft manuscript preparation: Md Sharikur Rahman, Mohamed Abdel-Aty, Ling Wang, Jaeyoung Lee. All authors reviewed the results and approved the final version of the manuscript.

References

1. Peng, Y., M. Abdel-Aty, Q. Shi, and R. Yu. Assessing the Impact of Reduced Visibility on Traffic Crash Risk Using Microscopic Data and Surrogate Safety Measures. *Transportation Research Part C: Emerging Technologies*, Vol. 74, 2017, pp. 295–305.
2. Abdel-Aty, M., A. A. Ekram, H. Huang, and K. Choi. A Study on Crashes Related to Visibility Obstruction Due to Fog and Smoke. *Accident Analysis & Prevention*, Vol. 43, No. 5, 2011, pp. 1730–1737.
3. Hassan, H. M., M. A. Abdel-Aty, K. Choi, and S. A. Algadhi. Driver Behavior and Preferences for Changeable Message Signs and Variable Speed Limits in Reduced Visibility Conditions. *Journal of Intelligent Transportation Systems*, Vol. 16, No. 3, 2012, pp. 132–146.
4. Hassan, H., M. Abdel-Aty, and A. Oloufa. Effect of Warning Messages and Variable Speeds in Different Visibility Conditions. Presented at 90th Annual Meeting of the Transportation Research Board, Washington, D.C., 2011.
5. Yu, R., and M. Abdel-Aty. An Optimal Variable Speed Limits System to Ameliorate Traffic Safety Risk. *Transportation Research Part C: Emerging Technologies*, Vol. 46, 2014, pp. 235–246.
6. Lee, C., B. Hellinga, and F. Saccomanno. Evaluation of Variable Speed Limits to Improve Traffic Safety. *Transportation Research Part C: Emerging Technologies*, Vol. 14, No. 3, 2006, pp. 213–228.
7. Hellinga, B., and M. Mandelzys. Impact of Driver Compliance on the Safety and Operational Impacts of Freeway Variable Speed Limit Systems. *Journal of Transportation Engineering*, Vol. 137, No. 4, 2011, pp. 260–268.
8. Al-Ghamdi, A. S. Experimental Evaluation of Fog Warning System. *Accident Analysis & Prevention*, Vol. 39, No. 6, 2007, pp. 1065–1072.
9. Abdel-Aty, M. A., A. Oloufa, Y. Peng, T. Shen, X. Yang, and J. Lee. *Real Time Monitoring and Prediction of Reduced Visibility Events on Florida's Highways.* Report No. BDV24 962-01. Florida Department of Transportation, FDOT, 2014.
10. Broughton, K. L. M., F. Switzer, and D. Scott. Car Following Decisions Under Three Visibility Conditions and Two Speeds Tested with a Driving Simulator. *Accident Analysis and Prevention*, Vol. 39, No. 1, 2007, pp. 106–116.
11. Caro, S., V. Cavallo, C. Marendaz, E. R. Boer, and F. Vienne. Can Headway Reduction in Fog be Explained by Impaired Perception of Relative Motion? *Human Factors*, Vol. 51, No. 3, 2009, pp. 378–392.
12. Brooks, J. O., M. C. Crisler, N. Klein, R. Goodenough, R. W. Beeco, C. Guirl, P. J. Tyler, A. Hilpert, Y. Miller, J. Grygier, B. Burroughs, A. Martin, R. Ray, C. Palmer, and C. Beck. Speed Choice and Driving Performance in Simulated Foggy Conditions. *Accident Analysis and Prevention*, Vol. 43, No. 3, 2011, pp. 698–705.
13. Pang, M. B., S. S. Zheng, and Z. H. Cai. Simulation of Three Lanes One-way Freeway in Low Visibility Weather by Possible Traffic Accidents. *Physica A: Statistical Mechanics and its Applications*, Vol. 433, 2015, pp. 161–170.
14. Lyles, R. W., W. C. Taylor, D. Lavansiri, and J. Grossklaus. Field Test of Variable Speed Limits in Work Zones. Presented at 83rd Annual Meeting of the Transportation Research Board, Washington, D.C., 2004.
15. Abdel-Aty, M., K. Haleem, R. Cunningham, and V. Gayah. Application of Variable Speed Limits and Ramp Metering to Improve Safety and Efficiency of Freeways. *Proc., 2nd International Symposium on Freeway and Tollway Operations*, 2009, pp. 1–13.
16. Abdel-Aty, M., J. Dilmore, and A. Dhindsa. Evaluation of Variable Speed Limits for Real-time Freeway Safety Improvement. *Accident Analysis and Prevention*, Vol. 38, No. 2, 2006, pp. 335–345.
17. Wang, L., M. Abdel-Aty, and J. Lee. Implementation of Active Traffic Management Strategies for Safety of a Congested Expressway Weaving Segment. *Transportation Research Record: Journal of the Transportation Research Board*, 2017. 2635: 28–35.
18. Rämä, P. Effects of Weather-controlled Variable Speed Limits and Warning Signs on Driver Behavior. *Transportation Research Record: Journal of the Transportation Research Board*, 1999. 1689: 53–59.
19. Perrin, J., P. T. Martin, and B. Coleman. Testing the Adverse Visibility Information System Evaluation (ADVISE): Safer Driving in Fog. Presented at 81st Annual Meeting of the Transportation Research Board, Washington, D.C., 2002.
20. Fyfe, M., and T. Sayed. Safety Evaluation of Connected Vehicles for a Cumulative Travel Time Adaptive Signal Control Microsimulation Using the Surrogate Safety

Assessment Model. Presented at 96th Annual Meeting of the Transportation Research Board, Washington, D.C., 2017.

21. Olia, A., H. Abdelgawad, B. Abdulhai, and S. N. Razavi. Assessing the Potential Impacts of Connected Vehicles: Mobility, Environmental, and Safety Perspectives. *Journal of Intelligent Transportation Systems*, Vol. 20, No. 3, 2016, pp. 229–243.
22. Paikari, E., S. Tahmasseby, and B. Far. A Simulation-based Benefit Analysis of Deploying Connected Vehicles Using Dedicated Short Range Communication. *Proc., of IEEE Intelligent Vehicles Symposium, 2014.*
23. Tian, D., J. Zhou, Y. Wang, Z. Sheng, H. Xia, and Z. Yi. Modeling Chain Collisions in Vehicular Networks with Variable Penetration Rates. *Transportation Research Part C: Emerging Technologies*, Vol. 69, 2016, pp. 36–59.
24. Rahman, M. S., and M. Abdel-Aty. Longitudinal Safety Evaluation of Connected Vehicles' Platooning on Expressways. *Accident Analysis and Prevention*, 2017. pii: S0001-4575(17)30450-5. doi: 10.1016/j.aap.2017.12.012.
25. Nezamuddin, N., N. Jiang, T. Zhang, and S. T. Waller. *Traffic Operations and Safety Benefits of Active Traffic Strategies on TxDOT Freeways.* Report No. FHWA. TX-12/0-6576-1. Texas Department of Transportation, USA. 2011.
26. Zhang, B. *Traffic Conflict Analysis Under Fog Conditions Using Computer Simulation.* University of Central Florida, 2015.
27. Abdelfatah, A., M. Marzouk, A. Garib, and A.-H. Hussain. Effect of Fog on Freeway Performance: Initial Results. *Proc., 13th World Conference on Transport Research*, Rio De Janeiro, Brazil, 2013, pp. 1–11.
28. Habtemichael, F., and L. Picado-Santos. Sensitivity Analysis of VISSIM Driver Behavior Parameters on Safety of Simulated Vehicles and their Interaction with Operations of Simulated Traffic. Presented at 92nd Annual Meeting of the Transportation Research Board, Washington, D.C., 2013.
29. Lownes, N., and R. Machemehl. Sensitivity of simulated capacity to modification of VISSIM driver behavior parameters. *Transportation Research Record: Journal of the Transportation Research Board*, 2006. 1988: 102–110.
30. Treiber, M., A. Hennecke, and D. Helbing. Congested Traffic States in Empirical Observations and Microscopic Simulations. *Physical review E*, Vol. 62, No. 2, 2000, pp. 1805–1824.
31. Li, Y., Z. Li, H. Wang, W. Wang, and L. Xing. Evaluating the Safety Impact of Adaptive Cruise Control in Traffic Oscillations on Freeways. *Accident Analysis & Prevention*, Vol. 104, 2017, pp. 137–145.
32. Kesting, A., M. Treiber, and D. Helbing. Enhanced Intelligent Driver Model to Access the Impact of Driving Strategies on Traffic Capacity. *Philosophical Transactions of the Royal Society of London A: Mathematical, Physical and Engineering Sciences*, Vol. 368, No. 1928, 2010, pp. 4585–4605.
33. Milanés, V., and S. E. Shladover. Modeling Cooperative and Autonomous Adaptive Cruise Control Dynamic Responses Using Experimental Data. *Transportation Research Part C: Emerging Technologies*, Vol. 48, 2014, pp. 285–300.
34. Gettman, D., and L. Head. Surrogate Safety Measures from Traffic Simulation Models. *Transportation Research Record: Journal of the Transportation Research Board*, 2003. 1840: 104–115.
35. Abdel-Aty, M., M. M. Ahmed, J. Lee, Q. Shi, and M. Abuzwidah. *Synthesis of Visibility Detection Systems.* Report BDK78, 977-11. University of Central Florida, Orlando, FL, 2012.
36. Oh, C., S. Park, and S. G. Ritchie. A Method for Identifying Rear-end Collision Risks Using Inductive Loop Detectors. *Accident Analysis and Prevention*, Vol. 38, No. 2, 2006, pp. 295–301.
37. American Association of State Highway and Transportation Officials. *A Policy on Geometric Design of Highways and Streets.* American Association of State Highway and Transportation Officials, Washington, D.C., 2004.

The Standing Committee on Intelligent Transportation Systems (AHB15) peer-reviewed this paper (18-00846).

Article

TRR
JOURNAL OF THE TRANSPORTATION RESEARCH BOARD

Transportation Research Record
2018, Vol. 2672(19) 102–111

DOI: 10.1177/0361198118791612
journals.sagepub.com/home/trr
SAGE

Real-Time Magnetic Length-Based Vehicle Classification: Case Study for Inductive Loops and Wireless Magnetometer Sensors in Oklahoma State

Walid Balid[1] and Hazem H. Refai[1]

Abstract

Providing reliable, real-time vehicle volume and classification information is vital for 21st century intelligent transportation systems (ITS). Large vehicles have a major impact on traffic flow, as well as road maintenance. Information about passenger car versus truck volume is crucial for all transportation agencies. The work presented here provides a comprehensive method to develop length-based vehicle classification (LBVC) techniques that could be implemented using inductive loops (IDL) or magnetometer sensors (MAG). Distinctive LBVC schemes were developed to bin vehicles into groups based on structural similarity and statistical characteristics. Data collection, including vehicle magnetic length (VML) estimates using IDL and MAG, was performed at different sites located on Oklahoma highways and rural roadways to capture various traffic characteristics. Video images were utilized as ground-truth for accurate data labeling. Extensive data analyses, including machine learning methods and probabilistic modeling, were conducted to define decision boundaries for developed LBVC schemes. Three scenarios were developed for determining optimal thresholding methods: total classification accuracy maximization, per-group classification error minimization, and equal classification error optimization. Evaluation revealed consistent and accurate performance for all developed schemes. Classification accuracies of 97.70% and 99.00% were reported using MAG and IDL, respectively. The developed classification models are computationally efficient and can provide real-time LBVC data. The models are intended to supplement or replace axle-based data collection methods used throughout Oklahoma. The methodology developed in this work will also benefit other states and territories interested in developing LBVC schemes.

An escalating need for freight transport has led to a significant increase in truck traffic on U.S. highways nationwide. The U.S. Federal Highway Administration (FHWA) predicts a 23% increase in vehicle miles traveled by 2032 (i.e., 1.04% annual growth) (*1*). This rapid increase would significantly influence traffic safety and conditions. Moreover, according to (*2*), about 4,000 fatalities and more than 95,000 injuries in 2013 involved large trucks. Compared with passenger vehicles, trucks (or commercial vehicles) have a significant impact on pavement service life. For example, a vehicle hauling a 40-ton load causes over 4,000 times more road damage than a truck hauling a 5-ton load. This type of impact requires appropriate and distinctive traffic planning strategies to mitigate the effect of anticipated increases in truck traffic. Adequate information about freight volume and vehicle classification is necessary to support geometric and structural design of roadways and bridges.

Existing Knowledge

The FHWA requires all states to report various traffic data, including vehicle count and class. Each year, the Oklahoma Department of Transportation (ODOT), not unlike other U.S. department of transportation (DOT) agencies, is responsible for collecting permanent and temporary vehicle counts and classification data from over a hundred weight-in-motion (WIM) and automatic vehicle classification (AVC) sites located throughout the state. The data are initially used to develop various transportation planning strategies and management processes. Both WIM and AVC sites currently use inductive loops (IDL)

[1]University of Oklahoma, Tulsa, OK

Corresponding Author:
Address correspondence to Walid Balid: walid@ou.edu

and piezoelectric sensors for reporting vehicle count and class data, according to FHWA Scheme F with 13 classes (*3*). This scheme was developed during the 1980s based on manual classification of visually identifiable vehicle characteristics. The scheme was later amended to provide computational classification based on number of axles and wheelbase axle spacing. The axle-based vehicle classification (ABVC) system requires the installation of intrusive sensors that are plagued with several drawbacks, chiefly their reliance on pavement geometry. Resultant pavement deterioration is cause for unreliable ABVC data. A study conducted by the University of Oklahoma reported classification inaccuracy rates resulting from sensor misconfiguration at the state's AVC sites (*4*). ABVC schemes have proven problematic because of vehicle class overlap, especially for vehicles pulling one-, two-, or three-axle trailers (*4*). For example, one study reported that 43% of Class 3 vehicles were falsely classified as Class 2, and 45% of Class 5 were falsely classified as Class 3. In fact, three- or four-axle Class 3 trucks often overlap with Class 8 trucks, which causes Class 3 to be classified as Class 8. Class 6 trucks also overlap with Class 4 three-axle buses. Significant overlap has been observed for three- or four-axle Class 5 with trailers and Classes 8 and 5, as well as buses (*4*).

Prohibitive costs associated with ABVC scheme technologies, in addition to high regular maintenance and calibration costs, limit spatial distribution of such systems. Often ABVC schemes are placed only at strategic points where heavy truck volume has been previously observed, limiting the availability of ABVC data. Additionally, the feasibility of collecting ABVC data in urban areas is significantly hindered by bumper-to-bumper traffic during peak hours. Vehicle acceleration/deceleration and stop-and-go scenarios at signals are also factors contributing to erroneous data.

Length-Based Vehicle Classification Schemes

Most traffic planning and engineering studies require information only about car versus truck volume, and not so much about detailed data such as number of axles. Traffic flow characteristics are directly affected by long vehicles (i.e., trucks), and trucks have a major impact on road service life (*5*). Therefore, it is crucial for U.S. DOT agencies to estimate percentage of roadway damage. Accordingly, the *FHWA Traffic Monitoring Guide* (*3*) provides a generic length-based vehicle classification (LBVC) scheme based on nationwide IDL data. However, it is practically not feasible to define LBVC boundaries for FHWA F Scheme with 13 classes. VML overlapping often occurs between multiple classes (e.g., Class 2 through 5 and Class 8 through 10). The proposed LBVC scheme (*3*) categorizes vehicles into four groups, namely passenger vehicles (PV), single-unit trucks (SU), combination-unit trucks (CU), and multi-unit trucks (MU). The scheme is inexpensive and works best with dual-IDL stations. Notably, the FHWA guide permits U.S. states to develop their own LBVC scheme and define length threshold boundaries that best fit vehicle characteristics that are consistent within the state.

Because there is no generic definition for LBVC boundaries, several states have begun to adopt an LBVC scheme to replace outdated ABVC based on local vehicular data collected from AVC and WIM sites located throughout a state's roadways (*6, 7*). Many states have defined different length boundaries for their developed schemes (e.g., a four-group LBVC scheme and a three-group scheme by combining groups 3 and 4). For example, the Ohio DOT has adapted a statewide three-group LBVC scheme that bins FHWA Scheme F vehicles into PV, SU, and MU (*8, 9*). Florida, Illinois, Washington, and Idaho have also adopted three-group LBVC schemes. Their systems classify Scheme F vehicles into PV, short trucks (ST), and long trucks (LT) (*10*). Minnesota's DOT (MDOT) has proposed a four-group LBVC scheme—motorcycles (MC), short (S), medium (M), and long (L)—and recognizes a fifth group, namely very long (VL), in areas with multi-trailer trucks (*6*). Table 1 summarizes LBVC schemes from several states and lists their boundaries, which were identified from data collected using IDLs. Given this variety of classification schemes, it is interesting that very few efforts have been made to assess the feasibility of defining magnetic length boundaries based on data collected using wireless magnetometer sensors (MAG).

Unlike vehicle magnetic signatures (VMS) produced by IDLs, vehicle signatures sampled using MAG have less consistent peak magnitude. This is because of several factors, such as differences in sensitivity, vehicle trajectory, and non-symmetry of detection zone. Moreover, IDLs are active magnetic sensors, generating a magnetic field in the IDL zone, and resonate at a constant frequency, which increases when the generated field is induced by a passing vehicle. When this occurs, a detection state is triggered. Unlike IDLs, MAG are passive sensors that rely on the disturbance of the Earth's magnetic field relative to the presence of a metallic object (i.e., vehicle). Therefore, the amount of preamble ferrous material in a vehicle structure plays a major role in a sensor's detection range (*11–13*). A vehicle with a large amount of steel can be detected from quite a distance away. Differences in the amount of metal create significant variations in vehicle magnetic length (VML) per class, making it extremely challenging to use MAG for LBVC.

Earlier Studies

A significant number of studies that leverage IDLs in various traffic surveillance applications (e.g., vehicle

Table 1. Threshold Boundaries Defined by Different States in USA

Three-group LBVC schemes							
Description of vehicles	FHWA-ABVC	Illinois	Washington	Idaho	Florida	Ohio	New York
Passenger vehicles	F1–F3	0→6.7 m	0→6.1 m	0→7 m	0→6.52 m	0→8.53 m	0→6.7 m
Short trucks	F4–F7	6.7→11.9 m	6.1→12.8 m	7→12.2 m	6.52→12.9 m	8.53→14 m	6.7→11.9 m
Long trucks	F8–F13	≥11.9 m	≥12.8 m	≥12.2 m	≥12.9 m	≥14 m	≥11.9 m

Four-group FHWA-LBVC scheme boundaries; MDOT-LBVC scheme boundaries					
Description of vehicles	FHWA-ABVC	FHWA-LBVC	Description of vehicles	FHWA-ABVC	Minnesota
Passenger vehicles	F1–F3	0→3.96 m	Motorcycle	F1	0→1.98 m
Single unit trucks	F4–F7	3.96→10.67 m	Short vehicle	F2—F3	1.98→6.55 m
Combination trucks	F8–F10	10.67→18.59 m	Medium vehicle	F4—F7	6.55→14.93 m
Multi-trailer trucks	F11–F13	18.59→36.58 m	Long vehicle	F8—F13	14.93→36.58 m

detection, speed estimation, vehicle re-identification, and classification) can be found in the literature (*14–17*). LBVC using IDLs can be traced to 2001 (*6*). In 2008, 10 states led by MDOT aimed to establish standards for vehicle length-based measurements, as well as to define specific groups and their boundaries for LBVC (*7*). Since then, several other LBVC studies have been published. Researchers proposed an algorithm to classify vehicles into regular- and long-cars by estimating time gaps between consecutive vehicles and per-vehicle occupation time (*14*). The dataset was limited to 2,547 records of which 96% were cars. A high false positive rate was observed when classifying long vehicles. The system was also tested for a limited time under free-flow conditions on a highway where vehicle speed was assumed at 50 mph. In another study, dual-IDL were used for LBVC based on a scheme developed by the Ohio DOT (*15*). Although classification rate was high, the dataset was limited to 5,675 records, 96% of which belonged to Group 1. Another LBVC used vehicle speed estimates and length from a single IDL (*16*). Vehicles were classified into three groups, again based on the Ohio DOT LBVC scheme. Although a high classification rate was reported for Group 1, classification rate was not accurate for Groups 2 and 3. An interesting vehicle classification study reported promising results by using a heuristic method combined with a decision tree and *K*-means clustering approaches for building a vehicle classification model and, subsequently, classifying vehicles into 15 classes based on features extracted from VMS that had previously been collected using a single IDL (*17*). The dataset was limited to 3,718 records of which 80% belonged to Class 1. Classes 8 through 15 were highly underrepresented. Additionally, this approach requires high processing power and is thought to be valid only for specific IDL detectors that are equipped to capture detailed VMS. Moreover, deterioration in pavement geometry is expected to change IDL waveform and result in classification errors.

Very few studies can be found in the literature for LBVC schemes utilizing MAG technology. Admittedly, estimating vehicle physical length (VPL) using MAG is very challenging. Although some studies (*18–22*) have reported promising classification results using MAG, study results do not reflect actual performance, as testing datasets were small and limited (e.g., trucks found in Classes 5 through 13 were highly underrepresented). Most proposed methods are computationally expensive and require transmission of big data to an access point for processing. This is a fundamental limitation in the proposed wireless sensor network, as most are battery-powered with networks operating in the unlicensed spectrum (i.e., bandwidth is shared among different technologies and devices). A significant amount of data loss is highly probable as a consequence of an unstable wireless link resulting from either interference or heavy trucks traveling over the detection zone. Moreover, most presented studies have been conducted in controlled environments with limited evaluation for measuring classification accuracy over a full range of speeds under various traffic conditions.

Contribution

Presently, the Oklahoma DOT relies only on FHWA Scheme F for vehicle classification. Data collected from AVC and WIM sites across the state have demonstrated classification errors only months after calibration (*4*). The objective of the work presented in this paper is investigating the feasibility of using MAG for accurate LBVC based on VML estimation. A method is proposed for defining length thresholds between groups by means of probabilistic modeling and machine learning (ML). LBVC schemes that bin ABVC schemes into groups were developed and evaluated to find optimal boundaries for LBVC schemes. Determined boundaries were evaluated against boundaries developed by other states (See Table 1). The developed schemes are computationally efficient and can be executed in real-time, either on server side or on a master MAG node. The developed method was also used to define length boundaries for per-vehicle record (PVR) data collected from calibrated WIM sites in Oklahoma. Again, this work focuses on the development and implementation of an LBVC. The MAG-LBVC study was implemented in real-time using an innovative, nonintrusive, inexpensive, and portable vehicular traffic monitoring sensor, namely the intelligent vehicle counting and classification sensor (*i*VCCS). Companion papers have previously reported detailed information on sensor design, algorithm development, and system integration (*11–13*). Evaluation of the developed LBVC schemes was conducted using data collected from vehicles traveling on highways and urban roads at a full range of speeds (i.e., 15–88 mph) under various traffic conditions, including free-flow, congested traffic, and stop-and-go at a signalized intersection. This same approach can be standardized and generalized for use by other DOT agencies to develop LBVC not only for IDLs or MAG, but also for other nonintrusive detection technologies, such as microwave radars and vision systems.

Paper Organization

The balance of this paper is organized as follows. Section II presents field tests and data collection methodologies. Development of a real-time LBVC scheme and various data analysis and modeling procedures are detailed in Section III. Discussion and a conclusion are provided in Section IV and Section V, respectively.

Field Tests and Data Collection Methodology

MAG Deployment Setup

Two field tests were conducted—one on highway U.S. 412, Chouteau, Oklahoma 74337, and the other on an urban road at South Yale Avenue, Tulsa, Oklahoma 74135. Several *i*VCCS nodes were installed on the lane center of each roadway surface. Data from two nodes—one in each lane—were employed. All nodes were positioned as follows: *x*-axis along traffic direction; *y*-axis perpendicular to traffic; and *z*-axis perpendicular to the ground, pointing upward. An intelligent access point (*i*AP) was installed on the road shoulder and used to configure all nodes. Video images were used as ground truth for vehicle count and class validation. A Road Runner 3 Kit by Diamond Traffic Inc. (*23*) was used as ground truth for vehicle speed and length validation. Camera, Road Runner kit, *i*VCCS nodes, and *i*APs were synchronized using a GPS-based clock. Vehicles classes were manually extracted from video images by counting the number of axles for each vehicle. The human error factor of the extraction process was significantly reduced by comparing extracted classes from video images with the number of axles counted by the Road Runner kit. Vehicular measurements by *i*VCCS were later evaluated against IDL station data.

MAG and IDL Data Collection

Dataset size plays a vital role in developing an accurate LBVC scheme. Most earlier studies on vehicle classification (*14–22*) relied on small training sets consisting of only a few hundred vehicles—the majority of which were passenger cars—to train an intelligent classifier. Doing so significantly affects the accuracy of any classification algorithm. Moreover, such a practice limits the number of distinguishable classes. In this study, three different datasets were employed. Two were short-term deployments collected using *i*VCCS throughout the course of the day on both highway and urban roads. The other was a long-term deployment (i.e., 73 days) obtained from the highway WIM station immediately following station calibration.

MAG Data Preprocessing

To accurately define boundaries for an LBVC scheme, training datasets must be recorded and preprocessed before inputting the data into a classification algorithm (e.g., data cleaning; data imputation; filtering and smoothing; data alignment and labeling; and features computation, such as occupancy, speed, and length). *i*VCCS data preprocessing was conducted as follows.

Assume a vehicle traveling at speed of $v = 85\,\text{mph}$ has minimal length $l_V = 1.8\,\text{m}$ (e.g., motorcycle). Using (1), it is possible to calculate the minimum number of samples S_{VSL} (i.e., worst case scenario) given a maximum sampling rate $f_s = 400\,\text{Hz}$. The leading and trailing detection edge of the detection zone are denoted as l_{lde} and l_{tde}, respectively. α is correction factor. Therefore, a record belonging to a signature with length less than S_{VSL} data points will be considered an outlier and will be dropped from the dataset.

$$S_{\text{VSL}} = 3.6 \times \left(\frac{l_{\text{lde}} + l_{\text{tde}} + l_V}{v} \right) \times f_s \times \alpha \tag{1}$$

Data alignment and labeling is necessary to train an intelligent classifier when using a supervised ML method. Ground-truth data (i.e., video) was used to assign class labels to the vehicle class dataset collected using MAG. Erroneous records caused by double detection, as indicated by comparing classifications against ground-truth data, are removed from the dataset. Having to do so is rare, however, as *i*VCCS has demonstrated outstanding performance and precision in vehicle detection and speed estimation (*11–13*), (*24*). In this study only VML was used to train the classifier. Classification based on VML requires precise vehicle detection and highly accurate speed estimation. The most accurate and computationally efficient method to estimate vehicle speed is calculating travel time between two longitudinally positioned nodes ($N_A \rightarrow N_B$) separated by a distance d using (2). Once vehicle speed is known, VML (i.e., disturbance in the Earth's magnetic field induced by a vehicle structure) can be found as the product of vehicle speed and sensor occupancy time T_{Occ} using (3).

$$\begin{aligned} \overline{v_i} &\approx \frac{d^{(N_A \rightarrow N_B)}}{T_A^{(N_B)} - T_A^{(N_A)}} \approx \frac{d^{(N_A \rightarrow N_B)}}{T_D^{(N_B)} - T_D^{(N_A)}} \\ &\approx 2\frac{d^{(N_A \rightarrow N_B)}}{T_A^{(N_B)} - T_A^{(N_A)} + T_D^{(N_B)} - T_D^{(N_A)}} \end{aligned} \quad (2)$$

$$\begin{aligned} \overline{\text{VML}} &= \bar{v} \times T_{\text{Occ}}^{(N_i)} = \bar{v} \times \left(T_D^{(N_i)} - T_A^{(N_i)}\right) \\ &= \bar{v} \times \frac{T_D^{(N_A)} - T_A^{(N_A)} + T_D^{(N_B)} - T_A^{(N_B)}}{2} \end{aligned} \quad (3)$$

IDL Data Preprocessing

Preprocessing IDL data was significantly different, as it was not feasible to video record ground truth for long-term deployments. Therefore, a method employing domain specific outlier removal, as well as statistical tests such as the generalized extreme studentized deviant (ESD) test, Grubbs test, and Tietjen-Moore test (*25*), was developed to clean IDL data. The domain specific method analyzes the WIM PVR data file and eliminates entries that do not meet specific conditional testing criteria. Four testing conditions were used, namely vehicle speed range; number of axles counted by upstream IDL versus downstream IDL; maximum axle spacing versus class; and gross vehicle weight (GVW) versus class. Testing conditions were based on FHWA's *WIM Data Analyst's Manual* (*26*). For example, records were removed for the following conditions: 1) Records with a speed equal to 0 or some other extreme value; 2) Number of counted axles differs between loops; 3) Maximum axle spacing is larger than VPL; 4) GVW for a specific class significantly exceeds limits specified in (*26*), which is a common error between Class 3 or 5 with a trailer and Class 8; and 5) Number of counted axles for a specific class does not match the standard specified in (*26*) (e.g., Class 9 with four axles). Statistical tests remove outliers in a univariate dataset under an assumption that the underlying distribution can be approximated as a normal distribution (*25*), which is the case with classes in the study's IDL dataset. The original IDL dataset was collected over 73 days and included 554,332 records. The preprocessed dataset included 515,580 records, indicating 7% of records were removed from dataset for one of the aforementioned reasons.

One method for validating the output of the statistical test methods is fitting each class length distribution from the original dataset into a Gaussian distribution (*27*). Only records that fall within 3.4σ from the mean are used; this represents the 99.9% confidence level. Statistical measurements of the fitted dataset can be individually compared with those generated by a Tietjen-Moore test to ensure no records belonging to the statistical population of the fitted distribution were removed.

Development of a Real-Time LBVC Scheme

According to the FHWA's traffic monitoring guide, "If length-based classification is used, it should accommodate motorcycle identification as one of the groups" (*3*). In this study, a four-group LBVC scheme is proposed. Vehicles were grouped in each bin based on structural similarity and statistical measurements. Group 1 (G1) includes monocycles. Passenger cars, pickups, and SUVs are in G2. G3—the short-trailer group—includes buses, light trucks, and single unit trucks. G4 includes long vehicles, such as single-trailer and multi-trailer trucks. Statistical measurements for MAG and IDL datasets based on the proposed LBVC schemes are presented in Table 2.

Table 2. Statistical Measurements of VML Datasets Based on Proposed LBVC Schemes

	MAG					IDL				
Class	Mean	SD	Var	Max	Min	Mean	SD	Var	Max	Min
G1	1.6863	0.7899	0.6239	2.9106	0.8112	1.7203	0.5430	0.2948	2.7686	0.7112
G2	6.1783	0.9720	0.9447	12.5161	2.7606	4.9038	0.5628	0.3168	6.8580	3.2258
G3	9.9333	2.2571	5.0944	17.9241	6.2894	8.5328	1.8649	3.4777	13.7414	4.0640
G4	21.2157	2.9961	8.9765	34.6607	12.0606	20.4586	2.5696	6.6027	39.2430	10.6426

Note: SD = standard deviation; Var = variance; Max = maximum; Min = minimum.

LBVC Model Using ML

Intelligent classification algorithms learn to classify vehicles into predefined classes by statistically modeling the relationship between a vehicle's class and probabilistic distributions of features set (or predictors) extracted from VMS. Several ML classification methods (e.g., decision trees [DT], support vector machine [SVM], *k*-nearest neighbor [*k*NN], and naïve Bayes Classifier [NBC]) were evaluated, investigating best practices to build a system classification model and to infer optimal length boundaries of a real-time LBVC using MAG or IDL.

Each of the two VML datasets (i.e., MAG and IDL) was divided into two subsets—training and testing. The first was used for training the intelligent classifier and the other for assessing its performance. Predictive accuracy of fitted models was examined using 10-fold cross-validation. This method provides three functions: *a*) it builds *k*-different test sets from the same training set; *b*) it trains all models on the remaining training dataset; and *c*) it takes the best performing hypothesis on average to avoid overfitting.

All classification methods attempt to maximize total classification accuracy. However, each method implements different techniques. Notably, ML methods and their techniques are not the focus of this paper. Interested readers can review details in (*28*). Of great importance to report is that the radial basis function (RBF) kernel SVM outperformed other kernel types. Because SVM is a binary classifier, a one-versus-all approach was used. This means that *m* class models are fitted to the training data—one model for every class. When fitting the model for the *m*th class, training dataset labels are changed. Thus, the model is fitted to classify between two classes at once. During validation with the test dataset, *m* individual scores are estimated—one for every model. Then, the class with the highest score is chosen as the predicted class. For DT, a J84 version was implemented. This algorithm optimizes the tree size using "pruning" (i.e., cease splitting if the number of objects in a branch is smaller than a predefined value). Optimal values for other parameters (e.g., *k*-value in *k*NN; size of the tree and number of leaves in DT; *C* and gamma in SVM) were found using hyperparameter optimization algorithms.

Classifier performance was assisted using nine commonly used metrics, namely classification rate (C_R), mean absolute percentage error (MAPE), root-mean-square error (RMSE), root relative squared error (RRSE), true positive rate (TPR), false positive rate (FPR), *F*-measure (*F*1), Matthews's correlation coefficient (MCC), and area under receiver operating characteristic (ROC) curve (AUC). Mean absolute error (MAE) is a linear average of classification error magnitude. RMSE measures how far a true class is from predicted class. RRSE tells how a predicted class differs relatively from itself. TPR is an indicator of how many vehicles were classified correctly, and FPR is an indicator of how many vehicles were classified incorrectly. *F*1 measures the harmonic mean of the precision or positive prediction value (PPV) and FPR. MCC is an important balanced measure of classification quality even if classes are imbalanced. AUC measures overall quality of the classifier. Classification matrix (CM) includes true classes in rows and predicted classes in columns. TPR, FPR, *F*1, MCC, and ROC will be reported from this point as a weighted average score instead of a per-class score.

Table 3 presents performance results of all classification methods applied to the proposed LBVC schemes using the MAG and IDL datasets. Of interest is that all classification methods showed comparable accuracy when assigning a vehicle to a class. Results can be attributed to parameter optimization (i.e., hyperparameter optimization), which was conducted for each ML classification method. DT models were adopted primarily to identify length boundaries using a generated decision tree, mainly because they are quite easy to implement

Table 3. LBVC Scheme Performance Results for MAG and IDL Datasets

Criteria		Classification Method MAG				Classification Method IDL			
		DT	k-NN	C-SVM	NBC	DT	k-NN	C-SVM	NBC
Average	C_R	97.69%	97.79%	97.78%	97.75%	98.99%	98.98%	98.98%	98.99%
	MAE	0.022	0.0171	0.0111	0.0173	0.0099	0.0087	0.0051	0.0089
	RMSE	0.1057	0.0963	0.1054	0.0932	0.0704	0.0678	0.0715	0.0668
	RRSE	54.94%	50.08%	54.80%	48.47%	23.01%	22.13%	23.35%	21.82%
Weighted average	TPR	0.977	0.978	0.978	0.977	0.990	0.990	0.990	0.990
	FPR	0.238	0.195	0.206	0.204	0.029	0.030	0.030	0.029
	PPV	0.974	0.975	0.974	0.974	0.990	0.990	0.990	0.990
	F1	0.971	0.975	0.974	0.974	0.989	0.989	0.989	0.989
	MCC	0.838	0.849	0.848	0.846	0.975	0.974	0.974	0.975
	AUC	0.864	0.963	0.886	0.987	0.981	0.996	0.980	0.997

Table 4. Decision Boundaries for Proposed LBVC Schemes

Group	FHWA-S_F	ϒ-MAG	ϒ-IDL
G1	F_1	0.7→2.984 m	0.7→2.7686 m
G2	F_2–F_3	2.984→10.971 m	2.7686→6.858 m
G3	F_4–F_7	10.971→14.727 m	6.858→12.243 m
G4	F_8–F_{13}	>14.727 m	>12.243 m

and memory efficient. Slightly better performance can be observed with an IDL dataset when compared with a MAG dataset. This expected outcome can be attributed to the large variability in preamble ferrous materials in vehicle structures belonging to the same class. Length boundaries (ϒ) found by DT models are presented in Table 4 relative to corresponding FHWA Scheme F (S_F) classes.

One important observation from examining Table 3 is that MCC was slightly degraded in all classifiers. As stated earlier, MCC is a balance measure of statistical correlation between a true class and predicted class. A correlation coefficient of + 1 represents perfect agreement between prediction and observation; a 0 indicates random prediction; and −1 indicates total disagreement between prediction and observation. Declining MCC value, notwithstanding extremely high classification rate, is attributed to class imbalance in the training set. The primary problem with ML algorithms is that they try to maximize total classification accuracy, but imbalanced class distributions increase sensitivity of ML algorithms toward overrepresented classes (*29*).

Imbalanced vehicle class distributions are not unique to Oklahoma, where more than 97% of vehicles belong to Classes 2, 3, 5, 6, 8, and 9. Similar results can be found in (*14–22*), although the majority of studies do not consider this fact and most report classification rate only as an overall measure of classifier performance.

LBVC Model Using Probabilistic Modeling

Separating two neighboring classes from each other can be treated as a binary problem. As such, probabilistic models can be employed to determine optimal boundary decisions to separate neighboring classes in which vehicles have overlapping lengths. These methods can be implemented in real-time, require no training sets, and improve classification accuracy by minimizing classification errors. This section describes the implementation of probabilistic models and optimization theory.

Let x be the VML, and C_1 and C_2 represent two vehicle classes. The probability of error between two distributions includes $P(C_1|C_2)$ and $P(C_2|C_1)$, given in (4). They respectively represent the probability of error classifying C_2 as C_1 and C_1 as C_2.

$$P(\text{error}) = \int_{-\infty}^{\alpha} P(x|C_2)P(C_1)dx + \int_{\alpha}^{+\infty} P(x|C_1)P(C_2)dx \quad (4)$$

Distributions $P(x|C_1)$ and $P(x|C_2)$ statistically describe random variable x, corresponding to VML for the vehicles traveling on Oklahoma roadways. In accordance with the central limit theorem and law of large numbers, these distributions can be approximated as Gaussian distribution models, as presented in (5) and (6), where k is the class number.

$$X_k \propto N\left(\mu_{C_k}, \sigma_{C_k}\right) \rightarrow p\left(x|\mu_{C_k}, \sigma_{C_k}\right) = \frac{1}{\sigma_{C_k}\sqrt{2\pi}} e^{-\frac{\left(x-\mu_{C_k}\right)^2}{2\sigma_{C_k}^2}}; \quad \mu_{C_k} = \frac{1}{n}\sum_{i=1}^{n} x_i, \sigma_{C_k} = \sqrt{\frac{1}{n}\sum_{i=1}^{n}\left(x_i - \mu_{C_k}\right)^2} \quad (5)$$

$$P(\text{error}) = P(C_1)\int_{-\infty}^{\alpha_T} \frac{1}{\sigma_{C2}\sqrt{2\pi}} e^{-\frac{(x-\mu_{C2})^2}{2\sigma_{C2}^2}} .dx + P(C_2) \int_{\alpha_T}^{+\infty} \frac{1}{\sigma_{C1}\sqrt{2\pi}} e^{-\frac{(x-\mu_{C1})^2}{2\sigma_{C1}^2}} .dx \quad (6)$$

The integration of a probability distribution function (PDF) is a cumulative distribution function (CDF) represented by the error function erf, as given in (7).

$$P(\text{error}) = P(C_1)\frac{1}{2}\left[1 + \text{erf}\left\{\frac{\alpha_T - \mu_{C2}}{\sigma_{C2}\sqrt{2}}\right\}\right] + P(C_2)\left(1 - \left(\frac{1}{2}\left[1 + \text{erf}\left\{\frac{\alpha_T - \mu_{C1}}{\sigma_{C1}\sqrt{2}}\right\}\right]\right)\right) \quad (7)$$

Optimal threshold α_T can be found by taking the derivative of (10) and equating the outcome to 0, as presented in (8). Note that not all groups had the same number of vehicle instances per unit of time. To ensure that classification error will not change, given that new values are added to either group, weights $P(C_1)$ and $P(C_2)$ are proportional to the number of instances in each class group that are calculated, as in (9). It is worth mentioning that the optimal threshold found by solving (8) is global minima, which can be verified using the second derivative test, as in (10), such that $\alpha_T > 0$.

$$\frac{dP(\text{error})}{d\alpha_T} = 0 \rightarrow \frac{P(C_1)}{\sigma_{C2}\sqrt{2\pi}} e^{-\frac{(\alpha_T-\mu_{C2})^2}{2\sigma_{C2}^2}} - \frac{P(C_2)}{\sigma_{C1}\sqrt{2\pi}} e^{-\frac{(\alpha_T-\mu_{C1})^2}{2\sigma_{C1}^2}} = 0 \quad (8)$$

Table 5. Comparison between Decision Boundaries by Different Thresholding Methods

Group	FHWA-S_F	MAG			IDL		
		γ	α_T	α_E	γ	α_T	α_E
G1	F_1	0.7→2.984 m	0.7→3.736 m	0.7→2.9107 m	0.7→2.7686 m	0.7→3.2872 m	0.7→2.7686 m
G2	F_2–F_3	2.984→10.971 m	3.736→7.7516 m	2.912→7.427 m	2.7686→6.858 m	3.2872→6.0528 m	2.7686→5.715 m
G3	F_4–F_7	10.971→14.727 m	7.7516→14.95 m	7.427→15.136 m	6.858→12.243 m	6.0528→13.676 m	5.715→12.7 m
G4	F_8–F_{13}	>14.727 m	>14.95 m	>15.136 m	>12.243 m	>13.676 m	>12.7 m

Table 6. Comparison between Classification Rates by Different Thresholding Methods

Group	FHWA-S_F	MAG			IDL		
		γ	α_T	α_E	γ	α_T	α_E
G1	F_1	91.30%	100%	100%	98.00%	100%	100%
G2	F_2–F_3	99.90%	93.557%	89.8938%	100%	97.323%	93.0584%
G3	F_4–F_7	26.30%	81.4103%	87.5%	77.90%	88.132%	92.9789%
G4	F_8–F_{13}	98.70%	98.0952%	97.5238%	99.60%	98.1516%	99.4481%
Overall C_R		97.6951%	93.4419%	90.2131%	98.9895%	97.0888%	94.2768%

$$P(C_1) = \frac{\text{Length}(C_1)}{\text{Length}(C_1 + C_2)};\ P(C_2) = \frac{\text{Length}(C_2)}{\text{Length}(C_1 + C_2)} \tag{9}$$

$$\frac{d2P(\text{error})}{d\alpha_T^2} = 0 \rightarrow P(C_1)\frac{(\mu_{C2} - \alpha_T)}{\sigma_{C2}^3\sqrt{2\pi}}e^{-\frac{(\alpha_T-\mu_{C2})^2}{2\sigma_{C2}^2}} - P(C_2)\frac{(\mu_{C1} - \alpha_T)}{\sigma_{C1}^3\sqrt{2\pi}}e^{-\frac{(\alpha_T-\mu_{C1})^2}{2\sigma_{C1}^2}} = 0 \tag{10}$$

The α_E uses equal weights ($P[C_1] = P[C_2]$) but implements a classification error minimization algorithm among groups. In other words, α_E is found by solving (11) such that the probability of error for C_1 classified as C_2 is equal to the probability of error for C_2 classified as C_1 and ε is a precision value (e.g., 0.001). Equating classification errors will result in a decreased classification error for one particular group (i.e., the one with fewer instances), but will increase classification error for the second group (i.e., the one with more instances). Therefore, overall system classification error will increase.

$$P(\alpha_E, C_1) - P(\alpha_E, C_2) \leq \varepsilon \tag{11}$$

Developed models can then be applied on all overlapping problematic LBVC schemes that have imbalanced data, as well as those that have a high overlap ratio among neighbored groups.

Problematic groups are G2 and G3, which have more significant overlap in the MAG dataset when compared with the IDL dataset, as seen in Table 2. Table 5 presents a comparison between decision boundaries determined by various proposed thresholding methods for both MAG and IDL datasets. A comparison between classification rates for various thresholding methods is presented in Table 6.

For G1 and G2, an insignificant overlapping rendered comparable high rates for all thresholding methods. However, G2 and G3 demonstrated a significant overlap, as G3 included SU (e.g., Class 5), which are highly overlapped in length with Class 3, included in G2. ML methods set γ at a value that maximizes total classification accuracy at the expense of misclassifying most instances in the other group (e.g., G3 classification error 73.7%) because the number of instances in G2 is considerably larger than G3. Classification error for G3 was significantly reduced to 18.5897% when using α_T. However, overall classification accuracy was decreased. Classification rate for G3 improved when using α_E to achieve balanced classification rates among all groups at the expense of reducing overall classification accuracy to 90.2131% when compared with 97.6951% and 93.4419% for γ and α_T, respectively. All scenarios achieved comparable classification performance for G3 and G4, as a minimum overlapping can be observed. Moreover, the number of data points in both groups is relatively comparable.

Discussion

Boundaries developed in this study pertaining to Oklahoma and based on IDL or MAG data are provided for comparison (See Table 1 and Table 6). Observations include the following. First, vehicle length boundaries detected in various U.S. states are not alike because vehicle and traffic characteristics vary per road type, region, and state, among others; therefore, LBVC boundaries developed in one state may need adjustment to be applicable in another. Second, boundaries defined for MAG detectors are longer in length than those detected by IDL detectors, primarily because MAG estimates VML. The difference can be found between VML and the length of IDL's detection zone. It remains challenging to define an accurate detection zone for MAG, as disturbance to the Earth's magnetic field depends on detection, which is proportional to vehicle length, height of vehicle chasse above ground, and vehicle structure composition of ferrous materials. Third, a higher overlap ratio among LBVC scheme groups was observed in the MAG dataset when compared with the IDL dataset, which is attributed to variations in magnetic length for vehicles from the same class. These can be identified by comparing variance for each class in MAG and IDL, as shown in Table 2. Likewise, MAG classes have a larger length variance. Fourth, analyses demonstrated the importance of investigating not only the classification rate for evaluating classification models, but also other performance metrics (e.g., CM and MCC) to identify bias in the model caused by imbalanced class distributions. This phenomenon increases sensitivity of ML algorithms toward overrepresented classes. And finally, three algorithms developed for identifying optimal length boundaries for LBVC schemes resulted in three very different sets of thresholds (γ, α_T, and α_E). First, γ maximizes classification accuracy at the expense of minority classes; α_T achieves balance among overlapped groups regardless of number of instances in each group; and α_E attempts to equate classification error between two overlapped groups despite group population size. Given this information, the question remains as to which algorithm is suitable. The answer depends exclusively on the objective of the LBVC system. If traffic engineers are more concerned with vehicles in higher classes (i.e., trucks that are generally highly underrepresented), then it is important to utilize a method that achieves balance among classes (i.e., α_T). However, if classification accuracy is paramount regardless of classification rate per group, then the γ method would be suitable. For balanced classification error among groups, α_E should be implemented.

Conclusion

The work reported in this paper provided a comprehensive method to develop computationally efficient, real-time LBVC schemes that could be implemented using IDL or MAG. LBVC schemes were developed, implemented, and evaluated using ML algorithms and probabilistic modeling of VML. Classification accuracies of 97.70% and 99.00% were reported for MAG and IDL, respectively. The proposed classification schemes are intended to supplement or replace the ABVC data collection method used in the state of Oklahoma. The methodology developed in this work will benefit other states and territories interested in developing LBVC schemes. The work can be extended and mimicked for gathering LBVC data using nonintrusive technologies, such as vision systems and microwave radars.

Acknowledgments

The authors would like to acknowledge Oklahoma Department of Transportation (ODOT) for funding provided to support the research activities.

Author Contributions

The authors confirm contribution to the paper as follows: study conception and design: Balid; data collection: Balid, Refai; analysis and interpretation of results: Balid; draft manuscript preparation: Balid, Refai. All authors reviewed the results and approved the final version of the manuscript.

References

1. Office of Highway Policy Information. *FHWA Forecasts of Vehicle Miles Traveled (VMT): May 2014.* Office of Highway Policy Information, Federal Highway Administration, Washington, D.C., 2014. http://www.fhwa.dot.gov/policyinformation/tables/vmt/vmt_forecast_sum.cfm.
2. *Truck Safety—Advocates for Highway and Auto Safety.* Advocates for Highway and Auto Safety, http://saferoads.org/issues/truck-safety/. Advocates for Highway and Auto Safety.
3. Federal Highway Administration. *Traffic Monitoring Guide.* Report No. FHWA-PL-13-015. FHWA, U.S. Department of Transportation, 2013.
4. Refai, H., N. Bitar, J. Schettler, and O. Al Kalaa. *The Study of Vehicle Classification Equipment with Solutions to Improve Accuracy in Oklahoma.* Oklahoma Department of Transportation, Oklahoma City, Okla., 2014.
5. Adams, T., E. Perry, A. Schwartz, B. Gollnik, M. Kang, J. Bittner, and S. Wagner. *Aligning Oversize/Overweight Fees with Agency Costs: Critical Issues.* Report No. CFIRE 03-17. Wisconsin, Department of Transportation, 2013.
6. Minge, E. D., S. Peterson, H. Weinblatt, B. Coifman, and E. Hoekman. *Loop- and Length-Based Vehicle Classificaion, Federal Highway Administration—Pooled Fund Program.* Report No. MN/RC 2012-33. Minnesota Department of Transportation, Research Services, 2012.
7. National Cooperative Highway Research Program (NCHRP), *Loop and Length Based Classification Pooled Fund.* http://www.pooledfund.org/details/study/416.
8. Wei, H., Q. Ai, D. Eustace, and B. Coifman. Length-Based Vehicle Classification Models Using Dual-Loop Data against Stop-and-Go Traffic Flow. Presented at 90th

Annual Meeting of the Transportation Research Board, Washington, D.C., 2011.

9. Coifman, B. *Vehicle Classification from Single Loop Detectors*. Report No. MRUTC 05-02. Department of Electrical and Computer Engineering, The Ohio State University, 2007.
10. Mauga, T. *Development of Florida Length Based Vehicle Classification Scheme Using Support Vector Machines*. Florida State University, Tallahassee, Fla., 2016.
11. Balid, W., H. Tafish, and H. H. Refai. Development of Portable Wireless Sensor Network System for Real-Time Traffic Surveillance. *Proc., 18th IEEE International Conference on Intelligent Transportation Systems, IEEE - ITSC 2015,* September 15–18, 2015, pp. 1630–1637.
12. Balid, W., H. Tafish, and H. H. Refai. Versatile Real-Time Traffic Monitoring System Using Wireless Smart Sensors Networks. *Proc., 2016 IEEE Wireless Communications and Networking Conference (WCNC), Track 4: Services, Applications, and Business,* Doha, Qatar, 2016, pp. 1–6.
13. W. Balid, H. Tafish and H. H. Refai, "Intelligent Vehicle Counting and Classification Sensor for Real-Time Traffic Surveillance," in *IEEE Transactions on Intelligent Transportation Systems*, vol. 19, no. 6, pp. 1784–1794, June 2018. doi: 10.1109/TITS.2017.2741507
14. Liu, H. X., and J. Sun. Length-Based Vehicle Classification Using Event-Based Loop Detector Data. *Transportation Research Part C: Emerging Technologies*, Vol. 38, 2014, pp. 156–166.
15. Wu, L., and B. Coifman. Improved Vehicle Classification from Dual-Loop Detectors in Congested Traffic. *Transportation Research Part C: Emerging Technologies*, Vol. 46, 2014, pp. 222–234.
16. Coifman, B., and S. Kim. Speed Estimation and Length Based Vehicle Classification from Freeway Single-Loop Detectors. *Transportation Research Part C: Emerging Technologies*, Vol. 17, No. 4, 2009, pp. 349–364.
17. Jeng, S.-T., and S. G. Ritchie. Real-Time Vehicle Classification Using Inductive Loop Signature Data. *Transportation Research Record: Journal of the Transportation Research Board*, 2009. 2086: 8–22.
18. Cheung, S. Y., and P. P. Varaiya. *Traffic Surveillance by Wireless Sensor Networks: Final Report*. California PATH Research Report, USA, 2007.
19. Yang, B., and Y. Lei. Vehicle Detection and Classification for Low-Speed Congested Traffic with Anisotropic Magnetoresistive Sensor. *IEEE Sensors Journal*, Vol. 15, No. 2, 2015, pp. 1132–1138.
20. Taghvaeeyan, S., and R. Rajamani. Portable Roadside Sensors For Vehicle Counting, Classification, and Speed Measurement. *IEEE Transactions on Intelligent Transportation Systems*, Vol. 15, No. 1, 2014, pp. 73–83.
21. Kaewkamnerd, S., J. Chinrungrueng, R. Pongthornseri, and S. Dumnin. Vehicle Classification Based on Magnetic Sensor Signal. *2010 IEEE International Conference on Information and Automation,* Harbin, China, 2010, pp. 935–939.
22. Wang, R., L. Zhang, K. Xiao, R. Sun, and L. Cui. EasiSee: Real-Time Vehicle Classification and Counting via Low-Cost Collaborative Sensing. *IEEE Transactions on Intelligent Transportation Systems*, Vol. 15, No. 1, 2014, pp. 414–424.
23. Diamond Traffi; *Road Runner 3 Counter and Classifier Kit*. http://diamondtraffic.com/product/RoadRunner3Kit.
24. Balid, W. *Fully-Autonomous Self-Powered Intelligent Wireless Sensor for Real-Time Traffic Surveillance in Smart Cities*. Doctoral dissertation. University of Oklahoma, 2016.
25. *NIST: National Institute of Standards and Technology/ SEMATECH e-Handbook of Statistical Methods-Detection of Outliers*. http://www.itl.nist.gov/div898/handbook/eda/section3/eda35h.htm.
26. *WIM Data Analyst's Manual*. Report No. FHWA-IF-10-018. FHWA, U.S. Department of Transportation, 2010. https://www.fhwa.dot.gov/pavement/pub_details.cfm?id=668.
27. Bitar, N., and H. H. Refai. A Probabilistic Approach to Improve the Accuracy of Axle-Based Automatic Vehicle Classifiers. *IEEE Transactions on Intelligent Transportation Systems*, Vol. 18, 2016, pp. 1–8.
28. Bishop, C. M. *Pattern Recognition and Machine Learning (Information Science and Statistics)*. Springer-Verlag Berlin, Heidelberg, 2006.
29. Seiffert, C., T. M. Khoshgoftaar, J. Van Hulse, and A. Napolitano. RUSBoost: A Hybrid Approach to Alleviating Class Imbalance. *IEEE Transactions on Systems, Man, and Cybernetics—Part A: Systems and Humans*, Vol. 40, No. 1, 2010, pp. 185–197.

The Standing Committee on Intelligent Transportation Systems (AHB15) peer-reviewed this paper (18-00098).

Article

TRR
JOURNAL OF THE TRANSPORTATION RESEARCH BOARD

Transportation Research Record
2018, Vol. 2672(19) 112–123

DOI: 10.1177/0361198118780710
journals.sagepub.com/home/trr
SAGE

Motorway Path Planning for Automated Road Vehicles Based on Optimal Control Methods

Konstantinos Makantasis[1] **and Markos Papageorgiou**[1]

Abstract

A path-planning algorithm for automated road vehicles on multi-lane motorways is derived from the opportune formulation of an optimal control problem. In this framework, the objective function to be minimized contains appropriate respective terms to reflect: the goals of the vehicle advancement; passenger comfort; prevailing traffic rules (e.g., overtaking only from left); and the avoidance of obstacles (other moving vehicles) and of the vehicle departing from the road. Each term is coupled with a weighting factor that reflects its comparative importance. For the numerical solution of the optimal control problem, a very efficient feasible direction algorithm is used. To avoid local minima, a simplified dynamic programming algorithm is also conceived to deliver the initial guess trajectory for the optimal control algorithm. With low computation times, the approach is readily executable within a model-predictive control frame. The performance of the proposed algorithm is illustrated using two typical driving scenarios.

In the last few decades, a strong research effort has focused on the development of fully automated driving for road vehicles. Fully automated driving could reduce accidents caused by driver error and improve driver and passenger comfort. Although vehicle automation has already led to great achievements in supporting the driver in various monotonous and challenging tasks, see, for example, Reimer et al. (*1*), increasing the level of automation to fully automated driving is an extremely challenging problem. This is mainly due to the complexity of real-world environments, including avoidance of obstacles, compliance with traffic rules and consideration of aspects of human driving behavior.

The development of fully automated driving algorithms is inherently related to planning and updating a vehicle path, which should be safe, collision-free, and acceptable to users. For the path generation, the vehicle dynamics, the road geometry, the presence of obstacles, and the respect of traffic rules (*2*) should be taken into account. Planning such a path can be seen as a trajectory generation problem, that is, creation of a quasi-continuous sequence of states that must be tracked by the vehicle over a specific time horizon. Path planning in the presence of obstacles has been widely studied in robotics (*3*). However, path planning for road vehicles is a much more critical task, since the safety of the passengers must be guaranteed.

Under certain assumptions, simplifications, and conservative estimates, heuristic rules can be used to develop path-planning methods for road vehicles (*4*). However, such methods may have difficulties coping with the complexity of the real-world environment and the diversity of driving situations (*5*). To overcome the limitations of rule-based path-planning strategies, approaches based on the careful design and exploitation of potential field (PF) and optimal control (OC) methods have also been proposed.

PF methods generate a field, based on the design of potential functions for obstacles, road structures, traffic regulations, and the goals to be achieved. A vehicle trajectory is then generated by moving toward the descent direction of the field (*6–8*). Ji et al. (*9*) proposed a trajectory generation method along with a model-predictive path-tracking controller. The trajectory generation takes place by the exploitation of a PF that includes potential functions for the road structure and the obstacles, while the path-tracking controller is designed based on a constrained model-predictive control (MPC) problem.

[1]Technical University of Crete, School of Production Engineering and Management, University Campus, Kounoupidiana, Chania, Greece

Corresponding Author:
Address correspondence to Konstantinos Makantasis: kmakantasis@isc.tuc.gr

However, due to the fact that the system dynamics are not considered during the trajectory generation, the planned path may in some situations turn out to be unfeasible for vehicle to track (*10*). PF methods are quite popular due to the intuitive problem formulation and their low computational cost (*11*). However, there are still open issues regarding the integration of system dynamics and constraints within such methods (*12*).

This drawback can be alleviated by formulating the trajectory generation problem as an OC problem. Such approaches allow for concurrent consideration of system dynamics and obstacles during the trajectory generation process (*13*). For example, Carvalho et al. used the distance between the vehicle and an obstacle to generate approximate linear constraints for obstacle avoidance (*14*), while Gao et al. considered the obstacles as ellipse-shaped constraints, in order to keep the automated vehicle robustly far from them (*15*). In the same spirit, Horvath et al. (*16*) include in the cost function factors dependent on the longitudinal distance between the automated vehicle and the obstacles, in order to generate a collision-free path. OC approaches usually map the OC problem to a nonlinear programming (NLP) problem that can be solved using numerical NLP solvers (see, for example, [*17–20*]). The main drawback of the aforementioned mapping is that it produces a locally optimal solution, which is very much dependent on the initial guess trajectory employed. Furthermore, solving an NLP is very time consuming and thus these methods may not satisfy the real-time operation requirements of a path-planning system.

Another attractive alternative for solving OC problems is by using dynamic programming (DP) techniques. Although DP techniques produce a globally optimal solution for the OC problem, unfortunately, due to the curse of dimensionality (*21*), they are restricted to small-scale problems. For this reason, they are sometimes combined with other optimization methods. For example, Tanzmeister used DP to generate a rough reference vehicle trajectory for a parking scenario, where all obstacles are considered static (*22*).

Contribution of This Study

This paper develops a path-planning algorithm for automated vehicles on motorways. It generates a path that avoids obstacles, observes road structure and traffic rules, and takes into consideration both the comfort of the passengers and the system dynamics. Specifically, the objective function to be minimized contains appropriate respective terms to reflect: the goals of the vehicle advancement; the passenger comfort; the prevailing traffic rules (e.g., overtaking only from left); the avoidance of obstacles (other moving vehicles) and of road structures. Potential-field-like penalty functions are used to represent the moving obstacles, the road structure, and some traffic rules. Each term is coupled with a weighting factor that reflects its comparative importance, and thus the method is very flexible in considering different goals, rules, or priorities in an explicit and transparent way. Exploiting the structure of the system dynamics (state equations), a reduced gradient may be readily obtained, and thereby the OC problem is mapped to an NLP problem in the reduced space of the control variables. This marks a substantial reduction of the problem dimension, as the state variables are eliminated, and the numerical solution is generated using a very efficient iterative feasible direction algorithm. DP is also employed to produce a rough initial guess control trajectory to start the iterations of the numerical solution, in order to avoid convergence to a local minimum. The proposed methodology is distinguished by very low computational cost. Specifically, the future trajectory of a vehicle for a time horizon of 15 s is generated in less than 500 ms. Therefore, this methodology can be efficiently incorporated into an MPC frame in order to provide an online path-planning strategy.

In summary, the proposed method has the merits of both PF and OC methods. In addition, it is explicit and transparent (in contrast to black-box approaches), and allows for the study of different vehicle behaviors (via modification of the objective function weights) and of the respective impacts on traffic flow efficiency. Finally, the method adopts a problem formulation that circumvents the need for discrete variables for lane-changing maneuvers, thus opening the way for the mapping of the OC problem to an NLP problem in the reduced space of control variables. This leads to a substantial reduction of the problem dimension, hence the generation of the solution is distinguished by very low computational time.

Information Flow during the Path-Planning Task

The present work develops a path-planning algorithm for automated road vehicles on motorways. It is assumed that the path-planning algorithm is executed in the vehicle, to produce a reference vehicle trajectory over a future time horizon of some 15 s. The length of time horizon determines the dimension of the search space during optimization and, hence, the computational effort. A time horizon that is too short may lead to myopic vehicle behavior; a time horizon that is too long may consume computational resources unnecessarily. In this work we use an indicative time horizon which is deemed reasonable; leaving its fine-tuning for the MPC-based testing. Data required to produce the reference trajectory include:

- The current vehicle state, that is, its current two-dimensional (lateral and longitudinal) position and velocity.
- The current two-dimensional relative position, speed and (rough) length of other vehicles moving on the road near the reference vehicle.

This information can be made available to the path-planning algorithm through the vehicle's sensors and potentially also via advanced technologies for vehicle-to-infrastructure and vehicle-to-vehicle communication (*23*, *24*).

Based on this information, the proposed algorithm solves the corresponding OC problem and produces optimal control (two-dimensional acceleration) and state trajectories extending from the current time to some 15 s in the future. Note that, since the future movement of other vehicles is unknown at the start time, it is assumed that they will be moving with constant (current) speed over the 15-s horizon; but more accurate predictions may be possible if the current acceleration of other vehicles is also known.

The vehicle trajectories produced correspond to an open-loop solution, given the initial vehicle state and other vehicle data. To increase the robustness and reliability of path planning, the optimal vehicle trajectory problem may be cast in an MPC frame, whereby the 15-s trajectory is generated repeatedly at consecutive short time periods (e.g., every 1 s), each time with updated current state and other vehicle data; each time, only the very initial part (e.g., 1 s) of the produced trajectory is actually implemented, before being updated by the next produced trajectory, and so forth. The MPC approach with short time periods is facilitated by the very short computation time required to produce the optimal vehicle trajectories.

The outlined path-planning algorithm, including its real-time MPC usage, aims to produce reference state trajectories on the motorway based on simple vehicle kinematic dynamics. These state trajectories may then be used as a reference by the vehicle throttle and brake controllers, which are designed on the basis of more elaborated vehicle dynamics, to produce the actual movement of the vehicle on the road.

Optimal Control Problem Formulation

Vehicle Motion Dynamics

For the kinematic equations of vehicle motion, we consider the road as a two-dimensional plane, and the vehicle's position on this plane is expressed in global Euclidean coordinates. This way, using discrete-time equations with time period T, we have

$$x_1(k+1) = x_1(k) + Tx_3(k) + \frac{1}{2}T^2u_1(k), \tag{1}$$

$$x_2(k+1) = x_2(k) + Tx_4(k) + \frac{1}{2}T^2u_2(k), \tag{2}$$

$$x_3(k+1) = x_3(k) + Tu_1(k), \tag{3}$$

$$x_4(k+1) = x_4(k) + Tu_2(k), \tag{4}$$

where the state variables x_1, x_2, x_3, and x_4 correspond to the longitudinal position, lateral position, longitudinal speed, and lateral speed, respectively, of the vehicle; while the control variables u_1 and u_2 are its longitudinal and lateral acceleration, respectively.$\cdot(k)$ denotes the value of the corresponding variable at time $t = kT$, where $k = 0, 1, \ldots, K-1$ is the discrete-time index, and K is the fixed time horizon. The system has a known (measured) initial state $x_i(0) = x_{i,0}, i = 1, 2, 3, 4$. Note that the accelerations may be bounded by upper and lower (negative) values if necessary.

The Objective Function

The objective function, to be minimized in the frame of formulation of the OC problem, includes a number of terms to reflect efficient, convenient, rule-compliant, and safe driving. Each term is described in the following.

Penalty Function for Obstacle Avoidance. Obstacles correspond to non-crossable structures, which in the motorway case are essentially other moving vehicles. The penalty function (similar to potential-field functions) for obstacle avoidance should feature high values at the gross obstacle space, so that the reference vehicle is repulsed, and potentially unsafe trajectories are suppressed; and low (or virtually vanishing) values outside of that space. To this end, we adopt the exponential penalty function employed in (*25*), which is best understood in its one-dimensional form $f(x) = \exp\left(-(x/A)^2\right)$. This smooth function has a maximum of 1 for $x = 0$; for $|x| < A$, the function retains a value close to 1, while for $|x| > A$, it reduces rapidly to very small or virtually vanishing values; and the transition between both domains is mathematically smooth but steep. (The exponential function is computationally expensive; therefore, one might use some other, e.g., polynomial, functions with similar properties.) Thus, x could reflect the distance between the automated vehicle and each obstacle i, and A should be selected based on the vehicle's dimensions, taking into account also the speeds of the vehicle and obstacles for safe car-following distances. Thus, in two dimensions, we use the following exponential ellipsoid penalty function for each obstacle i

$$F_{o_i} = e^{-\left[\left(\frac{\Delta x_1^{(i)}}{r^{(i)}}\right)^2 + \left(\frac{\Delta x_2^{(i)}}{al}\right)^2\right]}, \tag{5}$$

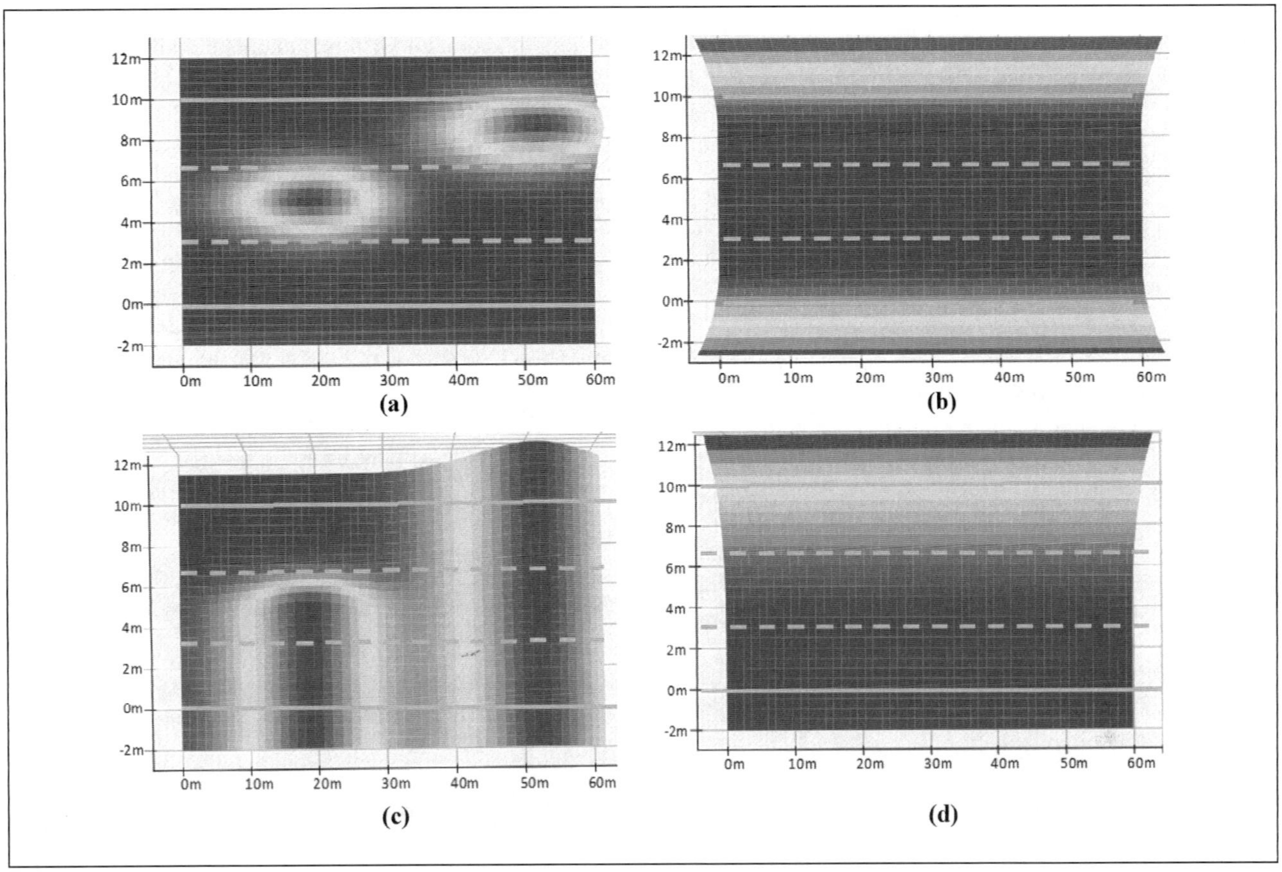

Figure 1. The form of the penalty functions of obstacles (*a*), road boundaries (*b*), and traffic rules (*c*) and (*d*). The green solid and dashed lines represent the road and lane boundaries respectively.

where $\Delta x_1^{(i)} = x_1^{(i)} - x_1$ and $\Delta x_2^{(i)} = x_2^{(i)} - x_2$ stand for the longitudinal and lateral distance between the *i*-th obstacle and the reference vehicle, with the convention that vehicle and obstacle positions refer to the position of the centers of the respective vehicles. We now need to specify the length and width of the ellipsoid, corresponding to the parameter A above.

Regarding the lateral safety distance, it is rational to assume that an obstacle and the vehicle must occupy different lanes, when their longitudinal distance is small. Thus, the lateral safety distance can be the same for all obstacles and be dependent on the lane width of a specific road. Thus, l in Equation 5 represents the lane width, and a is a positive constant that determines the exact width of the ellipsoid along the lateral direction.

Regarding the longitudinal safety distance, things are a little more complicated, since the safety distance should depend on the longitudinal speeds of the vehicle and the obstacle. For example, if the reference vehicle approaches an obstacle with high speed, then the longitudinal safety distance must be greater than approaching the same obstacle with lower speed. Thus, $r^{(i)}$ in Equation 5 is defined as

$$r^{(i)} = \begin{cases} cx_3 + L_i \; if \; x_1^{(i)} > x_1 \\ cv^{(i)} + L_i \; otherwise \end{cases}, \tag{6}$$

where $L_i = 0.5(\mu + \mu_i)$, and μ and μ_i are the lengths of the vehicle and obstacle, respectively; and $v^{(i)}$ is the longitudinal speed of the *i*-th obstacle. The upper case is activated when the vehicle is located upstream of the *i*-th obstacle and specifies the backward part of the ellipsoid. When the vehicle is located downstream of the *i*-th obstacle, the lower case, which specifies the forward part of the ellipsoid, is activated. For both cases c is the desired time gap. Figure 1*a* presents the penalty function generated when there are two obstacles on the road. The first obstacle occupies the middle lane, and its longitudinal position and velocity are 20 m and 8.0 m/s respectively; while the second obstacle occupies the left-most lane and its longitudinal position and velocity are 55 m and 15 m/s respectively; all vehicles are 5 m long.

Penalty Function for Road Boundaries. The penalty function of road boundaries must be designed to disallow the automated vehicle from departing from the road. To this end, we adopt a quadratic function, dependent on the width w of the road

$$F_r = \begin{cases} (x_2 - d)^2, x_2 \leq d \\ 0, d \leq x_2 \leq w - d \\ (x_2 + d - w)^2, x_2 > w - d \end{cases}, \tag{7}$$

where the positive parameter d is used to prevent the reference vehicle from moving close to the road boundaries. Figure 1b presents the penalty function generated for the road boundaries, with $w = 10$ m and $d = 1.5$ m.

Penalty Functions for Traffic Rules. Appropriate penalty functions may also be used to enable compliance with specific traffic rules. For example, European driving regulations do not allow overtaking of other vehicles from the right; also, when traffic is light, vehicles should use the right lane or return to it after overtaking.

Other vehicles that are present on the road are treated as obstacles. Therefore, the design of a penalty function that prevents the reference vehicle from overtaking from the right may be related to the penalty function for obstacle avoidance presented above. Specifically, we modify the penalty function in Equation 5 as follows

$$F_{o_i} = \begin{cases} e^{-\left[\left(\frac{\Delta x_1^{(i)}}{r^{(i)}}\right)^2 + \left(\frac{\Delta x_2^{(i)}}{al}\right)^2\right]} \textit{ if } x_2 \geq x_2^{(i)} \textit{ and } x_2^{(i)} < b \\ e^{-\left[\left(\frac{x_{1,new}^{(i)}}{r^{(i)}}\right)^2\right]} \textit{ otherwise} \end{cases}, \tag{8}$$

where b stands for the right-most point of the left-most lane. The upper case of Equation 8 is identical to Equation 5 and is activated when the i-th obstacle is not on the left-most lane and its lateral position is more to the right than the position of the vehicle. Otherwise, the lower case disallows the vehicle from overtaking the obstacle from the right. Figure 1c presents the corresponding penalty function when two obstacles are present on the road. The displayed obstacles and their states are identical as in Figure 1a.

Finally, the penalty function that attracts the reference vehicle to the right-most lane in light traffic must have its minimum in the middle of the right-most lane, be convex, and be very weakly rising across the road width, so that it only influences the vehicle movement if other objectives are virtually indifferent, for example,

$$F_t = \left(\frac{x_2 - \delta}{\beta}\right)^\gamma, \tag{9}$$

where δ corresponds to the middle of the right-most lane, β is bigger than the road width (so that $\beta > x_2 - \delta$ always holds), and γ is an even positive integer that controls the slope of the penalty. Figure 1d presents the generated penalty function in Equation 9, when $\delta = 1.7$ m, $\beta = 11$ m and $\gamma = 4$. The value of δ must coincide with the middle of the right-most lane, since Equation 9 presents its minimum when x_2 equals δ.

Vehicle Mission and Passenger Convenience. Beyond obstacles, road structure, and traffic rules, the proposed path-planning method takes into consideration the comfort of passengers and the vehicle mission. The comfort of the passengers is related to the magnitude of accelerations and lateral motion (*13*). Therefore, quadratic terms that penalize non-zero accelerations and lateral velocity are incorporated in the cost function. Note that the quadratic penalty for longitudinal acceleration is often used as a good proxy for the minimization of fuel consumption (*26*).

The vehicle mission is to advance with a speed close to a desired speed. Thus, a final term that penalizes the deviation between the vehicle speed and its desired speed is incorporated in the cost function to address the goal of longitudinal vehicle advancement.

Overall Objective Function. Based on the above, the overall cost function to be minimized in the frame of the OC problem has the following form

$$J = \sum_{k=0}^{K-1} \left[\frac{1}{2} w_1 u_1^2(k) + \frac{1}{2} w_2 u_2^2(k) + \frac{1}{2} w_3 (x_3(k) - v_d)^2 + \frac{1}{2} w_4 x_4^2(k) + w_5 \sum_{i=1}^{N} F_{o_i}(x_1(k), x_2(k)) + w_6 F_r(x_2(k)) + w_7 \left(\frac{x_2(k) - \delta}{\beta}\right)^\gamma\right]. \tag{10}$$

The overall objective function is a sum over the time horizon K of the individual contributions at each period k. The first, second, and fourth cost terms penalize non-zero accelerations and lateral motion and are related to passenger comfort. In the third cost term, the parameter v_d stands for the desired speed of the reference vehicle, and this term corresponds to the goal of vehicle advancement. The three last terms of the cost function integrate the penalty functions for obstacles, road structure, and traffic rules. Note that the obstacle term is a sum over all N obstacles that were detected to be near the reference vehicle at the start time.

Parameters w_i are constants that weight the contribution of each cost factor to the value of the overall objective function. These weights must be specified appropriately, for example, via trial and error, so as to lead to the desired vehicle behavior under all possible driving situations. This is a non-trivial endeavor, but it enables the design of a transparent and flexible vehicle behavior.

The proposed method traces a vehicle trajectory that essentially reflects the vehicle's longitudinal position and speed and its travelling lane. This trajectory needs to be tracked by the underlying vehicle control loops, based on high-definition maps. For the vehicle trajectory specification, possible curvatures may be aligned to form an equivalent straight section.

Problem Formulation

The path-planning problem may be mapped to an OC problem. The difference Equations 1–4 may be condensed in the following vector form

$$\boldsymbol{x}(k+1) = \mathbf{f}[\boldsymbol{x}(k), \boldsymbol{u}(k), k], k = 0, \ldots, K-1, \tag{11}$$

where x and u are the system states and control variable vectors, respectively. With known initial state $x(0) = x_0$, the OC problem consists of minimizing the objective function in Equation 10 subject to Equation 11 and possible acceleration bounds. Although expressing a dynamical process, the above minimization problem is, from a mathematical point of view, a static optimization problem due to the time-discrete nature of the process model involved (*27, 28*).

Problem Solution

The solution of a discrete-time OC problem corresponds to the calculation of optimal trajectories for the control variables, $\boldsymbol{u}(k)$, and states, $\boldsymbol{x}(k+1)$, $k = 0, 1, \ldots, K-1$, that minimize the cost function over the prespecified time horizon. Before describing the solution procedure, we derive the necessary optimality conditions for the formulated discrete-time OC problem.

Necessary Optimality Conditions

To present the necessary conditions of optimality, it is customary to use the Hamiltonian Function

$$\begin{aligned} H[\boldsymbol{x}(k), \boldsymbol{u}(k), \boldsymbol{\lambda}(k+1), k] &= \varphi[\boldsymbol{x}(k), \boldsymbol{u}(k), k] \\ &+ \boldsymbol{\lambda}(k+1)^T \mathbf{f}[\boldsymbol{x}(k), \boldsymbol{u}(k), k] \end{aligned} \tag{12}$$

where $\boldsymbol{\lambda}(k+1)$ stands for the co-state vector (equivalent to Lagrange multipliers) for the corresponding state equations conditions, and $\varphi[\boldsymbol{x}(k), \boldsymbol{u}(k), k]$ is the k-th summand of the objective function of Equation 10. For the problem at hand, we have the Hamiltonian Function

$$\begin{aligned} H &= \varphi[\boldsymbol{x}(k), \boldsymbol{u}(k), k] + \lambda_1(k+1) \\ &\left[x_1(k) + Tx_3(k) + \frac{1}{2}T^2 u_1(k)\right] \\ &+ \lambda_2(k+1)\left[x_2(k) + Tx_4(k) + \frac{1}{2}T^2 u_2(k)\right] \\ &+ \lambda_3(k+1)[x_3(k) + Tu_1(k)] \\ &+ \lambda_4(k+1)[x_4(k) + Tu_2(k)] \end{aligned} \tag{13}$$

then, we have the following conditions of optimality for the discrete-time OC problem that must be satisfied for $k = 0, \ldots, K-1$ (notation: $x_y = dx/dy$):

$$\boldsymbol{x}(k+1) = H_{\lambda(k+1)} = \mathbf{f}[\boldsymbol{x}(k), \boldsymbol{u}(k), k] \tag{14a}$$

$$\boldsymbol{\lambda}(k) = H_{x(k)} = \varphi_{x(k)}(k) + \boldsymbol{f}^T_{x(k)}\boldsymbol{\lambda}(k+1) \tag{14b}$$

$$H_{u(k)} = \varphi_{u(k)}(k) + \boldsymbol{f}^T_{u(k)}\boldsymbol{\lambda}(k+1) = 0. \tag{14c}$$

Equation 14*a* and *b* are the state and co-state difference equations, respectively, while Equation 14*c* specifies the OC variables. Furthermore, the following boundary conditions must be satisfied

$$\boldsymbol{x}(0) = \boldsymbol{x}_0 \tag{15a}$$

$$\boldsymbol{\lambda}(K) = 0. \tag{15b}$$

(for more details regarding the conditions of optimality, see [*27*]).

Feasible Direction Algorithm

The solution of complicated OC problems, like the one at hand, calls for an efficient numerical solver. Here, a feasible direction algorithm (*27*) is utilized. Extensive previous experience (*28*) with this algorithm suggests that it is the most efficient numerical solution approach for nonlinear OC problems of this kind. The feasible direction algorithm exploits the structure of the state equations and maps the OC problem into a NLP problem in the reduced space of control variables. Thus, the feasible direction algorithm attempts the calculation of a control trajectory $\boldsymbol{u}(k)$, $k = 0, 1, \ldots, K-1$, which corresponds to a local minimum of the cost function in the mK-dimensional space, where m corresponds to the number of control variables.

The feasible direction algorithm is an iterative procedure, whose algorithmic steps are presented below (superscripts indicate the iteration index);

Step 1: Guess an initial control trajectory $\boldsymbol{u}^{(0)}(k)$, $k = 0, \ldots, K-1$
Calculate states $\boldsymbol{x}^{(0)}(k)$ and co-states $\boldsymbol{\lambda}^{(0)}(k)$, $k = 0, \ldots, K-1$
Calculate the reduced gradient $\boldsymbol{g}^{(0)}(k) = \frac{dH}{d\boldsymbol{u}(k)}$, $k = 0, .., K-1$
Set iteration index $i = 0$.
Step 2: Specify a descent direction $\boldsymbol{s}^{(i)}(k)$, $k = 0, \ldots, K-1$.
Step 3: Specify a scalar step $\xi^{(i)} > 0$ through line optimization
Set $\boldsymbol{u}^{(i+1)}(k) = \boldsymbol{u}^{(i)}(k) + \xi^{(i)}\boldsymbol{s}^{(i)}(k)$, $k = 0, \ldots, K-1$
Calculate states $x^{(i+1)}(k)$, co-states $\boldsymbol{\lambda}^{(i+1)}(k)$, and gradient $\boldsymbol{g}^{(i+1)}(k)$, $k = 0, \ldots, K-1$.

Step 4: If a convergence test is satisfied, stop, otherwise start a new iteration $i := i + 1$, i.e., go to Step 2.

The algorithm starts with an initial guess control trajectory $\boldsymbol{u}^{(0)}(k)$, $k = 0, \ldots, K-1$. At each iteration i, it calculates the corresponding states, $\boldsymbol{x}^{(i)}(k)$ (from Equation 14*a*), and co-states, $\boldsymbol{\lambda}^{(i)}(k)$ (from Equation 14*b*). Then, it computes the reduced gradient $\boldsymbol{g}^{(i)}(k) = dH/d\boldsymbol{u}(k)$ (from Equation 14*c*), which is used to specify a descent direction $\boldsymbol{s}^{(i)}(k)$. Finally, the control trajectory is updated

$$\boldsymbol{u}^{(i+1)} = \boldsymbol{u}^{(i)}(k) + \xi^{(i)}\boldsymbol{s}^{(i)}(k), \tag{16}$$

where $\xi^{(i)}$ is a scalar step specified via line optimization. At the end of each iteration, if a convergence criterion (gradient close to zero) is satisfied, then all necessary conditions are satisfied, and the algorithm stops, otherwise it starts a new iteration. The algorithm may be readily extended to consider fixed bounds of the control variables, see (*27*, *28*) for all details.

Dynamic Programming (DP)

The trajectory produced by the feasible direction algorithm may correspond to a local minimum, which may depend on the initial guess trajectory used. In order to overcome this potential difficulty in the present case, DP is used to produce the initial guess trajectory. DP is characterized by computational cost, which increases exponentially with the problem dimensions and, therefore, cannot be used for efficient path planning. Instead, we employ DP to solve (with very low computation times) a simplified OC problem; the produced rough, but globally optimal, simplified-problem solution is then used as initial guess trajectory to be refined by the feasible direction algorithm.

To start with, the cost function

$$J_{DP} = \sum_{k=1}^{K-1} \left[\frac{1}{2}w_1u_1^2(k) + \frac{1}{2}w_2u_2^2(k) + \frac{1}{2}w_3(x_3(k) - v_d)^2 + \frac{1}{2}w_4x_4^2(k)\right] \tag{17}$$

that is used with DP is simpler than the corresponding cost function of the OC problem, since cost terms corresponding to obstacles, road structure, and traffic rules can be explicitly considered via non-feasible transitions. The problem simplifications introduced are:

- A larger time step T (e.g., 1 s) is used.
- The accelerations are only allowed to take three distinct values at each time step, corresponding to positive, zero, and negative acceleration, respectively, for each of them.

The second simplification implies that possible longitudinal and lateral transitions for the reference vehicle are

$$x_i(k+1) \in \{x_i(k) - \zeta T, x_i(k), x_i(k) + \zeta T\}, \tag{18}$$

for $i = 3$ (longitudinal movement) and $i = 4$ (lateral movement), where ζ denotes the utilized acceleration and decceleration along the longitudinal and lateral directions. Thus, at every time step, there are nine possible transitions, as visualized in Figure 2, where shaded regions indicate non-feasible transitions, the blue rectangle corresponds to the reference vehicle, while the green one to a moving obstacle.

On this basis, a forward DP algorithm is designed as follows. The initial state is branched into the nine possible transitions, whereby non-feasible transitions are ignored. The obtained branches are evaluated via the cost function, the most efficient one is branched, the new branches are evaluated, and so forth. The procedure ends when a branch has reached the final horizon, and its cost is lower than that of any of the other explored branches.

Illustration Examples

Among numerous scenarios used to select appropriate weights for the objective function terms and to evaluate the path-planning algorithm in different driving situations, two typical scenarios are presented to illustrate the features of the proposed method. In both scenarios, the reference vehicle moves on a three-lane motorway, each

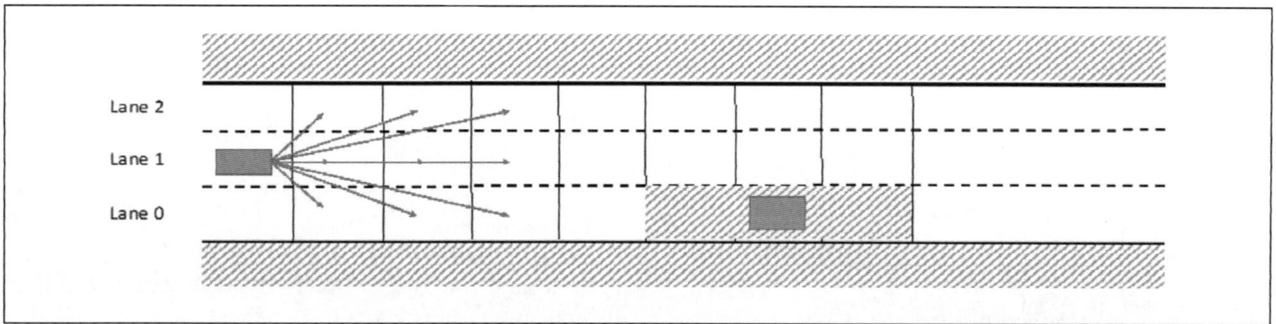

Figure 2. Schematic representation of the dynamic programming approach.

lane being 3.3 m wide, hence the middle of the left-most, middle, and right-most lane is located at 8.3, 5.0, and 1.7 m, respectively. In addition, for both scenarios the parameters of the path-planning algorithm are set to the following values:

- In Equation 10, $[w_1, w_2, w_3, w_4, w_5, w_6, w_7] = [1.0, 1.0, 5.0, 2.0, 150.0, 5.0, 10.0]$.
- In Equations 10 and 17, $v_d = 18$ m/s.
- In Equation 6, $c = 2$ stands for the desired time gap in seconds.
- In Equation 7, $d = 1.65$ m, that is, half of the lane width. When the distance between the vehicle and the road boundary is less than d the sixth penalty term of the cost function increases.
- In Equation 8, $a = 0.5$, which indicates that the fifth term of the objective function rapidly increases when the lateral distance between the vehicle and an obstacle is less than half of the lane width.
- In Equation 9, $\delta = 1.65$ m, $\beta = 11$ m and $\gamma = 4$. δ equals half the lane width. When x_2 is equal to δ, that is, the vehicle moves on the middle of the right-most lane, the seventh term of the objective function is minimized.
- In Equation 18, $\zeta = 3.0$ m/s^2, which corresponds to a reasonable acceleration value.
- Time horizon 15 s, $T = 100$ ms, hence $K = 150$.

The weights, w_i, were chosen empirically. Their values can be modified to create different emerging vehicle behaviors. In this work, safety was the main concern; for this reason, we selected the specific set of weights that penalizes the fifth (crashes) and the sixth (road departures) terms of the cost function the most. It should be noted that an update of the weights is likely to become necessary when tesing the vehicle in a more realistic environment with genuine feedback (MPC).

In the first scenario, the reference vehicle must follow the traffic rules and is therefore forced to move at a lower speed than its desired speed. In the second scenario, the reference vehicle accelerates to reach its desired speed, while at the same time it is overtaking in accordance with the traffic rules. For both scenarios, the path-planning algorithm utilizes the information about the longitudinal velocity and position of the obstacles at the start time and plans the path to be tracked by the reference vehicle for the next 15 s.

In the first scenario, there is only one moving obstacle downstream of the reference vehicle. The obstacle occupies the left-most lane, and its longitudinal speed is lower than the desired speed of the reference vehicle. In this case, the reference vehicle cannot overtake the obstacle, because this would violate the corresponding traffic rule. Instead, it moves to the right-most lane and drives with the same longitudinal speed as the obstacle. For this scenario, the initial state $\boldsymbol{x}(0)$ for the reference vehicle is $\boldsymbol{x}(0) = [0.0, 5.0, 10.0, 0.0]$, while the longitudinal speed for the obstacle is 10 m/s, and its initial position is $[x_1^1(0), x_2^1(0)] = [50.0, 8.3]$. Figure 3*a* presents the trajectories for the velocity and acceleration of the reference vehicle. The reference vehicle moves toward the right-most lane with very small lateral velocity and acceleration. Furthermore, it increases its longitudinal speed very smoothly in order to reach the obstacle's longitudinal speed. Figure 3*b* presents the positions of the reference vehicle (purple) and the obstacle (yellow) on the road plane.

In the second scenario, there are three moving obstacles downstream of the reference vehicle. All obstacles are moving on the first and second lanes with a longitudinal speed that is lower than the desired speed of the reference vehicle. However, in this scenario the left-most lane is free. Thus, the reference vehicle chooses the left-most lane and increases its longitudinal speed in order to reach its desired speed. For this scenario, the initial state $\boldsymbol{x}(0)$ for the reference vehicle is $\boldsymbol{x}(0) = [0.0, 5.0, 10.0, 0.0]$, while the respective longitudinal speeds of the three obstacles are 12 m/s, 12 m/s, and 10 m/s, and their respective initial positions are $[x_1^1(0), x_2^1(0)] = [60.0, 5.0]$, $[x_1^2(0), x_2^2(0)] = [30.0, 5.0]$, and $[x_1^3(0), x_2^3(0)] = [20.0, 1.7]$. Figure 4*a* presents the trajectories for the velocity and acceleration of the reference vehicle. The vehicle moves toward the left-most lane and accelerates in order to reach its desired speed. We have set the maximum allowed longitudinal acceleration to be 3 m/s^2. When it reaches its desired speed, its longitudinal acceleration reduces gradually to zero. This way it preserves its desired speed and at the same time it smooths the acceleration which affects the value of the cost function. Figure 4*b* presents the positions of the reference vehicle (purple) and the obstacles (yellow, cyan, and blue) on the road plane.

For both scenarios, the Quasi-Newton Broyden-Fletcher-Goldfarb-Shanno (BFGS) method for descent directions was used within the feasible direction algorithm. Both solutions were generated in less than half a second (DP using $T = 1$ s plus feasible direction algorithm) using a conventional laptop computer with i5 Intel CPU. This suggests that the proposed algorithm can be efficiently integrated into an MPC frame in order to cope with disturbances, since for a vehicle that moves at 100 km/h the control decision is applied every 2.8 m (0.1 s), while a whole new trajectory is produced in ~15 m (~0.5 s).

The aforementioned case study conditions were carefully selected to indicate the behavior of the vehicle under two general but important scenarios. In the first scenario there is no path for the vehicle to achieve its

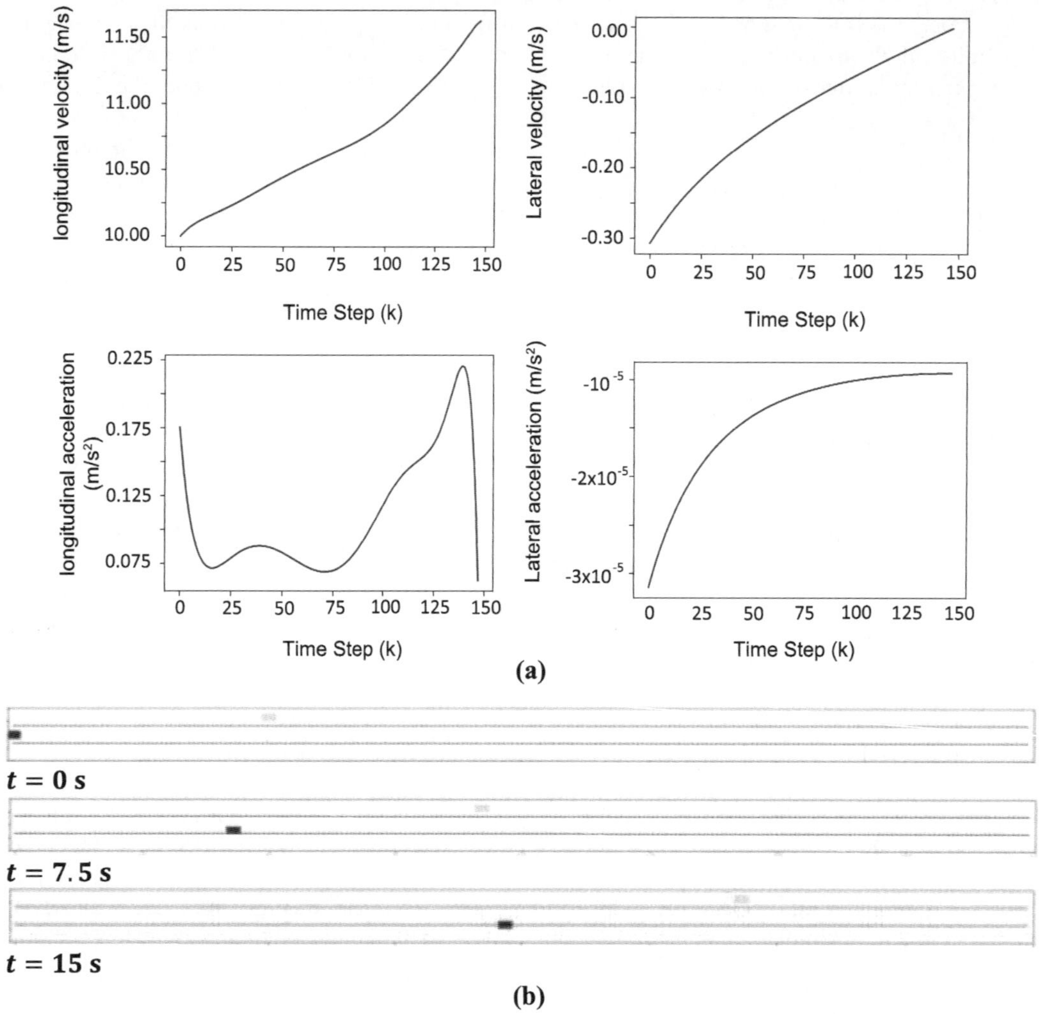

Figure 3. (*a*) The trajectories for longitudinal and lateral velocity and acceleration of the reference vehicle during the first scenario; and (*b*) The positions on road plane of the reference vehicle and the obstacle for three different timestamps (the purple rectangle represents the reference vehicle).

mission, while in the second scenario such a path exists. Evaluation on these two scenarios under a quasi-static environment showed that if a path exists for the vehicle to achieve its mission, the vehicle will find and follow it.

Conclusion

A path-planning algorithm for automated road vehicles was developed. The proposed method formulates the path-planning task as an OC problem, which is then efficiently solved via a combination of DP and NLP techniques. The method is transparent and flexible and allows for a variety of objectives, rules, and restrictions to be considered or modified as needs require. The method provides a very solid basis for planning the trajectory of automated vehicles, but it could also be used to interpret other, less transparent (e.g., deep-learning) methods or analyze their behavior or assess different driving policies and objectives.

The proposed solution corresponds to an open-loop system, but, thanks to the efficiency of the employed solution algorithms, it can be integrated in an MPC frame to provide an online path-planning algorithm that is robust to disturbances.

Ongoing and future work comprises:

- Embedding the algorithm in an MPC frame and using it for the automated vehicle to drive in an ordinary microscopic simulator. This would provide for a plethora of very realistic driving

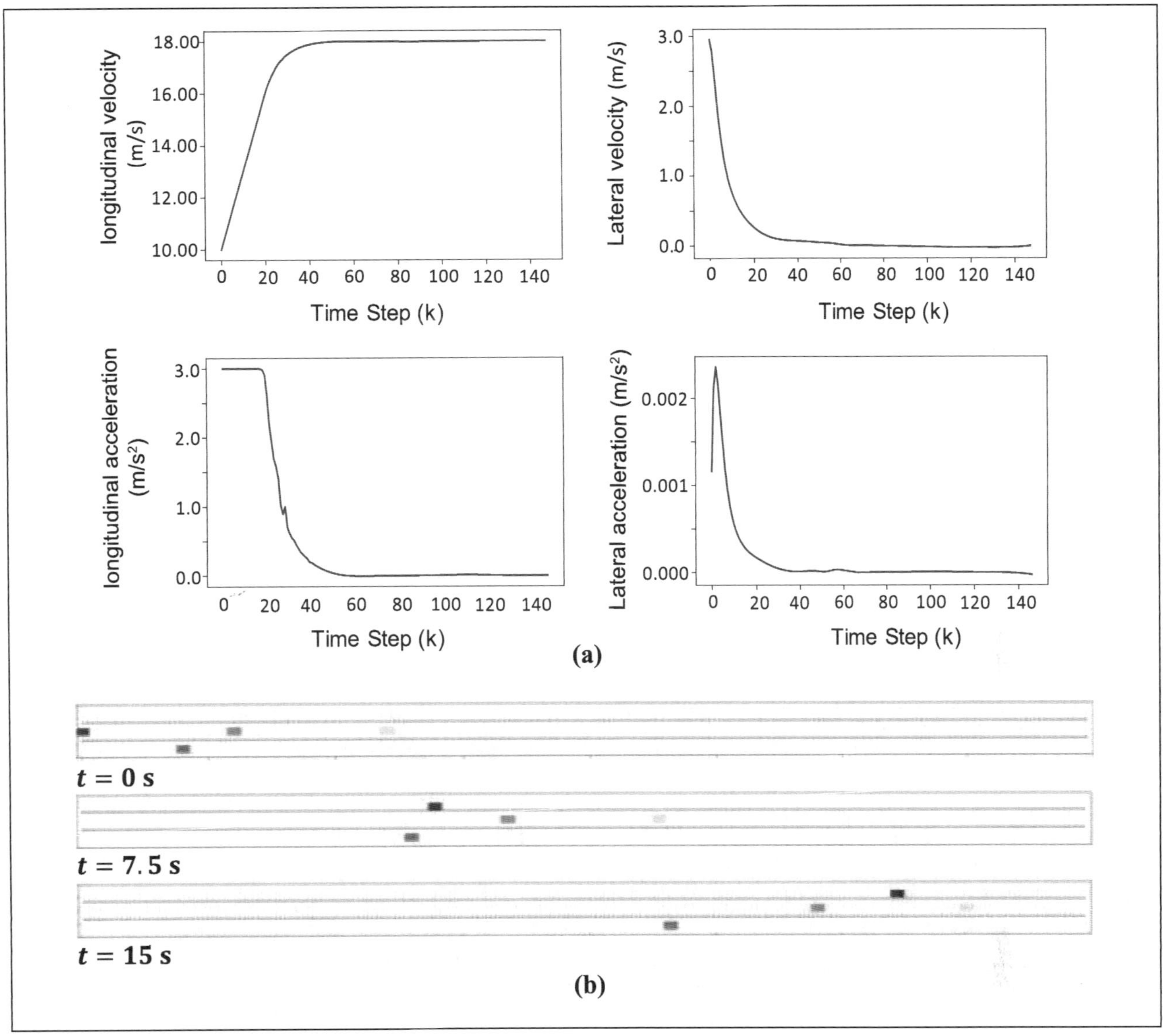

Figure 4. (*a*) The trajectories for longitudinal and lateral velocity and acceleration of the reference vehicle during the second scenario; and (*b*) The positions on road plane of the reference vehicle and the obstacles for three different timestamps (the purple rectangle represents the reference vehicle).

situations and thereby enable further fine-tuning and a very comprehensive evaluation of the induced driving behavior.

- Employ multiple vehicles driving with the same path-planning algorithm, so as to enable the study of their impact on the overall traffic flow efficiency.
- Formulate and solve a common OC problem for simultaneous path-planning of multiple automated vehicles and study the resulting synergistic effects and impact on the overall traffic flow.
- Attempt to apply objective functions that will not merely serve the selfish objectives of the automated vehicle, but will also contribute to a more efficient traffic flow.

Acknowledgments

The research leading to these results has received funding from the European Research Council under the European Union's Seventh Framework Programme (FP/2007-2013) / ERC Grant Agreement no. [321132], project TRAMAN21.

Author Contributions

The authors confirm contribution to the paper as follows: study conception and design: Konstantinos Makantasis, Markos Papageorgiou; data collection: Konstantinos Makantasis, Markos Papageorgiou; analysis and interpretation of results: Konstantinos Makantasis, Markos Papageorgiou; draft manuscript preparation: Konstantinos Makantasis, Markos Papageorgiou. Both authors reviewed the results and approved the final version of the manuscript.

References

1. Reimer, B., B. Mehler, and J. F. Coughlin. *An Evaluation of Driver Reactions to New Vehicle Parking Assist Technologies Developed to Reduce Driver Stress*. MIT Center for Transportation Logistics, Cambridge, Mass., 2010.
2. Zhang, S., W. Deng, Q. Zhao, H. Sun, and B. Litkouhi. Dynamic Trajectory Planning for Vehicle Autonomous Driving. *Proc., 2013 IEEE International Conference on Systems, Man, and Cybernetics*, Manchester, IEEE, New York, 2013, pp. 4161–4166.
3. Goerzen, C., Z. Kong, and B. Mettler. A Survey of Motion Planning Algorithms from the Perspective of Autonomous UAV Guidance. *Journal of Intelligent and Robotic Systems*, Vol. 57, No. 1–4, 2010, pp. 65–100.
4. Werling, M., T. Gindele, D. Jagszent, and L. Groll. A Robust Algorithm for Handling Moving Traffic in Urban Scenarios. *Proc., 2008 IEEE Intelligent Vehicles Symposium*, Eindhoven, Netherlands, IEEE, New York, 2008, pp. 1108–1112.
5. Fletcher, L., S. Teller, E. Olson, D. Moore, Y. Kuwata, and J. How. The MIT–Cornell Collision and Why It Happened. *Journal of Field Robotics*, Vol. 25, No, 10, 2008, pp. 775–807.
6. Wolf, M. T., and J. W. Burdick. Artificial Potential Functions for Highway Driving with Collision Avoidance. *Proc., 2008 IEEE International Conference on Robotics and Automation*, Pasadena, Calif., IEEE, New York, 2008, pp. 3731–3736.
7. Wang, J., J. Wu, and Y. Li. The Driving Safety Field Based on Driver - Vehicle - Road Interactions. *IEEE Transactions on Intelligent Transportation Systems*, Vol. 16, No. 4, 2015, pp. 2203–2214.
8. Schildbach, G., and F. Borrelli. Scenario Model Predictive Control for Lane Change Assistance on Highways. *Proc., 2015 IEEE Intelligent Vehicles Symposium (IV)*, Seoul, South Korea, IEEE, New York, 2015, pp. 611–616.
9. Ji, J., A. Khajepour, W. W. Melek, and Y. Huang. Path Planning and Tracking for Vehicle Collision Avoidance Based on Model Predictive Control With Multiconstraints. *IEEE Transactions on Vehicular Technology*, Vol. 66, No. 2, 2017, pp. 952–964.
10. Erlien, S. *Shared Vehicle Control Using Safe Driving Envelopes for Obstacle Avoidance and Stability*. PhD thesis. Stanford University, 2015.
11. Rasekhipour, Y., A. Khajepour, S. K. Chen, and B. Litkouhi. A Potential Field-Based Model Predictive Path-Planning Controller for Autonomous Road Vehicles. *IEEE Transactions on Intelligent Transportation Systems*, Vol. 18, No. 5, 2017, pp. 1255–1267.
12. Koren, Y., and J. Borenstein. Potential Field Methods and Their Inherent Limitations for Mobile Robot Navigation. *Proc., 1991 IEEE International Conference on Robotics and Automation Proceedings*, Sacramento, Calif., IEEE, New York, 1991, pp. 1398–1404.
13. Ntousakis, I. A., I. K. Nikolos, and M. Papageorgiou. Optimal Vehicle Trajectory Planning in the Context of Cooperative Merging on Highways. *Transportation Research Part C: Emerging Technologies*, Vol. 71, 2016, pp. 464–488.
14. Carvalho, A., Y. Gao, S. Lefevre, and F. Borrelli. Stochastic Predictive Control of Autonomous Vehicles in Uncertain Environments. *12th International Symposium on Advanced Vehicle Control*, 2014, pp. 1–8.
15. Gao, Y., A. Gray, H. E. Tseng, and F. Borrelli. A Tube-based Robust Nonlinear Predictive Control Approach to Semiautonomous Ground Vehicles. *Vehicle System Dynamics*, Vol. 52, No. 6, 2014, pp. 802–823.
16. Hrovat, D., E. Tseng, F. Borrelli, T. Lin, and Y. Gao. Predictive Control of Autonomous Ground Vehicles with Obstacle Avoidance on Slippery Roads. *Proc., ASME 2010 Dynamic Systems and Control Conference*, Cambridge, MA, 2010, pp. 265–272.
17. Gray, A., M. Ali, Y. Gao, J. Hedrick, and F. Borrelli. Semi-autonomous Vehicle Control for Road Departure and Obstacle Avoidance. In T. Stoilov (Ed.), *IFAC Control of Transportation Systems*, Sofia, Bulgaria, 2012, pp. 1–6.
18. Werling, M., and D. Liccardo. Automatic Collision Avoidance Using Model-predictive Online Optimization. *Proc., 2012 IEEE 51st IEEE Conference on Decision and Control (CDC)*, Maui, Hawaii, IEEE, New York, 2012, pp. 6309–6314.
19. Ziegler, J., P. Bender, T. Dang, and C. Stiller. Trajectory Planning for Bertha – A Local, Continuous Method. *Proc., 2014 IEEE Intelligent Vehicles Symposium Proceedings*, Dearborn, Mich., IEEE, New York, 2014, pp. 450–457.
20. Gao, Y., A. J. Gray, J. V. Frasch, T. Lin, H. E. Tseng, J. K. Hedrick, and F. Borrelli. Spatial Predictive Control for Agile Semi-autonomous Ground Vehicles. *Proc., 11th International Symposium on Advanced Vehicle Control*, Seoul, South Korea, 2012.
21. Bellman, R. *The Theory of Dynamic Programming*. Report No: RAND-P-550. Rand Corporation, Santa Monica, Calif., 1954.
22. Tanzmeister, G., M. Friedl, D. Wollherr, and M. Buss. Path Planning on Grid Maps with Unknown Goal Poses. *Proc., 16th International IEEE Conference on Intelligent Transportation Systems (ITSC 2013)*, The Hague, Netherlands, IEEE, New York, 2013, pp. 430–435.
23. Santa, J., A. F. Gómez-Skarmeta, and M. Sánchez-Artigas. Architecture and Evaluation of a Unified V2V and V2I Communication System Based on Cellular Networks. *Computer Communications*, Vol. 31, No. 12, 2008, pp. 2850–2861.
24. Papageorgiou, M., C. Diakaki, I. Nikolos, I. Ntousakis, I. Papamichail, and C. Roncoli. Freeway Traffic Management in Presence of Vehicle Automation and Communication Systems (VACS). In *Road Vehicle Automation 2* (G. Meyer, and S. Beiker, eds.), Springer, Cham, 2015, pp. 205–214. (Lecture Notes in Mobility).
25. Papageorgiou, M., and A. Steinkogler. Real-time Optimal Control of Moving Vehicles in Changing Environments. *IEEE Transactions on Control Systems Technology*, Vol. 2, No. 4, 1994, pp. 447–454.
26. Typaldos, P., I. Papamichail, and M. Papageorgiou. Minimization of Fuel Consumption for Vehicle Trajectories. Presented at 97th Annual Meeting of the Transportation Research Board, Washington, D.C., 2018.

27. Papageorgiou, M., M. Marinaki, P. Typaldos, and K. Makantasis. *A Feasible Direction Algorithm for the Numerical Solution of Optimal Control Problems – Extended Version*. Report No: 2016-26. Technical University of Crete, Dynamics Systems and Simulations Laboratory, Chania, Greece, 2016.

28. Papageorgiou, M., M. Leibold, and M. Buss. *Optimierung: Statische, Dynamische, Stochastische Verfahren für die Anwendung*, 4th ed. Springer, Berlin, 2015.

The Standing Committee on Intelligent Transportation Systems (AHB15) peer-reviewed this paper (18-01437).

Article

TRR
JOURNAL OF THE TRANSPORTATION RESEARCH BOARD

Transportation Research Record
2018, Vol. 2672(19) 124–134

DOI: 10.1177/0361198118776813
journals.sagepub.com/home/trr
SAGE

A Low-Cost Real-World Planning Strategy for Deploying a Dedicated Short-Range Communications Roadside Unit on a Highway Off-Ramp

Jonathan B. Walker[1] and Kevin Heaslip[2]

Abstract

The deployment of dedicated short-range communications (DSRC) roadside units (RSUs) allows a connected or automated vehicle to acquire information from the surrounding environment, such as a traffic light's signal phase and timing, using vehicle-to-infrastructure communication. Several scholarly papers exist on planning strategies for DSRC RSU deployments using simulation without accounting for wireless communication constraints and environmental changes. This paper proposes an empirical-based planning strategy for a highway off-ramp in a real-world environment. The research goal focuses on developing a low-cost and structured deployment plan for DSRC RSUs with the following objectives: use free planning tools; apply the deployment strategy in a real-world environment; utilize publicly available DSRC RSU data measurements; and leverage existing intelligent transportation systems infrastructure when possible. The proposed planning strategy includes three steps: (1) conduct a virtual site survey, (2) gather baseline performance data for the DSRC RSU equipment, and (3) generate a predictive radio frequency signal. The planning strategy was successfully applied on a highway off-ramp at exit 19A of the Capital Beltway, which encircles Washington, DC.

The deployment of dedicated short-range communications (DSRC) roadside units (RSUs) has the potential to revolutionize the world's transportation systems by optimizing highway capacity and accelerating the pathway to highly connected and automated vehicles (*1–3*). A DSRC RSU allows a connected or automated vehicle to acquire information from the surrounding environment, such as a traffic light's signal phase and timing, using vehicle-to-infrastructure (V2I) communication. Based on a review of the literature, several scholarly papers have been published on planning strategies for DSRC RSU deployments, both simulation- and empirical-based. This paper proposes an extension of these previous studies by detailing an empirical-based planning strategy on a highway off-ramp in a real-world environment that leverages existing intelligent transportation systems (ITS) infrastructure.

The main purpose of this strategy is to reduce initial deployment planning costs while accounting for wireless communication constraints and environmental changes. The process is two-fold:

1. Propose a novel process that can be applied to most DSRC RSU deployment use cases; and
2. Apply the novel process to a highway off-ramp located on the Capital Beltway in Prince George's County, MD.

Background

When leveraged with vehicle-to-vehicle (V2V) technology, V2I communication will enable a broader range of safety, mobility, and environmental applications since the communication technology has been designed to work as an interoperable unit for ITS (*4*). As noted in the National Highway Traffic Safety Administration (NHTSA) readiness report on V2V communication, V2I communication can be utilized in numerous safety applications such as red light violation warning, curve speed warning, reduced speed zone warning, and railroad crossing violation warning.

[1]Virginia Polytechnic Institute and State University, Falls Church, VA
[2]Virginia Polytechnic Institute and State University, Blacksburg, VA

Corresponding Author:
Address correspondence to Jonathan B. Walker: WalkerPE@vt.edu

The V2I communication standards have been designed around the use of an RSU that broadcasts via DSRC (*5*). DSRC is a description of a specific type of IEEE 802.11 standard in operation that is customized for ITS. In the United States, the Federal Communications Commission (FCC) has allocated 75 MHz (5.850–5.925 GHz) as a primary service under CFR 47 Parts 90 and 95. Three sets of standards are commonly used for DSRC operation:

1. One developed under the IEEE 802.11.p standard;
2. Two under SAE J2735 and SAE J2945; and
3. Several standards under IEEE 1609 (*4*).

Thus, the planning strategies for a DSRC RSU are vital to the success of a robust wireless connection and an adequate reaction time for the V2I application. If the wireless connection is poor or non-existent, the safety application will not obtain sufficient data to perform the operation services. Thus, there is a need to develop a low-cost and structured deployment plan for DSRC RSUs.

Approach Summary

This paper presents an approach for a low-cost and real-world planning strategy to deploy a DSRC RSU applied to a highway off-ramp in the Washington, DC, area. The approach is structured to achieve the following:

1. Employ free tools for planning purposes;
2. Identify a real-world site to deploy strategy;
3. Utilize publicly available DSRC RSU data measurements in the process; and
4. Leverage existing ITS infrastructure to deploy DSRC RSUs when possible.

The proposed planning strategy includes three steps: (a) conduct a virtual site survey, (b) gather baseline performance data for the DSRC RSU equipment, and (c) generate a predictive radio frequency (RF) signal. The paper's main contributions are as follows:

1. Introduce a low-cost, virtual, five-step method to identify potential locations for mounting a DSRC RSU on existing infrastructure;
2. Establish a free method to retrieve publicly available baseline performance data for the DSRC RSU without physically deploying the equipment in the region of interest;
3. Apply a novel technique that utilizes baseline DSRC RSU data to generate a predictive RF signal:
   - Using the free space path loss (FSPL) model;
   - Applying the reflective coefficient theory for a ground surface; and
   - Calculating the location of destructive or null points from the horizontal/vertical polarization reflection.

In this paper, the term predictive RF signal represents a key parameter that can be calculated using mathmetical forumulas. At times, the RF signal(s) can be measured, at a cost (e.g., equipment purchase/rental and hiring RF engineering experts). However, the research goal is focused on developing a low-cost development plan. The following sections describe the literature review, planning strategy approach, derivation of key RF propagation formulas, and results from a real-world use case.

Literature Review

The literature review showed that the planning strategies for DSRC RSU deployments can be segmented as *simulation-based* and *empirical-based*. Simulation-based strategies center on ideal conditions with no restrictions on equipment location; infrastructure limitations; or local, state, or federal regulations. Empirical-based strategies leverage existing ITS infrastructure to reduce deployment costs and account for wireless communication constraints and dynamic environments that could impact a DSRC signal.

The simulation-based strategies fall into three areas of scholarly research: (a) analyzing connectivity, (b) optimizing RSU deployments, and (c) load balancing RSUs. Recently, the empirical-based strategies were detailed in a series of field experiments that leveraged existing ITS infrastructure and analyzed constraints on DSRC transmission relative to distance (*6–11*).

The literature review revealed that the simulation- and empirical-based strategies both have advantages and disadvantages. The following subsections summarize the simulation- and empirical-based strategies. As mentioned previously, this paper uses an empirical-based strategy to apply the technique in a real-world environment for a highway off-ramp near a power-equipped pole.

Simulation-Based Planning Strategies

Analyzing Connectivity. Kafsi et al. conducted an analysis of the connectivity between RSUs and onboard units by computing the probability of two-vehicle nodes in a wireless coverage range (i.e., a signal strength that meets a given threshold) (*12*). Kafsi assumed the two-vehicle nodes communicated through a wireless link bounded by a critical distance. An extensive set of simulations was developed using a sophisticated radio model that accounted for vehicle density, traffic flow, congestion, and the penetration of V2V or V2I devices.

Khairnar et al. simulated the importance of a strong and uninterrupted wireless communication connection (*13*). Their simulation revealed that an onboard unit is forced to drop over 80% of the data packets/messages if the wireless connection is not maintained. Eventually, the onboard unit's software does not have enough data to perform a given operation (e.g., red light violation, work zone alert), which implies an adverse consequence for the V2V or V2I safety application(s).

Optimizing RSU Deployments. Ou et al. proposed a technique called connectivity-oriented maximum coverage scheme (CMCS), which optimized the wireless coverage in the RSU deployment (*14*). The CMCS was simulated for an urban environment in Chengdu, China—one of the three most populous cities in western China. Using the ns-2 and VanetMobiSim simulators, a clustering technique grouped road segments in an urban intersection. The road clusters were static and maintained an equal distance between road segments. Thus, the number of clusters on a road was equal to the ratio of the road length over the average communication range for a given onboard unit. The CMCS technique encompasses two objectives and two constraints, respectively: (a) maximize RSU connectivity, (b) maximize vehicle coverage, (c) minimize installation cost, and (d) include each RSU.

Liu et al. simulated three heuristic models to solve for the optimal RSU placement based on a transmission delay analysis in urban areas (*15*). The transmission delay analysis was used to calculate the maximum distance that a message could be disseminated within a bounded delay. Yan et al. studied the fundamental principles of roadside infrastructure planning and proposed a class of algorithms, called Tailor, to determine the minimum number of intersections that require roadside infrastructure (*16*). These works do not account for fluctuation in the signal strength from the RSU due to channel fading (e.g., reflection, diffraction, refraction, and/or scattering).

Load Balancing RSUs. Nawaz Ali et al. imitated an enhanced cooperative load balancing approach that utilized the residual bandwidth of neighboring RSUs to prevent network overload (*17*). Both the RSU and onboard unit have a limited transmission and reception range that affected the connection time and packet loss. In turn, the connection time and packet loss were directly related to the reaction time of a V2V or V2I application. During a simulated traffic peak-hour, the RSU was likely to experience a request overload from the onboard unit.

Empirical-Based Planning Strategies

In general, wireless communication is subject to fluctuation in signal attenuation, known as channel fading. Singya et al. demonstrated the multiple paths of an RSU's transmission signal being received by an onboard unit through: (a) reflection, (b) diffraction, (c) refraction, and/or (d) scattering. The multiple paths were mathematically derived and illustrated with a DSRC RSU mounted on a building rooftop and the onboard unit mounted on the ground (*18*).

The Crash Avoidance Metrics Partnership is a joint research venture by Ford and General Motors. Their DSRC field testing has shown zero packet loss with one interfering vehicle when there is direct-line-of-sight. The field test scenario(s) consisted of a stationary sending vehicle (i.e., acting as a DSRC RSU) and a receiving vehicle traveling at a distance of 350 meters at 60 mph. However, there is a need to conduct more field testing with multiple vehicle heights at various distances and speeds (*19*).

Recently, the empirical-based strategy was detailed in six notable papers by Shivaldova et al. on a series of field experiments conducted by the Institute of Telecommunications on IEEE 802.11p V2I and V2V communications via roadside equipment (*6–11*). The Institute of Telecommunications is part of the Faculty of Electrical Engineering and Information Technology at Vienna University of Technology (see their website at https://www.nt.tuwien.ac.at). The experiments were implemented in the field using V2I communication to and from a DSRC RSU with mobile vehicles. The six papers present results of: (a) using a statistical method to analyze the performance of V2I communication using a packet delivery ratio technique that was plotted against distance (*11*), and (b) comparing the performance of directional and omnidirectional antennas in terms of radio transmission coverage and the frame success ratio (*10*).

Based on the literature review, several simulation-based planning strategies assume a strong RF signal strength between the DSRC RSU and onboard equipment. In other words, the wireless connection is free of signal attenuation due to path loss, reflection, diffraction, refraction, or scattering. Miucic et al. (*20*) have documented DSRC signal strength drops (i.e., null areas) that should be accounted for in planning strategies. In a real-world environment, the DSRC signal strength drops must be accounted for in the planning strategy to determine factors such as: (a) technical requirements for a DSRC RSU, (b) the optimum location for mounting the DSRC RSU, and (c) the transmission power setting (i.e., if the power setting is adjustable as a software configuration). Overall, this literature review was utilized to identify gaps in the empirical planning strategies and introduce a low-cost, real-world planning strategy for deploying a DSRC RSU.

Planning Strategy Approach

The proposed planning strategy is a low-cost, structured approach to deploying a DSRC RSU with the following intent: (a) use free tools to plan the deployment process, (b) use publicly available data on the DSRC RSU to generate a predictive RF propagation model, and (c) leverage existing ITS infrastructure to deploy a DSRC RSU with a focus on low deployment cost.

The following outlines the planning strategy approach:

1. **Conduct Virtual Site Survey**
   - Conduct a virtual site survey to identify a potential mounting location for a DSRC RSU on existing infrastructure;
2. **Gather Baseline Performance Data on the DSRC RSU Equipment**
   - Retrieve baseline performance data of the DSRC RSU, without physically deploying the equipment in the region of interest, using the publicly available data submitted to the FCC by the manufacturer;
3. **Generate a Predictive RF Signal**
   - Antenna Polarization
   - Determine the FSPL
   - Analyze the reflective coefficient
   - Calculate the horizontal and vertical polarization critical distance.

Conduct Virtual Site Survey

Google Map™ and Google Earth™ are valuable tools in land-use planning and calculating the distance of road segments within a region of interest (*21*, *22*). Bodzin and Cirucci developed a land-use planning simulation that can be used in multiple geographic areas for real-world planning decisions. Real-world planning is an inexpensive approach to identify existing infrastructure (e.g., electrical power, mounting poles, garners) and study the foliage over a period of time (*21*). Janssen and Rosu introduced a step-by-step geographic information system (GIS) protocol for researchers to measure the distance and coverage of sidewalks along roadways (*22*).

A virtual site survey allows investigation into a region of interest and enables initial planning of the surrounding environment for wireless coverage prediction. Utilizing the time lapse feature in Google Street View, the works of Bodzin and Janssen (*21*, *22*) can be extended to survey the region of interest over a seven-year period. We propose a low-cost, virtual, five-step method to identify potential mounting locations of a DSRC RSU on existing infrastructure. The six steps are as follows:

1. Identify a potential DSRC RSU and retrieve the electrical network specifications/ constraints for the equipment;
2. Using Google Map or Google Earth, search for an existing infrastructure with electrical/solar power to mount the DSRC RSU;
3. Find the region of interest and record the essential road dimensions;
4. Document the location of foliage, buildings, and physical structures surrounding the region of interest over various weather seasons (e.g., spring, winter, summer); and
5. Obtain the latitude and longitude of the proposed mounting of the DSRC RSU;
6. Apply for a DSRC license with the FCC which is outlined in detail by FHWA-JPO-16-267 as well as limitations in the regulation, administrative practices, and deployment planning (*23*).

Google Maps was used to implement the virtual site survey on the Capital Beltway in Prince George's County, MD, with a focus on obtaining the latitude and longitude of an existing infrastructure with electrical/ solar power.

Figure 1 presents four images of a highway off-ramp at exit 19A of the Washington, DC, Capital Beltway, departing I-495 North (Outer Loop) onto Route 50 East of the John Hanson Highway. The bottom-right photo is the region of interest from August 2009. The bottom-left photo is an image from October 2012. The top-left photo is an image from November 2016, while the top-right photo is a top view over the region of interest. These images are available using Google Street View under the time lapse sliding bar. There is an automatic weather station with some form of power at latitude 38.941354 and longitude −76.856145 that was installed just before July 2012. This virtual site survey employs a free tool that can be used for planning purposes and provides a rapid visit to a region of interest when there is a need to identify existing infrastructure for mounting a DSRC RSU in a real-world environment.

Gather Baseline Performance Data on the DSRC RSU Equipment

FCC RF Test Report. The manufacturers of DSRC RSUs are responsible for demonstrating compliance with the FCC rules and requirements under an equipment certification process (*24*). This process contains empirical data regarding the RF propagation characteristics of the

Figure 1. Utilizing time lapse in Google Street View to conduct the virtual site survey.

DSRC RSU. The DSRC RSU test report is an equipment authorization issued by a recognized telecommunication certification body (TCB), and the testing is completed by an accredited, FCC-recognized testing laboratory. Moreover, the TCB is tasked with examining the test data, confirming the accuracy of the supporting documentation, ensuring the testing aligns with the appropriate FCC protocols, and ensuring the data validates the technical parameters (*25*, *26*).

This paper establishes a low-cost method of retrieving baseline performance data for the DSRC RSU without physically deploying the equipment at the region of interest. The baseline performance data is retrieved as follows:

1. Obtain the FCC identification (FCC ID) for the DSRC RSU, which is located on the outside of the hardware
   - The first element is a three-character code with the grantee code
   - The second element is the product code.
2. Using the FCC ID, retrieve the DSRC RSU test report
   - The DSRC RSU test report is available online under *Equipment Authorization Search* at https://apps.fcc.gov/oetcf/eas/reports/GenericSearch.cfm
3. Record the data under equivalent isotropically radiated power (EIRP) measurements from the test report, which the test procedure defines as the power emitted by the antenna after accounting for losses (or gains) between the output signal generator to the antenna.
   - Document the tested frequencies, polarization, height of transmitter/receiver, data rate, EIRP, transmitter/receiver power, bandwidth, margin, and correction factor.

The research for this paper was conducted with an Arada Systems DSRC RSU with an FCC ID of XZB-MAXR552-4 (here, RSU-XZB). Per the FCC Testing Guidelines (*26*), the Locomate Commando RSU was tested and measured with a spectrum analyzer. The DSRC RSU was placed on an 0.8 m height turntable that was rotated 360° to search for the maximum radiation power. In the meanwhile, the receiver antenna was rotated in a vertical and horizontal direction from 1 to 4 m for the maximum polar radiated power.

Figure 2 is a numerical plot and graph of data submitted by Arada Systems for their DSRC RSU under report number RF150120C04A for RSU-XZB. The plot shows a coalition between the EIRP measurements and data rate. There were other coalitions that were not as significant, so they were not presented on the figure to maintain a clear and concise graph. For example, the data sets were paired in the following channel/frequencies: test experiment numbers (or points) 1–2 = 172/5.860 GHz, points 3–4 = 174/5.870 GHz, points 5–6 =

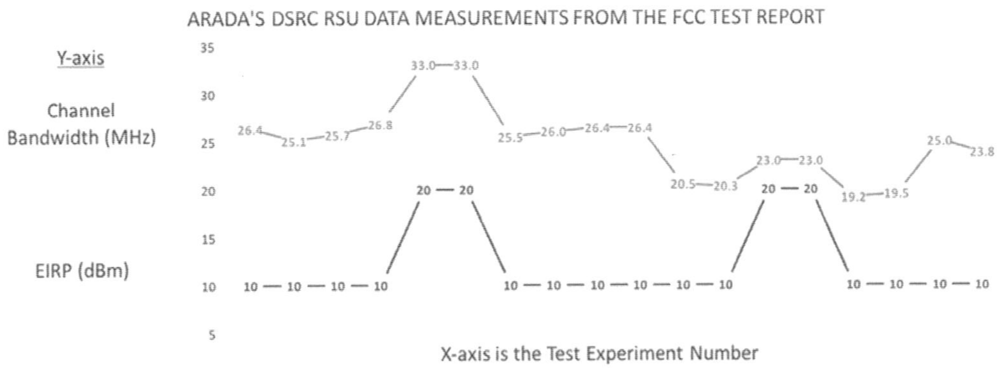

	1	2	3	4	5	6	7	8	9	10	11	12	13	14	15	16	17	18
Channel Bandwidth (MHz)	10	10	10	10	20	20	10	10	10	10	10	10	20	20	10	10	10	10
Data Rate (Mbps) is not plotted	3	27	3	27	6	6	3	27	3	27	3	27	6	6	3	27	3	27
EIRP (dBm)	26.4	25.1	25.7	26.8	33.0	33.0	25.5	26.0	26.4	26.4	20.5	20.3	23.0	23.0	19.2	19.5	25.0	23.8

Channel Bandwidth (MHz) Data Rate (Mbps) is not plotted EIRP (dBm)

Figure 2. Plot of the Arada Systems DSRC RSU data set from the FCC Test Report No. RF150120C04A.

175/5.875 GHz, points 7–8 points = 176/5.890 GHz, points 9–10 = 178/5.890 GHz, points 11–12 = 180/5.90 GHz, points 13–14 = 181/5.905 GHz, points 15–16 = 182/5.910 GHz, and points 17–18 = 184/5.920 GHz.

The DSRC RSU test report is a free data set that has been validated by an accredited testing laboratory. It can be used as a baseline data set without physically deploying the equipment in a region of interest.

Generate a Predictive RF Signal

This subsection focuses on identifying and explaining three fundamental equations to generate a predictive RF signal using the baseline DSRC RSU data. The three fundamental equations are: (a) FSPL, (b) reflective coefficient, and (c) horizontal/vertical polarization null points at critical distances. The objective is to account for various RF signal gains and losses between the DSRC RSU and onboard unit in the vehicle while accounting for wireless communication constraints and environmental change. This subsection provides examples of the fundamental principles of antenna polarization as a smooth transition into the reflection and transmission characteristics of a DSRC RSU.

Fundamental Principles of Antenna Polarization. In the simplest form, an electromagnetic wave may be modeled as two sinusoidal signals in the horizontal and vertical planes traveling through space, which are the electric and magnetic fields (*27, 28*). A traveling wave is considered vertically (or linearly) polarized if the electrical field is moving in the vertical direction through space. A polarized antenna is sensitive to the mounting position and provides the best transmission/reception when the electrical fields are on the same plane.

In Figure 3, the antenna of a sports utility vehicle (SUV) is linearly polarized with the electrical field, which is traveling perpendicular (i.e., the left signal moving north and south) to the Earth's surface. The RSU antenna is linearly polarized, but the electrical field is traveling parallel (i.e., the right signal moving east to west) to the Earth's surface. The DSRC RSU is mounted in the east direction and the SUV antenna is mounted in the north direction. Therefore, the electrical fields are not propagating on the same plane. In this scenario, a linearly polarized RSU antenna will receive the most energy from the SUV antenna when both antennas are physically mounted in the same direction and the electrical fields are traveling on the same plane.

Antenna Mounting. The polarization directly relates to the propagation direction (e.g., horizontal or vertical) of the electrical fields that depart the antenna. An antenna can be designed with a sensitivity to the mounting direction, which implies the transmitting and receiving antennas must retain the same polarization for the best transmission/reception path. If the mounting direction is not observed during installation, the target will experience at least a 3-dB loss because the electrical fields are not propagating on the same plane. The antenna polarization and mounting direction are important factors in the reflective coefficient and polarization critical distance.

FSPL Model. As an electromagnetic wave propagates from the DSRC RSU antenna in free space, the power density

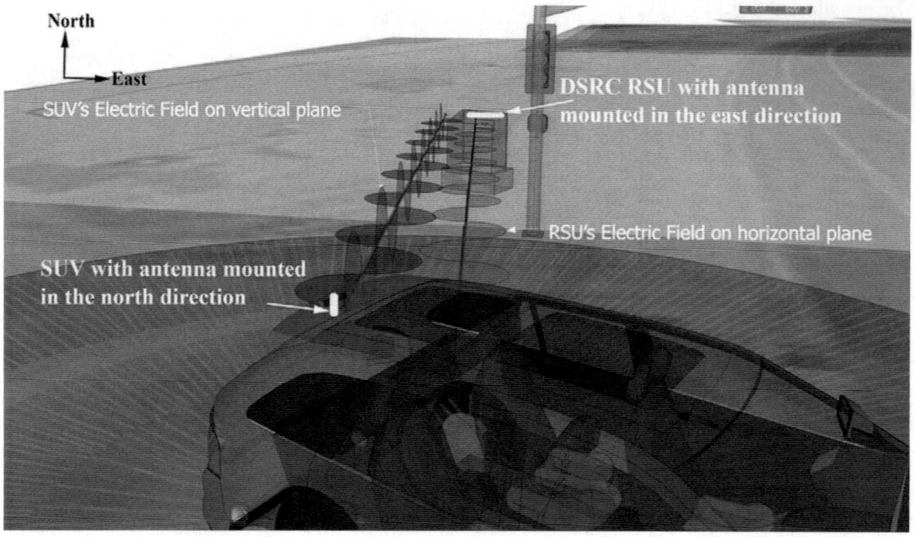

Figure 3. The antennas of the SUV and the RSU are linearly polarized; however, the electromagnetic waves are not propagating on the same plane.

per unit area decreases in proportion to the frequency and the square of the distance traveled (*29*, *30*). As a baseline reference, the FSPL model can be used as an estimate of the signal loss in a non-atmospheric and non-reflective environment. The IEEE standard 145-1993 states that the FSPL is "not due to dissipation, but rather to the fact that the power flux density decreases with the square of the separation distance" (*31*). The equation for FSPL is presented below in terms of dB (*27*):

$$\text{FSPL} = \left(\frac{4\pi df}{c}\right) \quad (1)$$

$$\text{FSPL(dB)} = 20\text{Log}_{10}\left(\frac{4\pi df}{c}\right) \quad (2)$$

$$\text{FSPL(dB)} = 20\text{Log}_{10}(d) + 20\text{Log}_{10}(f) + 20\text{Log}_{10}\left(\frac{4\pi}{c}\right) \quad (3)$$

$$\text{FSPL(dB)} = 20\text{Log}_{10}(d) + 20\text{Log}_{10}(f) + 92.45 \quad (4)$$

where c is the speed of light (3×10^8 m/s), d is the distance from the transmitter (in kilometers), and f is the signal frequency (in GHz).

Normally, a manufacturer will produce an antenna with gain in reference to an isotropic antenna. The antenna gain for the radio hardware transmitter (source) and receiver (target) must be accounted for in the FSPL such that

$$\text{FSPL(dB)} = 20\text{Log}_{10}(d) + 20\text{Log}_{10}(f) + 92.45 - \text{GTx} - \text{Grx} \quad (5)$$

where d is the distance from the transmitter (in kilometers), f is the signal frequency (in GHz), GTx is the transmitter gain (in dBi), and GRx is the receiver gain (in dBi). Once again, the FSPL is a baseline reference for a DSRC RSU antenna, and the equation will be used after determining a critical threshold, as discussed in a later subsection.

Reflective Coefficient: Conductive Properties of the Ground Surface. Because the DSRC RSU can be mounted at various heights, this subsection addresses the reflection and transmission coefficients of the ground surface surrounding the region of interest. The RF signal transmitted by a DSRC RSU is rarely the result of a traveling wave captured directly between the RSU and an onboard unit in a vehicle. More likely, the traveling wave is the result of multipath signals both directly and indirectly with different amplitudes and phases of an electromagnetic wave. Figure 4 is a realistic view with two sets of both vertical and horizontal waves between the RSU and onboard units in a vehicle. A traveling wave will be partially reflected from the surface and partially absorbed by the ground surface.

The angle and level of reflection are based on several factors such as the electromagnetic wave frequency, the relative permittivity of the surface, the thickness of the surface material(s), and the angle of a traveling wave that approaches the material. Wilson et al. conducted extensive measurements on 20 different materials (e.g., plexiglass, red brick, glass, and concrete) to estimate the reflectivity coefficients at 2.3 and 5.25 GHz (*32*).

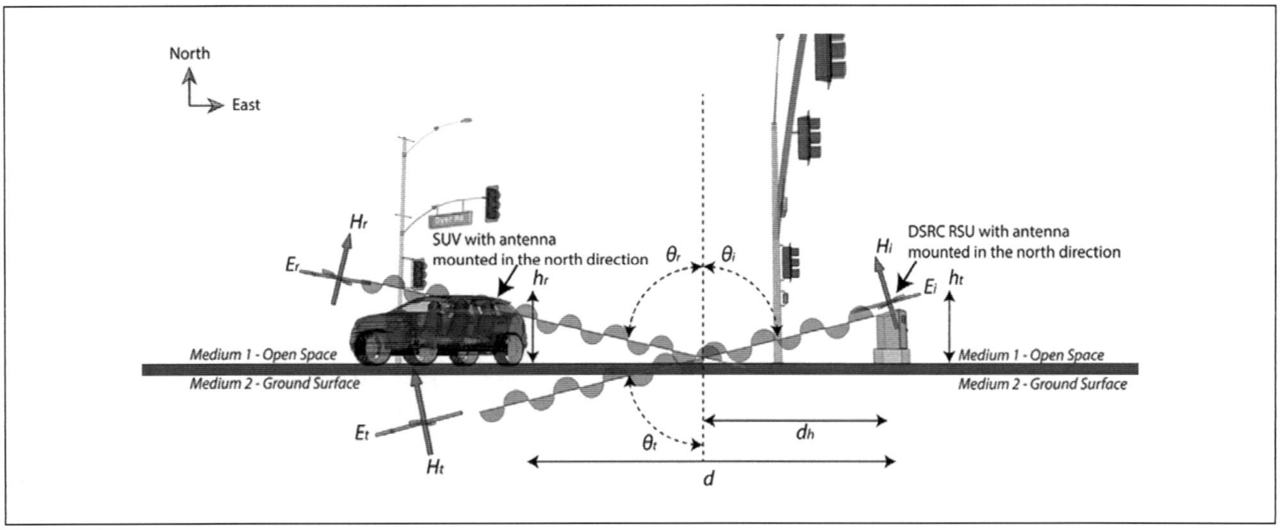

Figure 4. The reflection electric and magnetic fields (Er and Hr) and transmitted electric and magnetic fields (Et and Ht) at the ground surface (medium 2) in relation to open space (medium 1).

Figure 4 depicts the essential variables in wave reflection and transmission. The mounting height of the DSRC RSU antenna (or transmitter) is h_t and the mounting height of the vehicle antenna (or receiver) is h_r. The incident electric and magnetic fields are denoted as Ei and Hi, respectively. The reflective electric and magnetic fields are denoted as Er and Hr, respectively. The transmitted electric and magnetic fields are denoted as Et and Ht, respectively. At a single transmission frequency, the angle of incidence (θ_i) and angle of reflection (θ_r) are equal with the angle of transmission (θ_t). Between the DSRC RSU and vehicle antennas, there is a distance of d and the null distance for the horizontal polarization is d_h. The open space (or air) is medium 1 and the ground surface is medium 2.

When a DSRC RSU is mounted near the ground, the angle of incidence (θ_i) is close to 90° and termed a grazing angle (*21*). For an angle of incidence (θ_i) at 90°, the Fresnel reflection and transmission coefficients for both the perpendicular and parallel polarization are undefined since cosine of 90° is zero. Similarly, an angle of incidence (θ_i) at 85° and 80° will be 8.71% and 17%, respectively, of the ratio between the intrinsic impedances (η) of air and the street/road (or a given medium). Thus, the height of the antenna will affect the angle of traveling waves that graze at the street/road (or a given medium) and are reflected to the onboard unit in a vehicle from the DSRC RSU.

Horizontal and Vertical Polarization Null Points and Critical Distances. The reflection of the DSRC RSU traveling wave will have a destructive effect, which may produce a significant signal loss (or null point) on the resultant wave due to the law of reflection and Briggs Law (*27, 28*). This subsection focuses on calculating the critical distances of each null point of distributive traveling waves by a DSRC RSU. NHTSA's Vehicle Safety Communications project conducted field experiments on DSRC, during which null points were observed at various distances due to multipath reflections (*33*). Miucic et al. (*20*) also documented DSRC signal strength null points and derived the critical distance that should be accounted for in planning strategies.

As shown by Miucic et al. (*20*), the null points can be determined for the horizontal (d_h^k) and vertical (d_v^k) polarizations, whereby $k = 0$ is the first critical null point, $k = 1$ is the second critical null point, $k = 2$ is the third critical null point, and so on. Each of the critical null points depends on the RF spectrum (or the more commonly used wave length λ of a given frequency in meters). For a vertically polarized wave, the null point will occur when the direct and indirect waves are equal but opposite in magnitude (i.e., out of phase by 180° or $\frac{\lambda_i}{2} + k\lambda_i$). For a horizontal polarization wave, the null point will occur at $\lambda_i + k\lambda_i$:

$$d_h^k = \sqrt{\left(\frac{4h_rh_t - (\lambda_i + k\lambda_i)^2}{2(\lambda_i + k\lambda_i)}\right)^2 - (h_t - h_r)^2} \quad (6)$$

$$d_v^k = \sqrt{\left(\frac{4h_rh_t - (\lambda_i/2 + k\lambda_i)^2}{2(\lambda_i/2 + k\lambda_i)}\right)^2 - (h_t - h_r)^2} \quad (7)$$

Based on the null point formulas and as shown in Figure 4, the relative antenna heights between the DSRC RSU (h_t) and onboard unit in a vehicle (h_r) will govern the null points based on the path distance of d between the DSRC RSU and vehicle antennas. The direct and indirect waves contribute to the multipath signals when the relative difference in the antenna heights $|h_t - h_r|$ is much larger than the critical path distance d_c. The critical path distance d_c is calculated as follows (*34*):

$$d_c = \frac{4\pi h_t h_r}{\lambda} \quad (8)$$

On average, the path loss will follow the free space loss equation. When the path distance (d) between the DSRC RSU and vehicle antennas is much greater than the critical path distance d_c, the relative difference in the antenna heights $|h_t - h_r|$ becomes very small (or $h_t \cong h_r$). Thus, the horizontal d_h^k and vertical d_v^k null points must be calculated and multipath loss is calculated as follows (*34*):

$$PL_{multipath}(\text{dB}) = 40\text{Log}_{10}(d) - 20\text{Log}_{10}(h_t h_r) + 120 \quad (9)$$

where d is the path distance between the DSRC RSU and vehicle antennas (in kilometers), h_t is the height of the DSRC RSU in meters, and h_r is the height of the vehicle antenna in meters. As previously discussed, the height of the antenna will impact the amount of traveling waves that graze at the ground surface. This paper assumes the heights of the DSRC RSU and onboard unit in the vehicle are approximately the same ($h_t \cong h_r$).

Findings

Baseline Predictive RF Signal

Using the baseline performance data for the Arada Systems DSRC RSU, the FSPL and multipath loss were calculated based on a path distance d of 3 m between the DSRC RSU and onboard unit as described in the test report number RF150120C04A. The DSRC RSU and onboard unit were approximately the same height ($h_t \cong h_r$) at 1 m. As shown in Table 1, the critical path distance d_c was calculated between 245.46 and 247.98 m at 5.860 and 5.920 GHz, respectively. Based on the baseline data, the FSPL can be applied for an onboard unit that is stationary at a distance of 3 m and a height of 1 m.

Region of Interest Predictive RF Signal

Now, the same formulas can be applied at the site survey location of latitude 38.941354 and longitude −76.856145. Using the baseline performance data, the FSPL and multipath loss were calculated based on a path distance d of 7.62 m between the DSRC RSU and onboard unit. The path distance is 7.62 m to account for a lane width of 3.6576 m, a weather station setback of 3.9624 m, and a DSRC RSU height at 1.524 m ($h_t \cong h_r$) As shown in Table 1, the critical path distance d_c was calculated between 570.11 and 575.94 m at 5.860 and 5.920 GHz, respectively.

Predictive Horizontal Polarization Null Points for Region of Interest

Finally, the horizontal (d_h^k) polarization null points are calculated for six critical distances, whereby $k = 0–5$. In Figure 5, Google Maps Pro was used to depict the critical null points at:

$$d_h^0 = 91.17\,\text{m(red circle)}$$
$$d_h^1 = 45.55\,\text{m(yellow circle)}$$
$$d_h^2 = 30.32\,\text{m(orange circle)}$$
$$d_h^3 = 22.70\,\text{m(green circle)}$$
$$d_h^4 = 11.84\,\text{m(blue circle)}$$
$$d_h^5 = 9.82\,\text{m(white circle)}.$$

Table 1. FSPL and Multipath Loss from the Baseline Data and the Region of Interest at Latitude: 38.941354 and Longitude: −76.856145 for the Arada Systems DSCR RSU

Frequency (GHz)	EIRP (dBm)	FSPL (dB)	Multipath loss (dB)	Critical distance (m)
Baseline predictive RF signal				
5.860	26.35	57.35	19.08	245.46
5.870	25.70	57.37	19.08	245.88
5.880	25.50	57.38	19.08	246.30
5.890	26.40	57.39	19.08	246.72
5.900	20.52	57.41	19.08	247.14
5.910	19.16	57.42	19.08	247.56
5.920	25.00	57.44	19.08	247.98
Region of interest predictive RF signal				
5.860	26.35	65.45	27.96	570.11
5.870	25.70	65.46	27.96	571.08
5.880	25.50	65.48	27.96	572.05
5.890	26.40	65.49	27.96	573.03
5.900	20.52	65.51	27.96	574.00
5.910	19.16	65.52	27.96	574.97
5.920	25.00	65.54	27.96	575.94

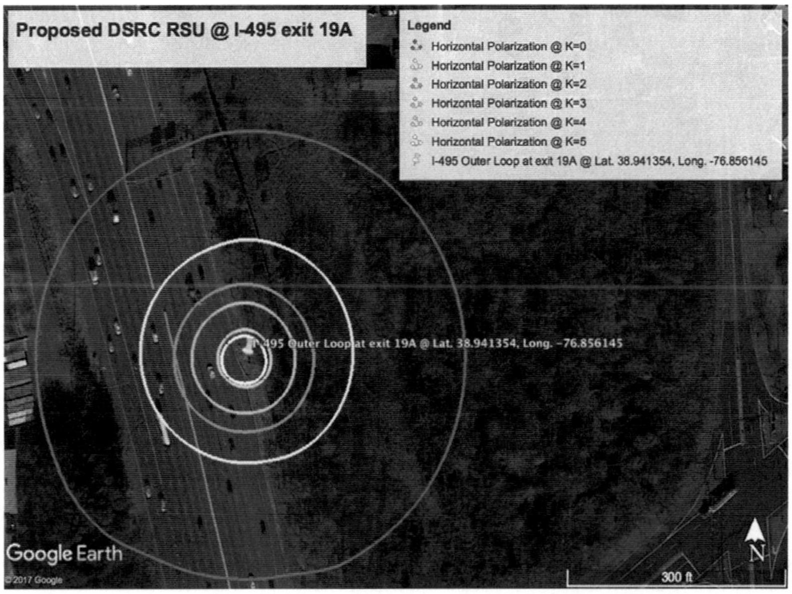

Figure 5. The predictive horizontal polarization null points for the virtual site survey location at latitude 38.941354 and longitude −76.856145.

Conclusion

The proposed planning strategy was successfully applied on a highway off-ramp at exit 19A of the Capital Beltway. We were able to demonstrate a low-cost planning approach for deploying a DSRC RSU without incurring equipment cost for testing equipment or hiring RF experts. Moreover, we utilized FCC test data, submitted by the DSRC RSU vendor, without purchasing the hardware to conduct initial calculations. The FCC test data was used to calculate critical distance and horizontal polarization null points for the potential DSRC RSU. There is a need for future field testing to compare the measured results against the theoretical calculations for critical distance and horizontal polarization (i.e., crucial field measurements).

Author Contributions

The authors confirm contribution to the paper as follows: study conception and design: JW, KH; data collection: JW; analysis and interpretation of results: JW, KH; draft manuscript preparation: JW, KH. Both authors reviewed the results and approved the final version of the manuscript.

References

1. Desiraju, D., T. Chantem, and K. Heaslip. Minimizing the Disruption of Traffic Flow of Automated Vehicles During Lane Changes. *IEEE Transactions on Intelligent Transportation Systems*, Vol. 16, No. 3, 2015, pp. 1249–1258.
2. Meissner, E., T. Chantem, and K. Heaslip. Optimizing Departures of Automated Vehicles from Highways While Maintaining Mainline Capacity. *IEEE Transactions on Intelligent Transportation Systems*, Vol. 17, No. 12, 2016, pp. 3498–3511.
3. Fishelson, J. *Platooning Safety and Capacity in Automated Electric Transportation*. Utah State University, 2013.
4. Harding, J., G. Powell, R. Yoon, J. Fikentscher, C. Doyle, D. Sade, M. Lukuc, J. Simons, and J. Wang. *Vehicle-to-Vehicle Communications: Readiness of V2V Technology for Application*. USDOT NHTSA, Washington, D.C., 2014, p. 327.
5. IEEE Vehicular Technology Society Sponsored. *IEEE Standard for Wireless Access in Vehicular Environments (WAVE)—Networking Services IEEE*. The Institute of Electrical and Electronics Engineers, Inc., New York, 2016.
6. Shivaldova, V., A. Winkelbauer, and C. F. Mecklenbrauker. Signal-to-Noise Ratio Modeling for Vehicle-to-Infrastructure Communications. *Proc., 6th International Symposium on Wireless Vehicular Communication WiVeC 2014*, Vancouver, Canada, IEEE, 2014.
7. Shivaldova, V., and C. F. Mecklenbrauker. Real-World Measurements-Based Evaluation of IEEE 802.11p System Performance. *Proc., 5th International Symposium on Wireless Vehicular Communication WiVeC 2013*, Dresden, Germany, IEEE, 2013.
8. Shivaldova, V., and C. F. Mecklenbräuker. Quantization-Based Complexity Reduction for Range-Dependent Modified Gilbert Model. *Proc., IEEE Sensor Array Multichannel Signal Processing Work*, A Coruna, Spain, IEEE, 2014, pp. 345–348.

9. Shivaldova, V., A. Winkelbauer, and C. F. Mecklenbräuker. Realistic Performance Model for Vehicle-to-Infrastructure Communications. *Proc., 2014 International Symposium on Wireless Personal Multimedia Communications (WPMC)*, Sydney, Australia, IEEE, 2014, pp. 557–561.
10. Shivaldova, V., A. Paier, D. Smely, and C. F. Mecklenbräuker. On Roadside Unit Antenna Measurements for Vehicle-to-Infrastructure Communications. *Proc., 23rd International Symposium on Personal Indoor and Mobile Radio Communications (PIMRC)*, Sydney, Australia, IEEE, 2012, pp. 1295–1299.
11. Shivaldova, V., A. Winkelbauer, and C. F. Mecklenbräuker. Vehicular Link Performance: From Real-World Experiments to Reliability Models and Performance Analysis. *IEEE Vehicular Technology Magazine*, Vol. 8, No. 4, 2013, pp. 35–44.
12. Kafsi, M., P. Papadimitratos, O. Dousse, T. Alpcan, and J. P. Hubaux. *VANET Connectivity Analysis*, IEEE AutoNet, New Orleans, 2009.
13. Khairnar, V. D., and S. N. Pradhan. Simulation Based Evaluation of Highway Road Scenario Between DSRC/802.11p MAC Protocol and STDMA for Vehicle-to-Vehicle Communication. *Journal of Transportation Technologies*, Vol. 3, No. 1, 2013, pp. 88–104.
14. Ou, D., Y. Yang, L. Xue, and D. Dong. Optimal Connectivity-Based Deployment of Roadside Units for Vehicular Networks in Urban Areas. *Transportation Research Record: Journal of the Transportation Research Board*, 2016. 2559: 46–56.
15. Liu, C., H. Huang, H. Du, and X. Jia. Optimal RSUs Placement with Delay Bounded Message Dissemination in Vehicular Networks. *Journal of Combinatorial Optimization*, Vol. 33, No. 4, 2017, pp. 1276–1299.
16. Yan, T., W. Zhang, G. Wang, and Y. Zhang. Access Points Planning in Urban Area for Data Dissemination to Drivers. *IEEE Transactions on Vehicular Technology*, Vol. 63, No. 1, 2014, pp. 390–402.
17. Nawaz Ali, G. G. M., P. H. J. Chong, S. K. Samantha, and E. Chan. Efficient Data Dissemination in Cooperative Multi-RSU Vehicular Ad Hoc Networks (VANETs). *Journal of Systems and Software*, Vol. 117, 2016, pp. 508–527.
18. Singya, P. K., N. Kumar, and V. Bhatia. Mitigating NLD for Wireless Networks: Effect of Nonlinear Power Amplifiers on Future Wireless Communication Networks. *IEEE Microwave Magazine*, Vol. 18, No. 5, 2017, pp. 73–90.
19. Shulman, M., and R. K. Deering. *Third Annual Report of the Crash Avoidance Metrics Partnership, April 2003–March 2004*. Washington, D.C., 2005.
20. Miucic, R., Z. Popovic, and S. M. Mahmud. Experimental Characterization of DSRC Signal Strength Drops. *Proc., 2009 12th International IEEE Conference on Intelligent Transportation Systems*, 2009, pp. 1–5.
21. Bodzin, A. M., and L. Cirucci. A Land-Use-Planning Simulation Using Google Earth. *Science Scope*, Vol. 32, No. 7, 2009, pp. 30–38.
22. Janssen, I., and A. Rosu. Measuring Sidewalk Distances Using Google Earth. *BMC Medical Research Methodology*, Vol. 12, No. 39, 2012.
23. Bayless, S., A. Guan, A. Shaw, M. Johnson, G. Pruitt, and B. Abernathy. *Recommended Practices for DSRC Licensing and Spectrum Management: A Guide for Management, Regulation, Deployment, and Administration for a Connected Vehicle Environment*, USDOT, ITS-JPO, Washington, D.C., 2015.
24. Federal Communications Commission. *Equipment Authorization Order & Accredited Testing Laboratories FCC Equipment Authorization Process*. 2017. https://transition.fcc.gov/oet/ea/presentations/files/may17/10-EA_Accredited-Labs-GT-Final.pdf. Accessed May 15, 2018.
25. Chan, K., R. F. Cleveland Jr, and D. L. Means. Evaluating Compliance with FCC Guidelines for Human Exposure to Radiofrequency Electromagnetic Fields. Supplement C to *OET Bulletin 65, Version 97-01*. Office of Engineering and Technology, Federal Communications Commission, Washington, D.C., 1997.
26. Cleveland, R. F., D. M. Sylvar Jr, and J. L. Ulcek. Evaluating Compliance with FCC Guidelines for Human Exposure to Radiofrequency Electromagnetic Fields. *OET Bulletin 65, Edition 97-01*. Office of Engineering and Technology, Federal Communications Commission, Washington, D.C., 1997.
27. Cheng, D. K. *Field and Wave Electromagnetics*, 2nd ed. Addison-Wesley, Reading, Mass., 1989.
28. Hayt, W. H., Jr. *Engineering Electromagnetics*, 5th ed. McGraw-Hill, Inc., New York, 1989.
29. Cisco Systems. *White Paper on Antenna Patterns and Their Meaning*, Cisco Systems, San Jose, California, 2007.
30. Linx Technology. *Application Note*. Linx Technologies, Merlin, Oregon, 2012.
31. Antenna Standards Committee of the IEEE Antennas and Propagation Society. *IEEE Standard for Definitions of Terms for Antennas*, IEEE Standards Board, Piscataway, New Jersey, 1993.
32. Wilson, R. *Propagation Losses Through Common Building Materials 2.4 GHz vs 5 GHz*. Magis Networks, Inc., 2002, pp. 1–28. https://www.am1.us/wp-content/uploads/Documents/E10589_Propagation_Losses_2_and_5GHz.pdf
33. National Highway Traffic Safety Administration (NHTSA). *Vehicle Safety Communications Project Final Report DOT HS 810 591*, USDOT NHTSA, Washington, D.C., 2006.
34. Campbell Scientific. *The Link Budget and Fade Margin (Application Notes)*, Campbell Scientific, Inc., Logan, Utah, 2016.

The Standing Committee on Intelligent Transportation Systems (AHB15) peer-reviewed this paper (18-02172).

Article

TRR
JOURNAL OF THE TRANSPORTATION RESEARCH BOARD

Transportation Research Record
2018, Vol. 2672(19) 135–144

DOI: 10.1177/0361198118787387
journals.sagepub.com/home/trr
SAGE

An Agent-Based Simulation Model for Parking Variable Message Sign Location Problem

Xun-You Ni[1,2] and Daniel (Jian) Sun[3]

Abstract

Parking spaces are often in short supply in urban areas. To balance the supply and demand and alleviate the overconsumption of public spaces, parking variable message signs (parking VMSs) are commonly used to release information on space availability to drivers en route. The aim of this study was to find the optimal positions for parking VMSs. To achieve the objective, we first define the major decision point (MDP) as the intersection where the newly generated path deviates from the previous one. When informed that the target parking lot is fully occupied, the driver would divert to an alternative one. The route to the alternative parking lot is indicated as the newly generated path, while the one leading to the original parking lot is denoted as the previous one. Quantitatively, MDPs with the highest frequency of occurrence are selected as the candidate positions. Then, an agent-based simulation is proposed to identify the MDPs induced by changes of space availability and the selection of routes. The results indicate that the proposed location algorithm slightly outperforms the scheme with the completed parking information in terms of average travel time and average travel distance. The algorithm can be further integrated into a simulation package, which may assist in the design and operation of an urban parking guidance and information system.

Parking spaces are generally an over-consumed public resource and are usually highly utilized or even fully occupied during peak hours (*1*, *2*). Traffic conditions are aggravated because of drivers cruising for vacant spaces; for example, Barone et al. illustrated that the cruising process may increase the traffic volume by as much as 25% to 40% (*3*, *4*). Parking guidance and information systems (PGIS) have become widely used to facilitate parking in urban areas (*5*). Parking variable message signs (parking VMSs), as a commonly used form of PGIS display terminal, mainly release information on space availability to drivers en route (*6*). This could divert excess demand from the popular parking lots to the less utilized ones and thus reduce travel time. With the assistance of parking VMSs, drivers would alter their parking choice when they received information that there were no vacant spaces within the target parking lot. The drawbacks are that by providing all drivers with the same information parking VMSs would induce competition for the best parking lots (*7*); and the construction and operation costs of parking VMSs are higher than internet PGIS. In recent years, drivers have been able to receive parking information from smart phones or other in-vehicle terminals (*8*, *9*), and even to reserve a parking space at pre-trip or en-route stage. However, safety issues arise when drivers are paying attention to updates of parking information in-vehicle (*10*), and only a few drivers are interested in parking reservation services (*11*). Another advantage of parking VMSs is the release of parking information to all drivers passing by, while in-vehicle terminals only provide proprietary information to the particular vehicles with on-board installations. Unfortunately, although parking VMSs have been in use for more than 20 years, no systematic and practical models have yet been developed to solve the problem of where best to locate them. In the practical implementation, planners should determine where it is suitable to

[1]School of Civil Engineering and Architecture, Zhejiang Sci-Tech University, Hangzhou, China
[2]Center for ITS and UAV Applications Research, School of Naval Architecture, Ocean & Civil Engineering, Shanghai Jiao Tong University, Shanghai, China
[3]State Key Laboratory of Ocean Engineering, School of Naval Architecture, Ocean & Civil Engineering, Shanghai Jiao Tong University, Shanghai, China

Corresponding Author:
Address correspondence to Daniel (Jian) Sun: danielsun@sjtu.edu.cn

locate parking VMSs, aiming to reduce costs and satisfy projected demand.

Chen and Shi (*12*) classified parking VMSs into three levels in terms of travel distance, and illustrated that the location of the second-level parking VMS needs to be optimized. Based on this, a model with an objective of maximizing parking guidance utilities was proposed, in which a genetic algorithm (GA) was adopted to obtain the solutions. However, this does not conform precisely to the actual situations, in which the utility decreases with increasing number of intersections passed by from the parking VMS position to the target parking lot. Other problems of the model are that traffic flow on the network was assumed to be fixed, and even the solution method GA is heuristic in nature. Taking into account the reassigning function of parking guidance, Ji et al. (*13*) assigned parking demands on the network based on stochastic user equilibrium (SUE) and formulated a location model to maximize parking guidance utilities. The utility was defined as traffic flow multiplied by the weighted summation of parking space, fee charged, and attraction ability. However, SUE is a static traffic assignment algorithm, which cannot reflect drivers' responses to the parking information dynamically. By investigating various parking choice models as well as their implementation, potential savings in travel time of three typical location scenarios were quantified (*14*): close to the parking lot entrance, at major decision point (MDP), and intensive coverage. The MDP should be where the majority of drivers make decisions, especially conducting parking choices and route selection simultaneously. The results indicate that the MDP location scenario demonstrates a large potential in reducing travel time, which unfortunately was not illustrated explicitly for lack of detailed solving steps. Ni et al. (*15*) proposed one possible form of MDP, defined as the intersections where the newly generated path deviated from the previous one. Then, an incremental assignment (IA) was implemented to identify the spatial positions of the MDPs. However, IA is still a static traffic assignment algorithm, and changes of space availability and the cruising process were not modeled dynamically. In summary, the existing research does not take the dynamic cruising process into account, and consequently the results of their practical implementation may be worse than the theoretical values.

Agent-based simulation is appropriate for representing the complex spatial–temporal relations of participants in transportation modeling (*16*, *17*) and describing drivers' responses to parking information dynamically. To describe the drivers' behaviors in a spatially explicit environment, Benenson et al. (*18*) proposed an agent-based model named PARKAGENT. The model was implemented as an ArcGIS application which integrates multiple visual tools to select simulation area, establish model scenarios, and store simulation results. Levy et al. (*19*, *20*) extended the model to describe cruising for on-street and off-street parking spaces. The other agent-based model for parking research is extended from the software MATSim, which is a platform for large-scale simulations. Waraich and Axhausen (*21*) developed the model into two hierarchical stages: cruising for different types of parking lots and choosing a desirable one with the highest utility. By integrating the cellular automaton approach into the software package, cruising for vacant parking spaces and parking reservation applications were further investigated (*22*, *23*). The two models were implemented by ArcGIS or Java; the re-development on ArcGIS is complex and the flowchart of Java programming is discrete. Sun et al. (*24*) proposed a more detailed agent-based model and used MATLAB to program the entire flowchart. However, it is not reasonable to ignore traffic congestion and assume that all the vehicles would queue at the fully occupied parking lots reached. In this paper, the model was extended to describe the cruising process more precisely, and then adapted to identify the MDPs where it was suitable to locate parking VMSs. Compared with the existing location research, the cruising process was reflected dynamically and the location algorithm may perform better in the practical implementation.

The remainder of this paper is structured as follows. The next section discusses how to determine the candidate positions where it would be suitable to locate the parking VMSs, from both spatial and quantitative perspectives. In the third section, each vehicle was treated individually as an autonomous agent with the attributes divided into four categories and an agent-based simulation framework was introduced to identify the MDPs induced by changes of space availability and selection of routes. The fourth section takes Xujiahui CBD, Shanghai, as a case study to demonstrate the performance of the proposed location algorithm. Finally, conclusions and recommendations for future work are presented.

Determining Principles for the Candidate Positions

To identify the candidate positions, spatial and quantitative rules were provided. The spatial positions of the MDPs are defined as where drivers make parking choices and route selection simultaneously, while the quantitative rule is to prevent locating parking VMSs at the MDPs with a low frequency of occurrence. The frequency of occurrence is the number of the specified MDPs occurring in the MDP set which is used to record the MDPs.

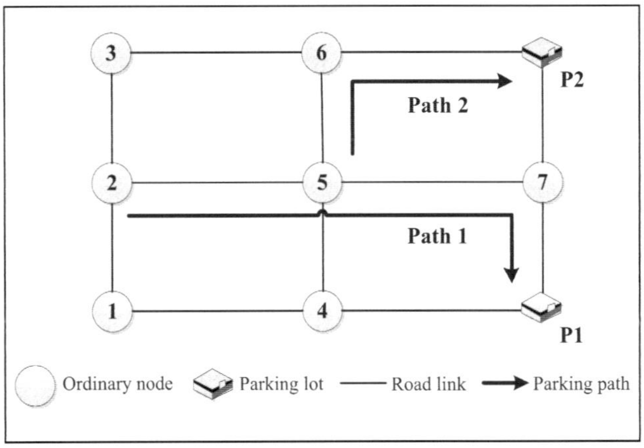

Figure 1. Spatial positions of the MDPs.

Spatial Positions of the MDPs

With the parking information, drivers make parking choice decisions according to the actual conditions. When informed that the target parking lot is fully occupied, a driver would divert to an alternative one (*25*). The route to the alternative parking lot is indicated as the newly generated path, while the one leading to the original parking lot is denoted as the previous one. The spatial positions of the MDPs are further defined where drivers learn that the target parking lot is fully occupied and select the new path deviating from the previous one (*15*). The spatial positions of the MDPs are presented in Figure 1.

In Figure 1, *P1* is the target parking lot and *P2* is a less utilized one. *Path 1* is the previous path from node 2 to *P1*, and *Path 2* is a new path from node 5 to *P2*. Drivers travel along *Path 1* from node 2 to node 5; if informed that the target parking lot *P1* is fully occupied, drivers would make parking choice decisions on link 2→5 (the inbound link of intersection 5) and try to access *Path 2* leading to *P2*. Since the second intersection (node 6) on the new path is not the third intersection (node 7) on the previous path, the new path deviates from the previous one. Link 2→5 is one MDP, which is generated by changes of space availability and selection of routes to an alternative parking lot with vacant spaces.

Quantitative Rule for the Candidate Positions

The MDPs mentioned were determined from the spatial perspective, however, it is almost impossible to locate parking VMSs at all the MDPs. To satisfy the drivers' information requirements, a quantitative rule was proposed. If an MDP does not satisfy the rule, it will be removed from the candidate positions for locating parking VMSs. The quantitative rule is represented as:

$$\Psi_a \geq Z \quad (1)$$

where Ψ_a is the frequency of link a occurring in the MDP set, and Z is the lower limit for the frequency of occurrence, which can be obtained according to the maximum number of parking VMSs allowed or the other objectives of the parking guidance.

Agent-Based Simulation Framework to Identify the MDPs

For each vehicle trip, drivers must park their vehicles within a parking lot, and then walk to their destinations. Since the information about availability of spaces affects parking choice behavior significantly (*26*), and the information changes with time (*27*, *28*), the cruising process is indispensable in congested urban areas. As stated by Chaniotakis and Pel (*29*), most drivers start cruising for vacant spaces when close to or even arriving at their destinations. It is essential to describe drivers' responses to the parking information. The agent-based simulation represents individual behavior by setting the agents with a series of attributes (*30*), which are extensible according to the actual conditions.

Agent Design

Each vehicle was treated individually as an autonomous agent with ten attributes, which can be divided into four categories: basic, spatial, parking, and statistic. The basic attributes are fixed and known before the vehicle enters into the parking zone; the spatial attributes are used to describe the vehicle moving on the network dynamically; the parking attributes are applied to describe the process of cruising for vacant parking spaces; and the statistic attributes are to export the performances. The parking zone is generally defined as a district with various parking lots, such as a Central Business District (CBD). Details of the attributes are designed as follows. For the basic attributes: ID indicates the sequence number of entering into the network, O denotes the origin, and D denotes the destination. For the spatial attributes: A indicates the intersection has been passed by (the first intersection on the parking path), B denotes the downstream intersection (the second intersection on the parking path), C denotes the third intersection on the parking path, and F denotes the current position on the road link. For the parking attributes, P indicates the target parking lot. For the statistic attributes, T1 and T2 indicate the time of entering into and departing from the network for each vehicle respectively.

Agent-Based Simulation Framework

By implementing the detailed individual behaviors and the travel time varying with the traffic flow, the agent-

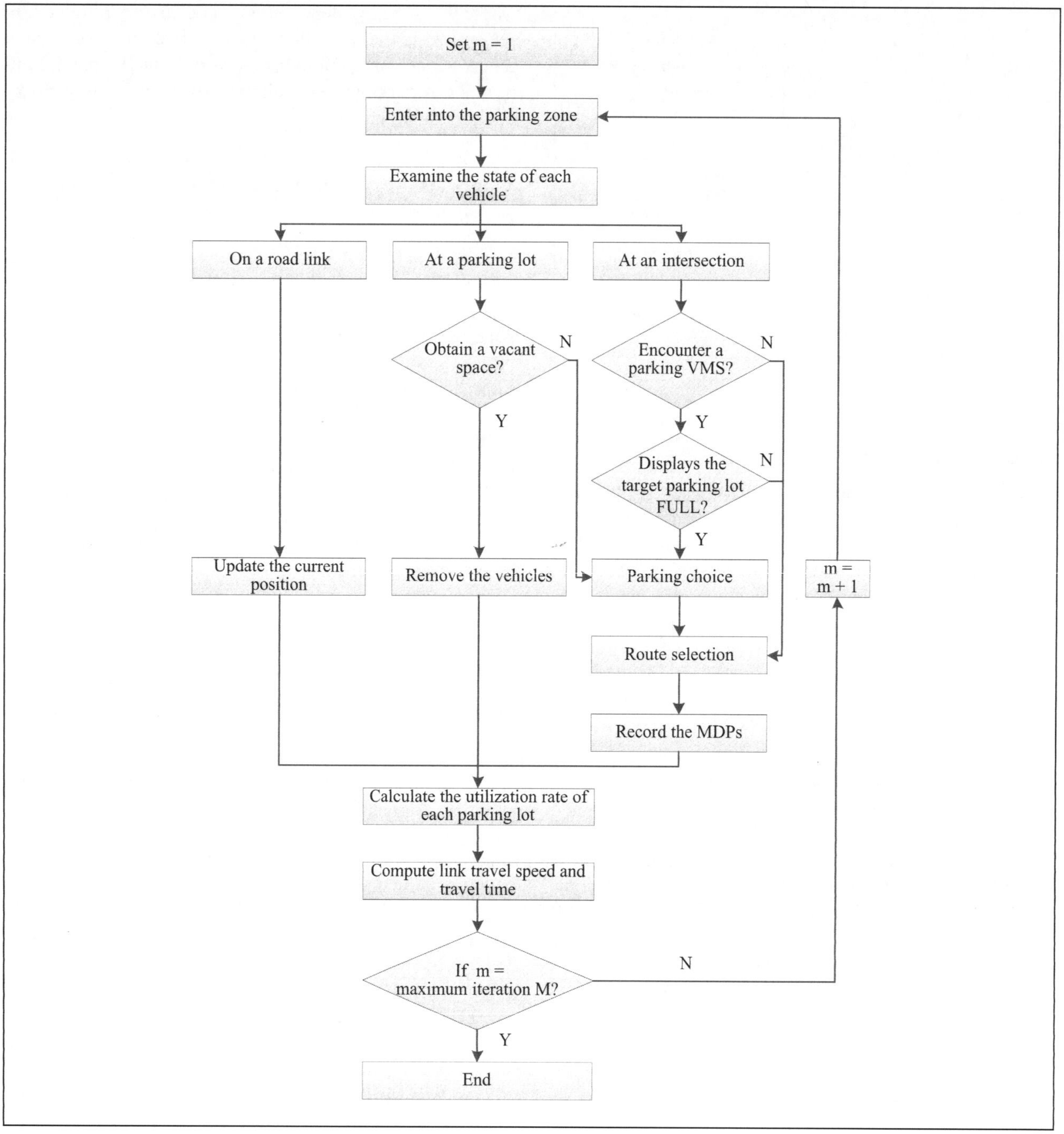

Figure 2. Framework of the agent-based simulation.

based simulation presented in Sun et al. (*24*) was extended to reflect the actual traffic conditions and involve the surplus of vehicles in the next iteration. To capture link travel speed varying with the traffic flow, the traffic flow model was introduced. Since the parking VMS is located some distance away from the target parking lot, some vehicles may get vacant spaces while others cannot obtain a vacant space upon arrival. The surplus of vehicles is defined as vehicles that cannot obtain a vacant space when arriving at the target parking lot.

Figure 2 demonstrates the framework of the agent-based simulation, which describes the vehicles from entering the parking zone, moving toward the target parking lot, and departing from the network dynamically. The overall process is summarized as follows.

Step 0: Initialization. Set the interval iteration $\Delta t = 1$ second, the current iteration $m = 1$, the maximum iteration $M = 7500$, the total travel time $\Gamma^m = 0$, and the total walking time $\Upsilon^m = 0$. Then, initialize link flow $x_a^m = 0$, link density $g_a^m = 0$, link travel time t_a^m to the free-flow travel time t_{af}, and the number of vacant spaces q_j^m to the parking lot capacity φ_j, where x_a^m indicates the traffic flow on link a at iteration m, g_a^m is the vehicular density on link a at iteration m, t_a^m is the travel time on link a at iteration m, t_{af} is the free-flow travel time on link a, q_j^m is the number of vacant spaces in parking lot j at iteration m, and φ_j is the capacity of parking lot j. Then, two empty sets **X** and **Y** are introduced to denote the vehicles parked in parking lots and on the network respectively. Moreover, all the MDPs are recorded in an empty set R.

Step 1: Vehicles enter into the parking zone. Initialize the vehicular attributes: ID, O, D, A, B, C, F, P, T_1, and T_2. Here, ID is initialized equal to the sequence number of entering into the network. O and D are known for each vehicle. Set $A = O$, $F = \varepsilon$, and $T_1 = m$. In order to bring the vehicle at the head of the downstream link, introduce a small constant ε, such as 0.01. P is obtained according to a parking choice model, which is illustrated in Sun et al. (*24*) and Thompson et al. (*31*). Then, B and C are updated as the second and third intersections on the shortest path leading to P respectively. Finally, add the vehicles into **Y**.

Step 2: Examine the state of each vehicle. This step refers to (*24*).

Step 3: Update attributes of vehicles in **Y**. For each vehicle on a road link, update the current position F according to the link speed:

$$F_a^m = \begin{cases} F_a^{m-1} + v_a^m \Delta t, if F_a^{m-1} + v_a^m \Delta t < L_a \\ 0, \qquad otherwise \end{cases} \tag{2}$$

where v_a^m is the travel speed on link a at iteration m, L_a is the length of link a, and F_a^m and F_a^{m-1} are the current positions on link a at iterations m and $m-1$ respectively. For each vehicle at a parking lot that obtains a vacant space, set $T_2 = m$. Then, all these vehicles parked are removed from **Y** and added into **X**. The number of vehicles arriving at the parking lot is computed as:

$$h_j^m = h_j^{m-1} + 1 \tag{3}$$

where h_j^m and h_j^{m-1} are the number of vehicles arriving at parking lot j at iterations m and $m-1$, respectively. The total travel time is calculated as:

$$\Gamma^m = \Gamma^{m-1} + T_2 - T_1 \tag{4}$$

where Γ^m and Γ^{m-1} are the total travel time at iterations m and $m-1$ respectively.

Then, the total walking distance is computed as:

$$\Upsilon^m = \Upsilon^{m-1} + w_{PD} \tag{5}$$

where w_{PD} is walking distance for the vehicle from the target parking lot P to destination D, and Υ^m and Υ^{m-1} are the total walking distance at iterations m and $m-1$ respectively. For each vehicle at a parking lot without vacant spaces, denote E and G as the second and third intersections on the shortest path leading to P respectively. Parking choice and route selection are conducted as follows: update P according to the parking choice model and recalculate E and G. Moreover, if $E \neq C$, it is one MDP where the new path deviates from the previous one. Record the MDP $A \to B$ (the inbound link of intersection B) into **R**. Then, the frequency of occurrence for MDPs is computed as:

$$\Psi_a^m = \Psi_a^{m-1} + 1 \tag{6}$$

where Ψ_a^m and Ψ_a^{m-1} are the frequency of link a occurring in MDP set at iterations m and $m-1$ respectively.

Then, set $A = B$, $B = E$, $C = G$, and $F = \varepsilon$.

For each vehicle at an intersection, if it encounters a parking VMS and receives information that the target parking lot P has no vacant spaces, the driver makes the parking choice and route selection and records the MDPs; otherwise, the driver only makes a route selection. Then, set $A = B$, $B = E$, $C = G$, and $F = \varepsilon$.

Note: for each vehicle in this step, link a can be identified with $A \to B$. The shortest path is computed according to the widely used Dijkstra algorithm (*32*).

Step 4: Calculate the utilization rate of each parking lot. The current number of vacant spaces is calculated as:

$$q_j^m = max\left(q_j^{m-1} - h_j^{m-1} + \rho_j^{m-1}, 0\right) \tag{7}$$

where ρ_j^{m-1} is the number of vehicles departing from parking lot j at iteration $m-1$ and q_j^m and q_j^{m-1} are the number of vacant spaces in parking lot j at iterations m and $m-1$.

The current space occupancy percentage (SOP) is calculated as follows:

$$r_j^m = \frac{q_j^m}{\varphi_j} \tag{8}$$

where, r_j^m is SOP for parking lot j at iteration m and φ_j is the capacity of parking lot j.

Step 5: Compute link travel speed and travel time. The density is defined as the number of vehicles within a length unit, and computed as:

$$g_a^m = \frac{x_a^m}{\eta_a L_a} \tag{9}$$

where η_a is number of lanes of link a, x_a^m is calculated as the number of the vehicles loaded. If one vehicle has the attributes A and B, then the vehicle should be loaded on link $A \rightarrow B$.

To capture link travel speed varying with the traffic flow, Li et al. (*33*) proposed a traffic flow model as follows:

$$v_a^m = \begin{cases} v_{af}, & if\ g_a^m \leq g_{ab} \\ v_{a0} + (v_{af} - v_{a0})\left(1 - \dfrac{g_a^m - g_{ab}}{g_{aj} - g_{ab}}\right)^{\gamma}, & if\ g_{ab} < g_a^m \leq g_{aj} \end{cases} \tag{10}$$

where v_{af} is the free-flow speed on link a, g_{ab} is the density breakpoint on link a, v_{a0} is the minimal travel speed on link a, g_{aj} is the jam density on link a, and γ is the shape parameter.

Link travel time is calculated as:

$$t_a^m = \frac{L_a}{v_a^m} \tag{11}$$

Step 6: Stopping rule. If $m = M$, stop and output the performances; otherwise, set $m = m + 1$ and then go to Step 1. Moreover, the average travel time is expressed as:

$$\bar{\Gamma} = \frac{\Gamma^M}{N} \tag{12}$$

where $\bar{\Gamma}$ is the average travel time, Γ^M is the total travel time at iteration M, and N is the number of vehicles in $\mathbf{X}$.

The average walking distance is calculated as:

$$\bar{Y} = \frac{Y^M}{N} \tag{13}$$

where $\bar{Y}$ is the average walking distance, and Y^M is the total walking distance at iteration M.

Case Study

The road network of Xujiahui CBD, Shanghai, China was used as a case study. The core area is the target region, and the surrounding road network was adopted to let drivers make decisions in a large spatial dimension. Six major roads: West Huaihai Rd, Middle Huaihai Rd, South Urumchi Rd, Dong'an Rd, No. 2 South Zhongshan Rd, and Kaixuan Rd, set the boundaries of the network, as presented in Figure 3. It consists of 101 nodes, 159 road links, five origins (from 1 to 5), two destinations (from 6 to 7), and six major parking lots (from 8 to 13), with 2,768 parking spaces in total. Labels along road links are their lengths, measured by the meter. The origins are the main entrances of the parking zone, and the destinations are the centroids where travelers were attracted. Parking demands have 10 origin–destination (OD) pairs. To simplify, the arrival rate for each OD pair was set as four vehicles per min. and the departing ratio for each parking lot was two vehicles per min. The bold lines denote the arterials, and the ordinary lines are the branches.

Capacities and rates of fees charged for each parking lot were obtained from the field surveys (*34*) and are presented in Table 1.

To describe the vehicles moving on the network, free-flow speed and minimum travel speed were initialized. The values of the free-flow speed on the arterials and the branches are 60 km/h and 30 km/h, respectively. Moreover, the minimal travel speed on all road links is set as 5 km/h. By implementing the agent-based simulation with the completed parking information, the spatial positions of the MDPs and the corresponding frequency of occurrence were obtained and are presented in Table 2.

In Table 2, 6→18 with 303 denotes the frequency of MDP 66→18 appearing in the MDP set. If parking VMSs were located at all the MDPs, drivers would get the completed parking information and the parking lots should be utilized uniformly.

Taking into account the quantitative rule, MDPs with the largest frequency of occurrence were selected as the candidate positions. The lower limit for the frequency of occurrence was determined according to the objectives of parking guidance, for example, setting Z equal to 15. Then, the candidate positions were determined. The scheme with the completed parking information was named as Scenario 1, while that with the frequency of occurrence equal to or greater than Z was denoted as Scenario 2.

Changes of SOP are presented in Figure 4. For the two scenarios, it is obvious that the SOP of each parking lot grows with time, so that the excess parking demands were diverted to the less utilized ones. Since the number of vacant spaces was not taken in the parking choice model, the SOP of the parking lot with large capacity is small. Compared with Scenario 1, the most popular P10 is less utilized and together with P12 and P13 these become the best parking lots in Scenario 2, which

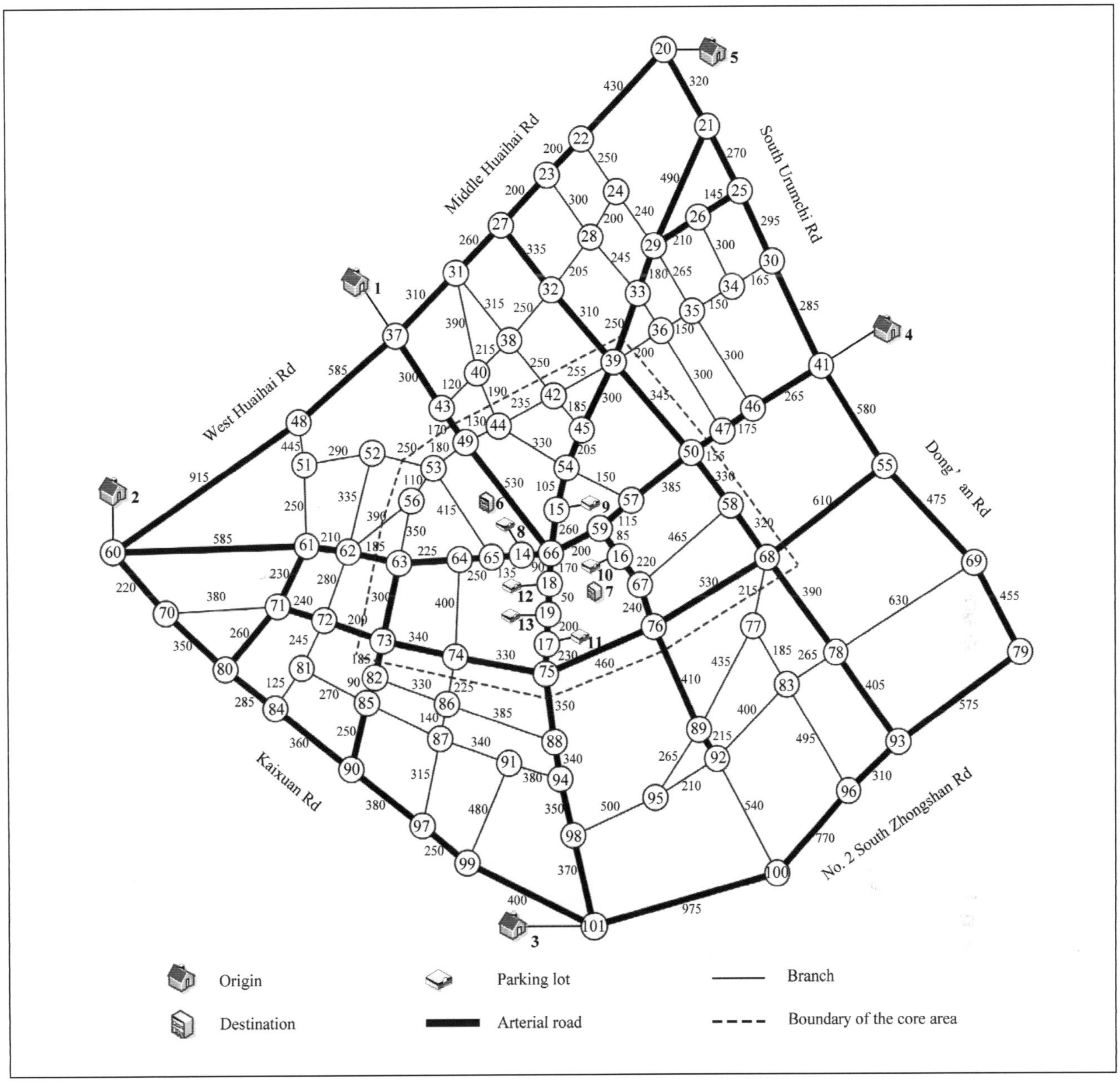

Figure 3. Network of Xujiahui CBD, Shanghai, China (101 nodes, 159 road links, 5 origins, 2 destinations, and 6 major parking lots).

Table 1. Capacities and Rates of Fees Charged at Parking Lots

Parking lot	8	9	10	11	12	13
Capacity	1260	500	191	333	250	234
Fee (RMB/h)	10	9	15	10	15	10

indicates the parking lots are utilized more equably in the latter scheme. In Scenario 2, it is obvious that the popular parking lots reach saturation more slowly, while the SOP of the less utilized ones increases. Consequently, too many parking VMSs may induce competition for the best parking lots and increase travel time.

In Scenario 1, average travel time and average walking distance are 1,153.62 seconds and 443.04 meters,

Table 2. MDPs Obtained from the Scheme with the Completed Parking Information

MDP	FO[a]	MDP	FO	MDP	FO
66 → 18	303	16 → 67	8	71 → 72	2
66 →59	242	67 → 76	7	59 → 57	2
49 → 66	143	66 → 49	6	72 → 73	2
18 → 19	102	74 → 75	6	37 → 43	1
57 → 59	94	45 → 39	6	49 → 43	1
59 → 66	82	39 → 50	6	47 → 50	1
14 → 66	70	60 → 61	5	32 → 39	1
50 → 57	64	58 → 68	5	45 → 42	1
88 → 75	42	45 → 54	5	49 → 44	1
19 → 17	30	68 → 76	4	67 → 58	1
15 → 54	29	64 → 74	4	65 → 53	1
59 → 16	26	16 → 59	4	73 → 74	1
75 → 76	18	50 → 39	3	37 → 31	1
15 → 66	17	20 → 21	3	42 → 45	1
76 → 75	16	61 → 71	3	53 → 56	1
18 → 66	14	43 → 49	3	63 → 73	1
76 → 67	13	58 → 50	3	44 → 40	1
17 → 75	12	44 → 54	3	60 → 48	1
65 → 64	12	39 → 45	2	44 → 49	1
50 → 58	11	57 → 50	2	56 → 53	1
14 → 65	10	58 → 67	2	55 → 68	1
75 → 17	10	57 → 54	2	67 → 16	1
17 → 19	8	19 → 18	2	53 → 49	1
54 → 57	8	54 → 44	2		
54 → 45	8	64 → 63	2		

[a]FO denotes the frequency of occurrence.

respectively. For Scenario 2, the values become 1,134.12 seconds and 441.17 meters. Because too many parking VMSs may induce competition, Scenario 2 slightly outperforms Scenario 1 with the average travel time reduced by 1.69% and the average walking distance decreased by 0.42%. Considering the majority of drivers' information requirements, the proposed location algorithm is feasible and the final location scenario with the economical number of parking VMSs is determined.

Conclusions

The parking VMS location problem is critical in the design and operation of urban PGIS. This study proposes a location algorithm considering the process of cruising for vacant spaces dynamically. Then, an agent-based simulation was introduced to identify the MDPs induced by changes of space availability and selection of routes. Conclusions are drawn as follows:

(i) Taking into account the process of cruising for vacant spaces, this paper proposes a location algorithm which performs better in the practical implementation. By providing the lower limit for the frequency of occurrence, the location scenario with the economical number of parking VMSs would be determined.

(ii) To satisfy the drivers' information requirements, the quantitative rule was proposed. If the indicator frequency of occurrence is small, the corresponding MDPs should be removed from the candidate positions for locating parking VMSs.

(iii) To describe the process of cruising for vacant spaces explicitly, all vehicles were treated as agents with the attributes divided into four categories: basic, spatial, parking, and statistic. Moreover, the vehicular attributes are easily extensible according to the actual conditions.

(iv) An agent-based simulation was introduced by integrating the agent design, in which, drivers' responses to the parking information were reflected dynamically and the MDPs induced by changes of space availability and selection of routes were recorded.

While the results are promising, further studies may be conducted to improve the performance of the proposed algorithm. Parking demand influences the spatial positions of the MDPs and the corresponding frequency of occurrence, which directly determines the candidate

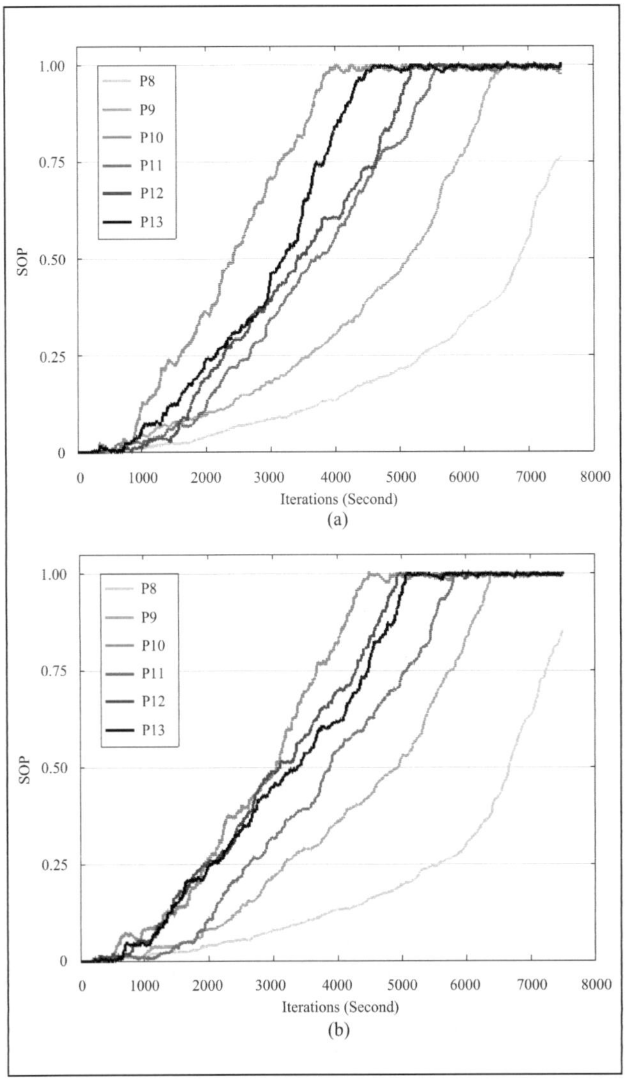

Figure 4. Changes of SOP with the two scenarios: (*a*) Scenario 1, (*b*) Scenario 2.

positions for locating the parking VMSs. During implementation, the peak-hour parking demands would be chosen for the location problem. Moreover, other location indicators, such as drivers' behaviors and traffic safety, would be quantified and assessed further. Additionally, as a foundation of cruising for vacant spaces is the parking choice model, it is critical to adjust the parameters to describe the cruising process accurately. The parameters of the traffic flow model used in this study may be further investigated. With large amounts of field data, the agent-based simulation may be studied to promote the parking level of service more practically. Further studies may implement the location algorithm in a real environment, and then the performances with more traffic data can be assessed.

Acknowledgments

The authors express their appreciation to Professor Zhong-Ren Peng from the Department of Urban and Regional Planning, University of Florida, for his valuable suggestions during the conduct of this study. Additionally, the support of the Humanities and Social Science Research Project, Ministry of Education, China (18YJCZH011), and the Shanghai Municipal Natural Science Foundation (17ZR1445500) are greatly appreciated.

Author Contributions

The authors confirm contribution to the paper as follows: study conception and design: Xun-You Ni and Daniel(Jian) Sun; data collection: Xun-You Ni; analysis and interpretation of results: Xun-You Ni and Daniel(Jian) Sun; draft manuscript preparation: Xun-You Ni and Daniel(Jian) Sun. All authors reviewed the results and approved the final version of the manuscript.

References

1. Broaddus, A. Tolling Heavy Goods Vehicles: Overview of European Practice and Lessons from the German Experience. *Transportation Research Record: Journal of the Transportation Research Board*, 2009. 2066: 106–113.
2. Lam, W. H., Z. C. Li, H. J. Huang, and S. C. Wong. Modeling Time-Dependent Travel Choice Problems in Road Networks with Multiple User Classes and Multiple Parking Facilities. *Transportation Research Part B*, Vol. 40, No. 5, 2006, pp. 368–395.
3. Liu, W., and N. Geroliminis. Modeling the Morning Commute for Urban Networks with Cruising-for-Parking: An MFD Approach. *Transportation Research Part B*, Vol. 93, 2016, pp. 470–494.
4. Barone, R. E., T. Giuffrè, S. M. Siniscalchi, M. A. Morgano, and G. Tesoriere. Architecture for Parking Management in Smart Cities. *IET Intelligent Transportation Systems*, Vol. 8, No. 5, 2014, pp. 445–452.
5. Ji, Y. J., D. N. Tang, P. Blythe, W. H. Guo, and W. Wang. Short-term Forecasting of Available Parking Space Using Wavelet Neural Network Model. *IET Intelligent Transportation Systems*, Vol. 9, No. 2, 2014, pp. 202–209.
6. Bagloee, S. A., A. Ceder, and C. Bozic. Effectiveness of En Route Traffic Information in Developing Countries Using Conventional Discrete Choice and Neural-Network Models. *Journal of Advanced Transportation*, Vol. 48, No. 6, 2012, pp. 486–506.
7. Ni, X. Y., and D. J. Sun. Agent-based Modelling and Simulation to Assess the Impact of Parking Reservation System. *Journal of Advanced Transportation*, Vol. 2017, Article ID 2576094, 2017, pp. 1–10.
8. Wu, E. H. K., C. Y. Liu, J. Sahoo, M. H. Jin, and S. H. Lin. Agile Urban Parking Recommendation Service for Intelligent Vehicular Guiding System. *IEEE Intelligent Transporation Systems Magazine*, Vol. 6, No. 1, 2014, pp. 35–49.
9. Abou-Senna, H. Congestion Pricing Strategies to Investigate the Potential of Route Diversion on Toll Facilities

Using En-route Guidance. *Journal of Traffic and Transportation Engineering (English Edition)*, Vol. 3, No. 1, 2016, pp. 59–70.
10. Geng, Y., and C. G. Cassandras. New "Smart Parking" System Based on Resource Allocation and Reservations. *IEEE Transactions Intelligent Transportation Systems*, Vol. 14, No. 3, 2013, pp. 1129–1139.
11. Ji, Y., W. Guo, P. Blythe, D. Tang, and W. Wang. Understanding Drivers' Perspective on Parking Guidance Information. *IET Intelligent Transportation Systems*, Vol. 8, No. 4, 2013, pp. 398–406.
12. Chen, Q., and F. Shi. Locating of Variable Message Signs in Parking Guidance Systems. *Proceedings of the ICE - Transport*, Vol. 163, No. 3, 2010, pp. 111–118.
13. Ji, Y., W. Deng, and W. Wang. A Planning Model for Determining Optimal Locations of Parking Guidance Sign Boards Based on Utility Maximization. *Proc., 11th International Conference of Chinese Transportation Professionals*, Nanjing, China, 2011.
14. Waterson, B. J., N. B. Hounsell, and K. Chatterjee. Quantifying the Potential Savings in Travel Time Resulting from Parking Guidance Systems – A Simulation Case Study. *Journal of the Operation Research Society*, Vol. 52, No. 10, 2001, pp. 1067–1077.
15. Ni, X. Y., D. J. Sun, and Z. R. Peng. An Improved Incremental Assignment Model for Parking Variable Message Sign Location Problem. *Journal of Advanced Transportation*, Vol. 49, No. 7, 2015, pp. 817–828.
16. Guo, L., S. Huang, and A. W. Sadek. A Novel Agent-based Transportation Model of a University Campus with Application to Quantifying the Environmental Cost of Parking Search. *Transportation Research Part A*, Vol. 50, 2013, pp. 86–104.
17. Chen, S., and D. J. Sun. An Improved Adaptive Signal Control Method for Isolated Signalized Intersection Based on Dynamic Programming. *IEEE Intelligent Transportation Systems Magzine*, Vol. 8, No. 4, 2016, pp. 4–14.
18. Benenson, I, K. Martens, and S. Birfir. PARKAGENT: An Agent-based Model of Parking in the City. *Computers, Environment and Urban Systems*, Vol. 32, No. 6, 20018, pp. 431–439.
19. Levy, N., K. Martens, and I. Benenson. Exploring Cruising Using Agent-based and Analytical Models of Parking. *Transportmetrica A*, Vol. 9, No. 9, 2012, pp. 773–797.
20. Levy, N., M. Render, and I. Benenson. Spatially Explicit Modeling of Parking Search as a Tool for Urban Parking Facilities and Policy Assessment. *Transport Policy*, Vol. 39, 2015, pp. 9–20.
21. Waraich, R. A., and K. W. Axhausen. Agent-based Parking Choice Model. *Transportation Research Record: Journal of the Transportation Research Board*, 2012. 2319: 39–46.
22. Horni, A., L. Montini, R. A. Waraich, and K. W. Axhausen. An Agent-based Cellular Automaton Cruising-for-Parking Simulation. *Transportation Letters*, Vol. 5, No. 4, 2013, pp. 167–175.
23. Sana, B., H. Riadh, and M. Rafaa. Intelligent Parking Management System by Multi-agent Approach: The Case of Urban Area of Tunis. *Proc., IEEE International Conference on Advanced Logistics and Transport*, Hammamet, Tunisia, May 2014.
24. Sun, D. J., X. Y. Ni, and L. H. Zhang. A Discriminated Release Strategy for Parking Variable Message Sign Display Problem Using Agent-based Simulation. *IEEE Transactions on Intelligent Transportation Systems*, Vol. 17, No. 1, 2016, pp. 38–47.
25. Yan, G., W. Yang, D. B. Rawat, and S. Olariu. SmartParking: A Secure and Intelligent Parking System. *IEEE Intelligent Transportation Systems Magzine*, Vol. 3, No. 1, 2011, pp. 18–30.
26. Tang, S., T. Rambha, R. Hatridge, S. Boyles, and A. Unnikrishnan. Modeling Parking Search on a Network by Using Stochastic Shortest Paths with History Dependence. *Transportation Research Record: Journal of Transportation Research Board*, 2014. 2467: 73–79.
27. Schlote, A., C. King, E. Crisostomi, and R. Shorten. Delay-tolerant Stochastic Algorithms for Parking Space Assignment. *IEEE Transactions on Intelligent Transportation Systems*, Vol. 15, No. 5, 2014, pp. 1922–1935.
28. Rajabioun, T., and P. Ioannou. On-street and off-street Parking Availability Prediction Using Multivariate Spatiotemporal Models. *IEEE Transactions on Intelligent Transportation Systems*, Vol. 16, No. 5, 2015, pp. 2913–2924.
29. Chaniotakis, E., and A. J. Pel. Drivers' Parking Location Choice Under Uncertain Parking Availability and Search Times: A Stated Preference Experiment. *Transportation Research Part A*, Vol. 82, 2015, pp. 228–239.
30. Sun, Q., and S. Wu. A Configurable Agent-based Crowd Model with Generic Behavior Effect Representation Mechanism. *Computer-Aided Civil and Infrastructure Engineering*, Vol. 29, No. 7, 2014, pp. 531–545.
31. Thompson, R. G., K. Takada, and S. Kobayakawa. Optimisation of Parking Guidance and Information Systems Display Configurations. *Transportation Research Part C*, Vol. 9, No. 1, 2000, pp. 69–85.
32. Sheffi, Y. *Urban Transportation Networks: Equilibrium Analysis with Mathematical Programming Methods*. Prentice-Hall, Inc., Englewood Cliffs, New Jersey, United States, 1985.
33. Li, A. C., L. Nozick, R. Davidson, N. Brown, D. A. Jones, and B. Wolshon. Approximate Solution Procedure for Dynamic Traffic Assignment. *Journal of Transportation Engineering*, Vol. 139, No. 8, 2013, pp. 822–832.
34. Sun, D. J., J. M. Cheng, and Y. Zhang. An AHP-Fuzzy Comprehensive Evaluation Model of Parking Lots in Urban CBD Area. *Advances in Transportation Studies: An International Journal*, Vol. 37, 2015, pp. 141–156.

The Standing Committee on Intelligent Transportation Systems (AHB15) peer-reviewed this paper (18-05592).

Article

TRR
JOURNAL OF THE TRANSPORTATION RESEARCH BOARD

Transportation Research Record
2018, Vol. 2672(19) 145–156

DOI: 10.1177/0361198118783598
journals.sagepub.com/home/trr
SAGE

Assistive System to Improve Pedestrians' Safety and Mobility in a Connected Vehicle Technology Environment

Sara Khosravi[1], Byungho Beak[1], K. Larry Head[1], and Faisal Saleem[2]

Abstract

Conflicts between vehicles and vulnerable road users (VRUs) often result in injuries and fatalities. This paper presents a vehicle-to-infrastructure and Wi-Fi based pedestrian conflict avoidance system to improve VRUs' safety. Compared with the vision based systems, the proposed system can improve VRUs' safety in non-line-of-sight situations. In particular, it can be helpful in a situation when drivers are making a right or left turn where there is a crosswalk and visibility conditions are poor. The system, called Smart Walk Assistant (SWA), was implemented as an application on a smartphone and server on a road-side unit. It includes two wireless communication pathways: pedestrian-to-infrastructure (P2I) and pedestrian-to-vehicle (P2V). The first communication pathway enables users to send a pedestrian signal request to the traffic signal controller and receive traffic signal status. The second communication pathway enables pedestrians and vehicles to exchange information including location, speed, and heading that can be used to provide conflict alerts. The SWA system may be especially beneficial to differently abled pedestrians, including blind or visually impaired pedestrians, who would benefit from active support to safely cross streets at signalized intersections. The P2I application was designed, developed, and field demonstrated. The P2V application was designed and developed, and will be field tested in the near future.

Preventing vehicle-to-vulnerable-road-user accidents are common topics in the intelligent transportation systems (ITS) community. Creating more walkable cities requires ensuring pedestrian safety and reducing conflicts between pedestrians and vehicles. Road traffic risk (*1*) is defined by four different elements: exposure, the probability of a crash, the probability of injury, and the outcome of the injury. Compared with other road users, vulnerable road users (VRUs) are particularly exposed to injury as they are not protected by a vehicle shell. According to the World Health Organization (*2*) the risk of death per kilometer traveled for pedestrians is higher than vehicles. Based on the Global Status Report on Road Safety 2013 (*2*), nearly half (46%) of those killed in road traffic crashes are pedestrians, cyclists, or users of motorized two-wheeled vehicles.

Predicting potential conflicts can prevent some vehicle-to-pedestrian accidents. However, safety systems for pedestrians are challenging since they are not protected by a vehicle shell that could serve as a platform for sensors and other system components. The lack of information about potential dangers and the inability of pedestrians to detect and perceive oncoming dangers can be the primary reason for accidents (*3*). Using sensors and computer vision techniques to detect pedestrians has been studied (*4*). However, these techniques do not perform well when pedestrians are not in the line of sight of the vehicle. Also, the accuracy of these systems decreases in poor visibility conditions, such as at night and in bad weather conditions. Wireless communications can help overcome these limitations.

Connected vehicles systems are emerging as the next generation vehicle safety system. These systems include wireless communication between vehicles, the infrastructure, and vulnerable road users. The U.S. Department of Transportation estimates that approximately 82% of all crash types involving unimpaired drivers in the U.S. and around 40% of all crashes occurring at intersections can be addressed by vehicle-to-vehicle (V2V) communication (*5*). These statistics show that connected vehicle technology can have a significant impact on improving road

[1]Department of Systems and Industrial Engineering, University of Arizona, Tucson, AZ
[2]Maricopa County, MCDOT, Phoenix, AZ

Corresponding Author:
Address correspondence to Sara Khosravi: sarakhosravi@email.arizona.edu

users' safety. The majority of research has been focused on V2V and vehicle-to-infrastructure (V2I) communications. Developing safety applications from the point of view of pedestrians can help reduce crashes between pedestrians and vehicles. A pedestrian and vehicle conflict can be avoided by alerting the vehicle driver of the existence of the pedestrian or by informing the pedestrian of the existence of the oncoming vehicle. Current research focuses on informing the driver about the existence of the pedestrian (*4*), but the pedestrian can react for his/her own safety too. Informing, or alerting, the pedestrian can have an important impact on reducing pedestrian fatalities.

Connected vehicle technology and wireless communications between road users can improve mobility, especially for visually impaired and disabled pedestrians. Crossing signalized intersections can be a dangerous activity for these users. The tasks required for visually impaired and blind pedestrians to utilize a signalized crosswalk are more challenging than they are for other pedestrians. They have to detect the street, locate the crosswalk or crossing location, locate and press the pedestrian actuation button (if needed), determine the appropriate time to cross, and travel in a straight path while crossing (*6*). It is difficult for blind pedestrians to be aware of the need to push the pedestrian actuation button without previous knowledge about the crossing area (*6*). They usually cannot find and push the pedestrian actuation button, and they generally begin to cross when they hear parallel vehicles starting to move. The time they have to cross the street is likely to be shorter than if there is pedestrian actuated phase at the intersection (*6*). Providing pedestrian-to-infrastructure (P2I) communication can improve mobility for visually impaired and disabled pedestrians by reducing the complexity and assisting them in executing the required tasks.

The goal of this paper is to present a system to assist pedestrians at the signalized intersection and provide a situational awareness system to improve pedestrians' safety by using connected vehicle technology.

Literature Review

The lack of information about oncoming traffic, traffic signal status, and geometry at an intersection makes crossing decisions difficult for visually impaired and blind pedestrians (*7*). Barlow et al. (*8*) showed that only a few (0–16.3%) of visually impaired pedestrians were able to find the pedestrian pushbutton. In addition, visually impaired pedestrians began walking during the "Walk" interval only 24.6% of the time and on 50% of these crossings they failed to complete their crossing before the onset of perpendicular traffic. Providing useful real-time information for visually impaired pedestrians can improve their mobility and safety. DeRuwe and Wall (*9*) developed the Advanced Accessible Pedestrian System (AAPS) to enhance pedestrian push buttons to provide tactile and audio feedback for visually impaired pedestrians.

Providing precise and real-time information for drivers can help them to perceive and identify dangerous situations so they can take proper actions to avoid conflicts. Developing a situational awareness system to alert drivers and pedestrians to possible threats can have a significant impact on safety. Recent research has focused on driver support systems to help prevent traffic accidents.

Gandhi and Trivedi (*4*) classify pedestrian collision avoidance systems into three main categories: improving infrastructure, active systems, and passive systems. Infrastructure improvement focuses on physical treatment such as separation of pedestrians and vehicles as well as control of vehicle speed. Passive pedestrian systems involve vehicle design and focus on reducing the impact of the collision with the pedestrians and include treatments such as intelligent bumpers, pop-up bonnets, and windscreen airbags. Active pedestrian collision avoidance systems focus on designing pedestrian detection systems by employing different types of sensors and computer vision algorithms. These systems involve predicting the possibility of collisions and providing warnings for the driver. Technologies that are currently used in active pedestrian collision avoidance systems can be divided into three main types: vision based, radar based, and wireless based (*10*). Each of these techniques has some advantages and disadvantages.

Vision-based pedestrian detection techniques and algorithms have been used to predict pedestrian trajectories and the possibility of collisions (*11*, *12*). Chen et al. (*13*) developed a vision-based system utilizing a camera to detect pedestrians and an extended Kalman Filter to predict the motion of a pedestrian. Based on the trajectory prediction, the time to collision was calculated and if it is less than a threshold, the driver is alerted to the possible collision. One of the main advantages of using vision-based techniques to detect pedestrians is the low cost. They also have some disadvantages, including degraded performance in poor visibility conditions, such as bad weather, at night, long recognition and processing latency, and background color and lighting.

Wang et al. (*14*) developed a vision-based application, which uses the phone's embedded camera. They utilized machine learning algorithms on the phone to recognize the front views and back views of moving vehicles to detect vehicles approaching the user and to provide an alert to the pedestrian. In this system, there is no communication between pedestrians and drivers. This system cannot be used if the smartphone is in a pocket or if the camera is facing the ground or the sky (*3*).

Wireless-based communication is one solution that can overcome vision-based systems limitations. Wireless communications can perform well in all weather conditions and non-line-of-sight situations. Ho and Chen (*10*) developed a Wi-Fi based application to provide situational awareness for pedestrians and vehicles. In their system, the pedestrian's smartphone is set to access point (AP) mode and vehicles connect to the AP to receive pedestrian position information which is sent as the service set identifier (SSID). They used a mathematical model based on the vehicle and pedestrian information to calculate the probability that the vehicle will successfully stop before running into the pedestrian. In their system, there is no communication between the pedestrian and the traffic controller. Also, vehicles should connect to the pedestrian mobile device to scan SSIDs. This platform may raise some security and privacy concerns.

Connected vehicle systems use dedicated short-range communication (DSRC) in the 5.9 GHz spectrum. Latency in DSRC is much less than in Wi-Fi, and the range of the DSRC is much greater than Wi-Fi, so drivers can receive the pedestrian message sooner and have more time to make proper decisions to avoid conflicts. Different vehicle-to-pedestrian (V2P) applications have been developed. Wu et al. (*15*) used a smartphone embedded with DSRC technology to communicate with vehicles, but currently there is no smartphone embedded with DSRC available.

Methodology and Implementation

The NHTSA (*16*) identified three categories of pedestrian crashes with vehicles, as illustrated in Figure 1. Half of the accidents happen on straight roads without a visual obstruction of the pedestrian. Approximately 37% of accidents happen if the pedestrian is not visible or if the vehicle turns into a street.

The vision-based systems for pedestrians' and bicyclists' protection may not perform well in these situations, while a wireless-based system can be used to detect hidden pedestrians (*17*). The advantage of wireless communications in the connected vehicle environment can provide a valuable solution for the development of future pedestrian protection systems.

In the connected vehicle environment, each intersection is equipped with a Roadside Unit (RSU) and vehicles are equipped with an On Board Units (OBUs). These devices use DSRC to communicate V2V and V2I. V2P and P2I communications need additional capabilities since there are no DSRC pedestrian devices available. Wireless media, including cellular (4G/LTE/5G) or Wi-Fi (802.11), can be used to provide P2I communications and V2I can be used to relay information to the pedestrian to create a V2P pathway.

The Society of Automotive Engineers (SAE) has established a set of DSRC message standards (SAE J2735) and performance requirements for V2P applications (SAE J2945/9). Some of the standard messages include:

- Basic safety message (BSM): vehicles broadcast position and motion information.
- Signal phase and timing (SPaT): current traffic signal status of signalized intersections.
- MAP message (MAP): digital description of the geometry of the roadway.
- Personal safety message (PSM): vulnerable road users broadcast a type of BSM that includes type of VRU (e.g., pedestrian, bicycle), position, and motion information.

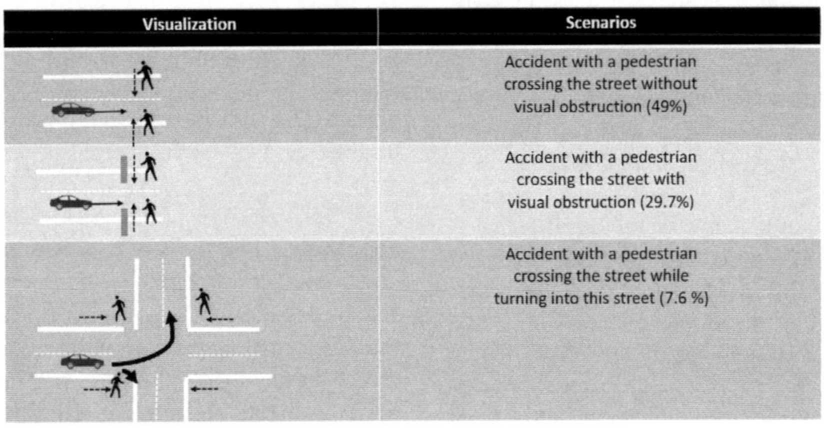

Figure 1. Crash scenarios between pedestrian and vehicle.

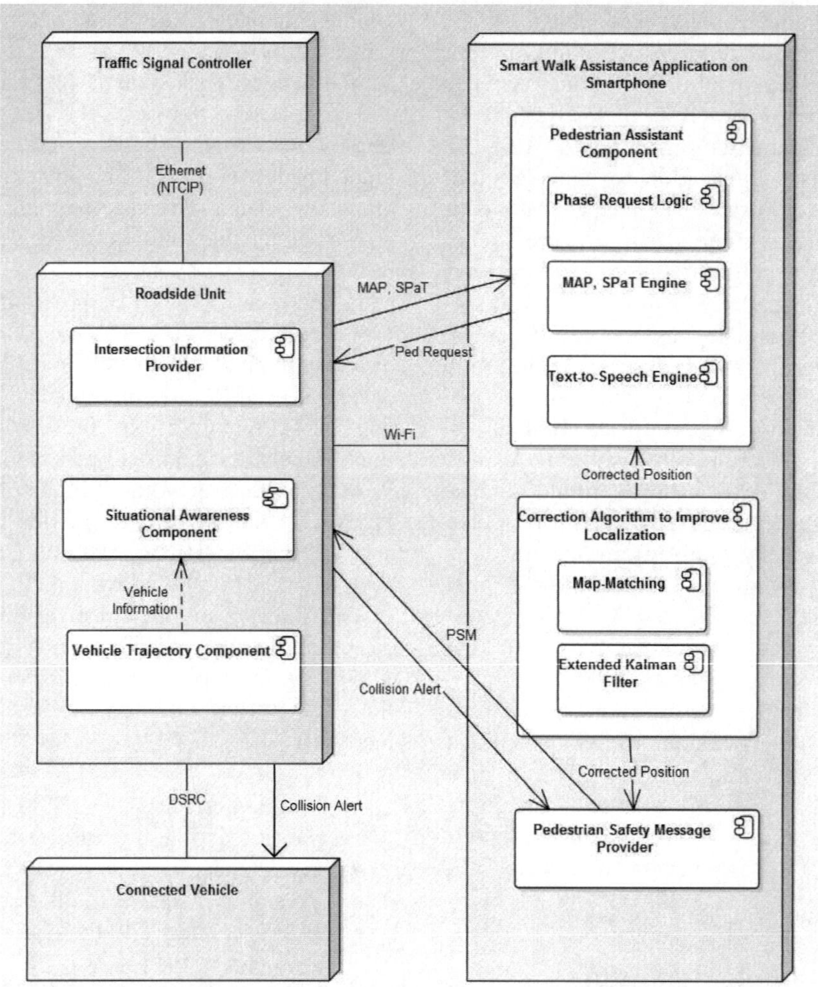

Figure 2. SWA component diagram.

The idea of the connected pedestrian application, which is called Smart Walk Assistant (SWA), is to assist pedestrians at signalized intersections and improve pedestrians' safety by using connected vehicle technology. Figure 2 shows the basic system components. The pedestrian assistant component is installed on the smartphone and communicates with the intersection information provider component on the RSU to assist pedestrians at intersections. The situational awareness component is installed on the RSU and communicates with the PSM provider on the smartphone. The situational awareness component detects collisions between pedestrians and vehicles. A position correction algorithm is implemented on the smartphone to improve pedestrian location estimation and overcome GPS positioning challenges. Each of these components is explained in the following sections.

Pedestrian Assistant Component

The goal of this component is to provide communication between pedestrians and a traffic signal controller to help pedestrians access different services through their smartphone. This can have a significant effect on improving disabled and visually impaired pedestrians' mobility and safety.

The SWA application was developed as an Android application. The smartphone communicates to the local intersection infrastructure equipment RSU using Wi-Fi. The intersection information provider on the RSU sends the MAP to the application. The MAP data are used to determine the location of crosswalks and sidewalks. The MAP is based on accurate GPS measurements, which were obtained using survey grade RTK-GPS equipment (Leica Geosystem RTK 1.30). Figure 3 shows MAP

Figure 3. MAP points.

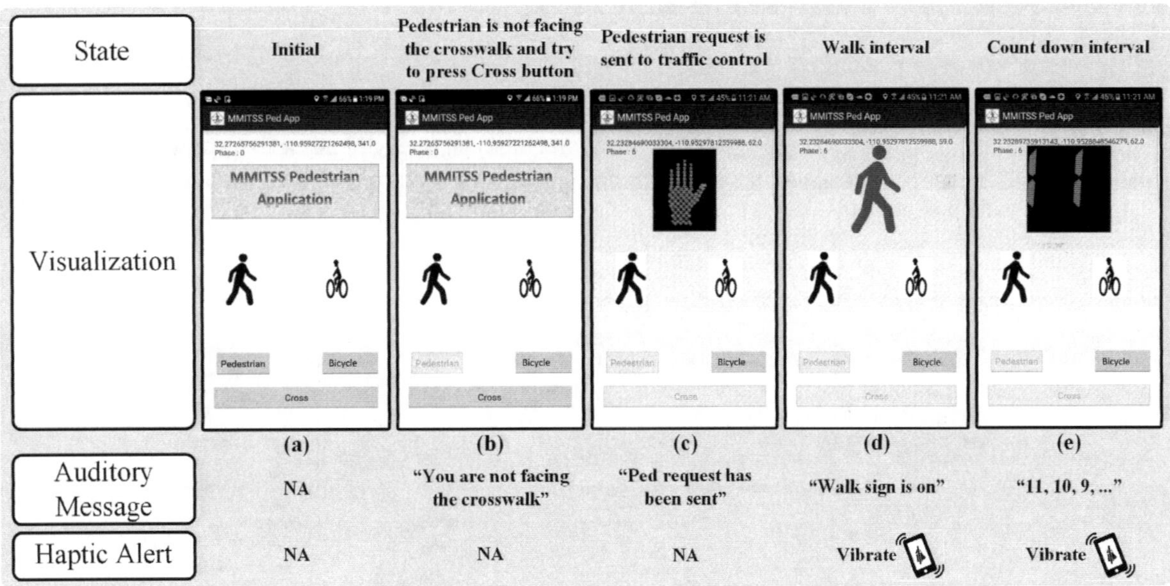

Figure 4. Signal status on the SWA application.

points that include the sidewalks and crosswalks on a Google Earth image.

Figure 4*a* shows the initial screen of the SWA application. A user must first select pedestrian or bicycle. For this paper, only the pedestrian features are discussed. The application uses the smartphone's GPS and inertial sensors to estimate the location and heading of the pedestrian (smartphone). In this paper it is assumed the user orients the phone in their direction of travel. Based on this location estimate, the application will determine which crosswalk in the MAP the pedestrian is facing. The compass orientation of the smartphone is very important and if it is not oriented in the same direction as the candidate crosswalk, it will not allow the pedestrian to request service when they press the cross button (but will provide feedback to the user that they are not aligned with the crosswalk), as shown in Figure 4*b*. With the phone oriented toward the crosswalk, the pedestrian can press the cross button and a pedestrian request will be sent to the RSU, and an audible message will be played to the pedestrian. The request will be forwarded to the traffic signal controller.

The intersection information provider on the RSU broadcasts the SPaT messages which the SWA application can receive over Wi-Fi. The smartphone will display pedestrian signal information using visual, haptic, and auditory cues based on the SPaT status associated with the identified crosswalk. The visual display uses the standard pedestrian signal head graphics and a combination of auditory and haptic alerts. During the "Don't walk" interval, a standard "Don't Walk" graphic is displayed. If the pedestrian presses the cross button during the "Don't Walk" interval, a message will warn him/her that the signal is red, as shown in Figure 4*c*. During the "Walk" interval a standard walk graphic is displayed, as shown in Figure 4*d*. During the pedestrian clearance interval (flashing "Don't Walk"), the standard pedestrian graphic and a countdown timer are displayed, including an auditory countdown, as shown in Figure 4*e*.

Situational Awareness Component

The situational awareness system is made of three components: the OBU deployed on the vehicle, the RSU at the intersection, and the smartphone of the pedestrian. Figure 5 shows the system architecture.

The situational awareness component on the RSU (Figure 2) is responsible for receiving vehicle information messages and pedestrian information messages to detect possible collisions. The RSU receives BSMs that are broadcasted by all connected vehicles within the radio range through DSRC. An application on the RSU, called the vehicle trajectory component, receives this information. On the smartphone, the pedestrian safety message provider creates a PSM. According to the SAE J2945/9 standard, the PSM includes:

- VRU type: the type of vulnerable road user.
- Temporary ID: identifier for each VRU.
- Location information: GPS location information.
- Position accuracy: GPS location accuracy.
- Speed: speed of VRU in m/s.
- Heading: compass heading of the VRU in degrees.
- Personal crossing request: VRU crossing request status.

Figure 6 shows an activity diagram for this component. Whenever a new message is received on the socket server, the component first determines the message type. If the received message is a vehicle information message, it stores vehicle information in a list of vehicles and waits to receive pedestrian information. When pedestrian information is received, the situational awareness component unpacks and stores this information in a list of pedestrians, then the situational awareness component utilizes the collision prediction algorithm to detect any possible collision between pedestrian and vehicles. If

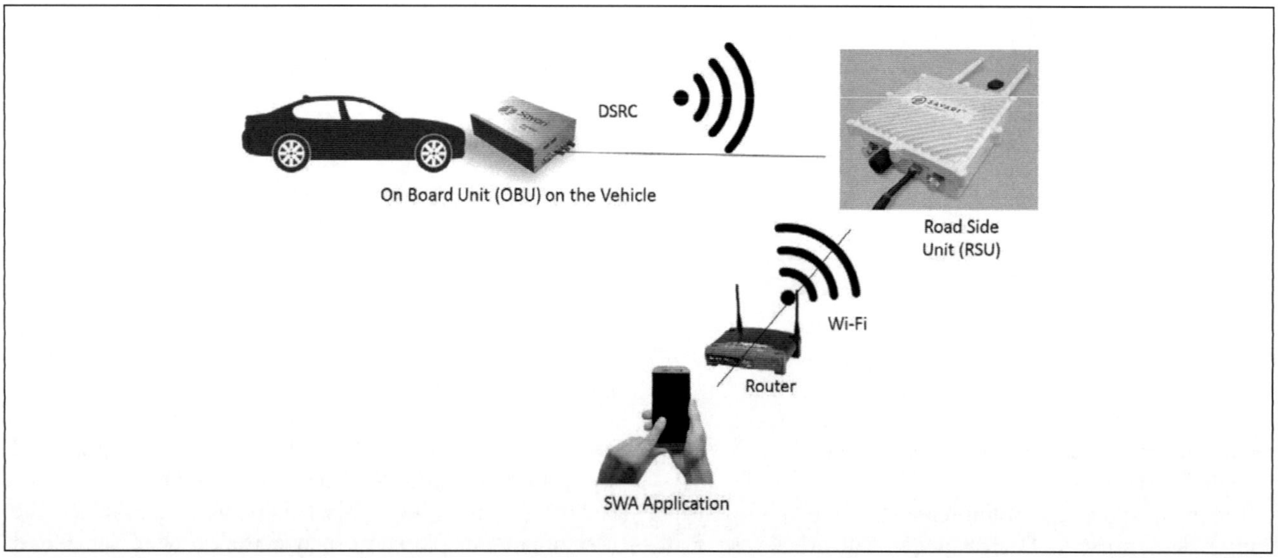

Figure 5. V2P communication components.

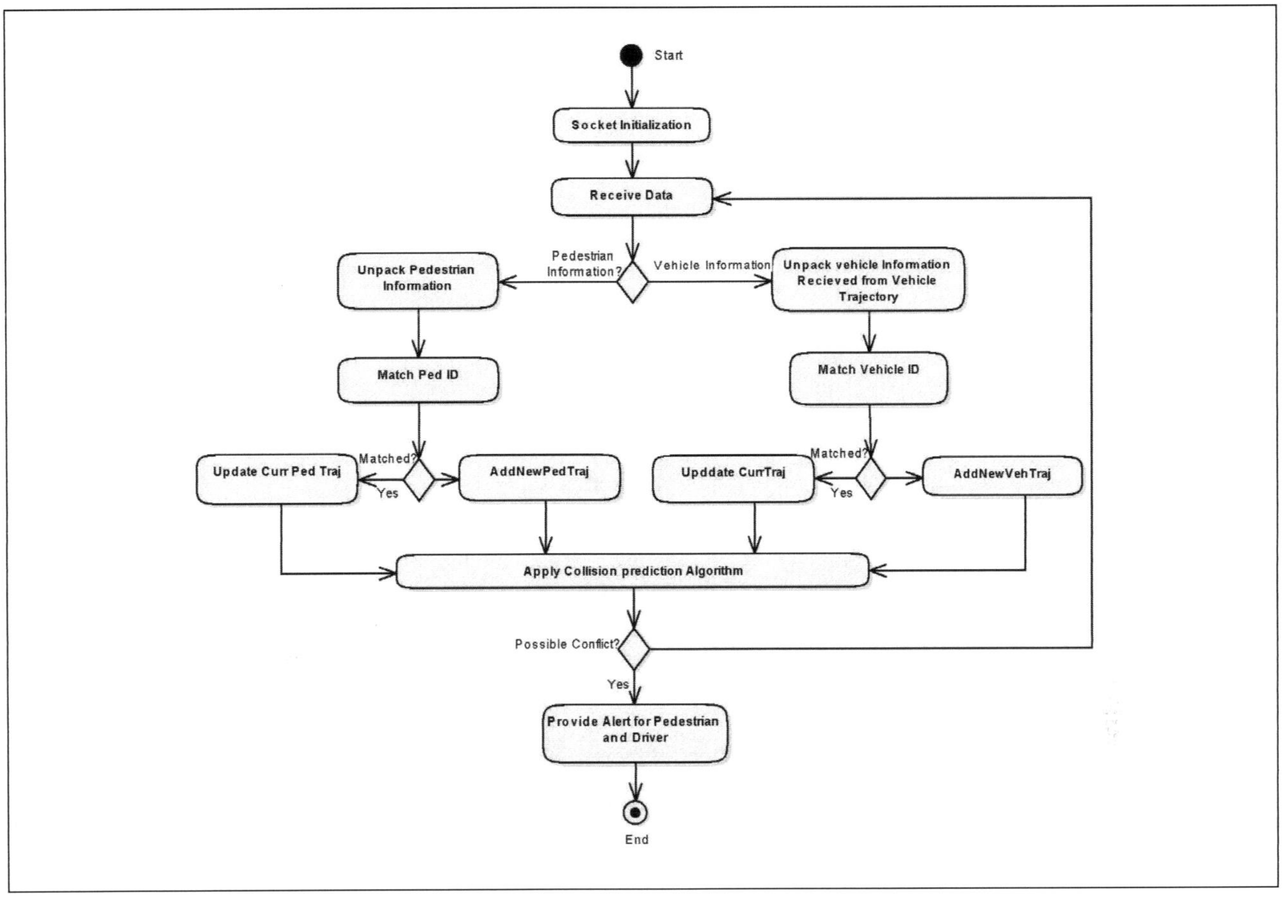

Figure 6. Activity diagram: situational awareness component.

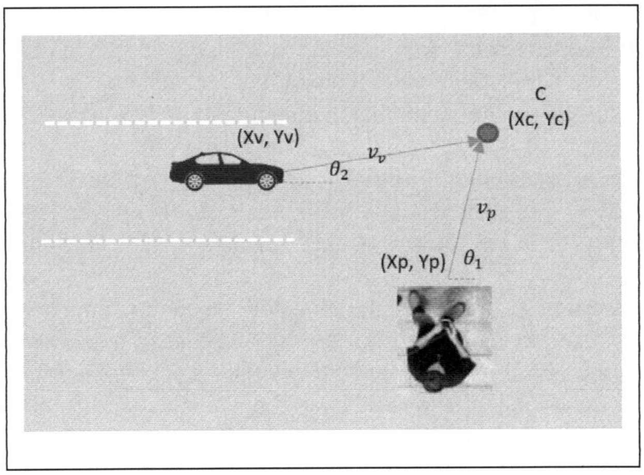

Figure 7. Collision prediction algorithm modeling.

there is a possible collision, it alerts the vehicle driver and pedestrian.

Situational Awareness Algorithm. The SWA application can address a variety of dangerous situations at the intersection, including the scenarios in Figure 1. Although, some of these scenarios would only result if there was a traffic violation at the signalized intersection, such as a pedestrian entering a crosswalk during the Don't Walk interval or a vehicle violating a red indication and entering a crosswalk. The only requirement to use this system is to have an equipped intersection with RSU and Wi-Fi.

An application on the RSU is responsible for retrieving information from vehicles and pedestrians to anticipate the possible collision point and estimate the time to collision (TTC). The TTC is one of the safety indicators in collision analysis (*18*, *19*). TTC is defined as the time the vehicle and pedestrian require to reach the conflict point based on the pedestrian's trajectory and the vehicle's trajectory. The collision prediction approach is based on the algorithm presented in (*18*, *19*). After receiving pedestrian PSMs and vehicle BSMs, the GPS coordinates of the pedestrian and vehicles are transformed to local x and y coordinates, respectively (x_v, y_v) and (x_p, y_p). The conflict point can be calculated by Equations 1 and 2. Where θ_i is the compass heading of pedestrian (1), vehicle (2), and x_c and y_c are collision point C, as shown in Figure 7:

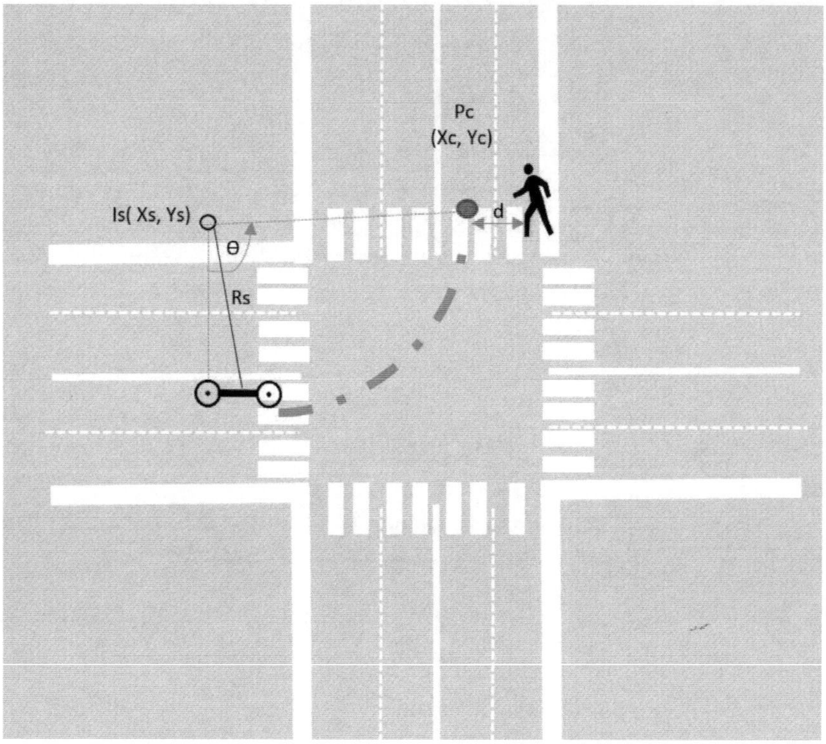

Figure 8. Bicycle model to calculate conflict point and TTC.

$$x_c = \frac{y_v - y_p - (x_v{*}\tan\theta_2 - x_p{*}\tan\theta_1)}{(\tan\theta_1 - \tan\theta_2)} \quad (1)$$

$$y_c = \frac{x_v - x_p - (x_v{*}\cot\theta_2 - x_p{*}\cot\theta_1)}{(\cot\theta_1 - \cot\theta_2)} \quad (2)$$

The time to reach to the collision point, from the perspective of the pedestrian (TTC_p), is different from the vehicle, TTC_v. However, if the difference is less than a specific threshold (δ), the situation is considered potentially unsafe. TTC_p and TTC_v can be calculated and compared by Equations 3, 4, and 5 where V_p and V_v are the speeds of the pedestrian and vehicle, respectively:

$$TTC_p = \frac{\sqrt{(x_p - x_c)^2 + (y_p - y_c)^2}}{V_p} \quad (3)$$

$$TTC_v = \frac{\sqrt{(x_v - x_c)^2 + (y_v - y_c)^2}}{V_v} \quad (4)$$

$$TTC_d = |TTC_p - TTC_v| < \delta \quad (5)$$

The value of the δ parameter depends on the driver reaction time, breaking time of the vehicle, pedestrian reaction time, and alert message transmission delay. The study of the impacts of these parameters is not in the scope of this project and values have been selected based on the literature (*18*) where 6 seconds is assumed. This collision prediction algorithm can address the first and second dangerous scenarios in Figure 1. In order to address the third scenario, the trajectory of the vehicle needs to be estimated. A algorithm proposed by Jiménez et al. (*19*) considered linear motion for the vehicle. To develop a system that can perform well in all scenarios, a new prediction algorithm was developed based on Anaya et al. (*3*). In their study, a bicycle model was used to predict vehicle trajectories during turning maneuvers (*20*). In the bicycle model, the left and right front wheels are represented by one single wheel at one point. Similarly the rear wheels are represented by one central rear wheel at one point (*21*). Figure 8 shows the bicycle model that is used to model the vehicle trajectory (blue dashed line) and a conflict point based on a vehicle trajectory. R_S, d, and (X_s,Y_s) are the radius to the gravity center, distance between conflict point and the pedestrian, and the coordinate of instantaneous center of rotation (I_s), respectively. (x_c, y_c) is the conflict point of the pedestrian (Pc).

Equations 6 to 9 describe how to calculate the conflict point and TTC. V_v, w, ε_r, V_p, and L are vehicle velocity, yaw rate, the vehicle steering angle, pedestrian velocity, and the length of the vehicle, respectively (SAE J2735). (x_p, y_p) is the position of the pedestrian:

Figure 9. Pedestrian GPS trajectory vs true path.

$$R_s = \frac{V_v}{w} \varepsilon_r = \frac{L}{R_s} X_s = 0 Y_s = R_s \times \cos\left(\frac{\varepsilon_r}{2}\right) \quad (6)$$

$$\theta = a\tan\left(\frac{Y_p - Y_s}{R_s}, \frac{X_p - X_s}{R_s}\right) \quad (7)$$

$$TTC = \frac{\theta - \frac{\varepsilon_r}{2} + \frac{\pi}{2}}{w} \quad (8)$$

$$X_c = R_s \times \cos\theta + \frac{\varepsilon}{2} - \frac{\pi}{2} + X_s$$
$$Y_c = R_s \times \sin\theta + \frac{\varepsilon}{2} - \frac{\pi}{2} + Y_s \quad (9)$$

Based on (x_c, y_c) the distance between the conflict point and the pedestrian *(d)* can be calculated by Equation 10. If the pedestrian will reach the conflict point before the vehicle arrives (i.e., $d < V_p \times TTC$), the pedestrian is in the conflict area and the pedestrian and the driver should be alerted to a potential conflict:

$$d = \sqrt{(x_p - x_c)^2 + (y_p - y_c)^2} \quad (10)$$

The SWA application addresses all dangerous scenarios in Figure 1 by using two conflict prediction algorithms which were presented above. The SWA application uses PSM information to monitor the phase that the pedestrian has requested. The situational awareness component on the RSU can create a warning message for the requested phase and start to alert vehicles that may share the same road path with the pedestrian based on the vehicle trajectory.

Correction Algorithm to Improve Localization

Developing a reliable situational awareness system for pedestrians is much more challenging than for vehicles because a vehicle's movement is more predictable and usually remains in the lane in the road. Tracking pedestrian trajectories can be effected by GPS positioning error. In order to study GPS position errors, field measurements were conducted to assess the accuracy of the smartphone position estimate. In this experiment, a pedestrian's trajectory was collected as he/she traveled along a sidewalk and crosswalk. Location data were collected every 1 second. Figure 9 shows the pedestrian location collected by the smartphone (green dots), and the true path (RTK-measured GPS points) is shown in yellow markers using Google Earth. As Figure 9 shows, the GPS points collected by the smartphone have a significant bias and vary over time. In order to provide better location-based services for pedestrians, a position accuracy not greater than the width of a crosswalk or sidewalk is needed. The Federal Highway Administration guideline for sidewalk design specifies a width of 1.52 meters.

The third component in the SWA application provides a correction algorithm to improve a pedestrian's location estimation. The algorithm uses map-matching to project the pedestrian's location to the MAP and an Extended Kalman Filter to correct for random measurement errors as the pedestrian moves along the MAP. Details of this algorithm are presented in (*22*). Different test scenarios were conducted to validate the algorithm. In all test scenarios, a pedestrian used the application and the location trajectory data was collected via the phone GPS.

In the first scenario, the pedestrian started walking from the sidewalk at the west end and south side of the intersection as shown by the red arrow in Figure 10*a* (32 meters from the crosswalks) on a path to the northeast side of the intersection. The route included two

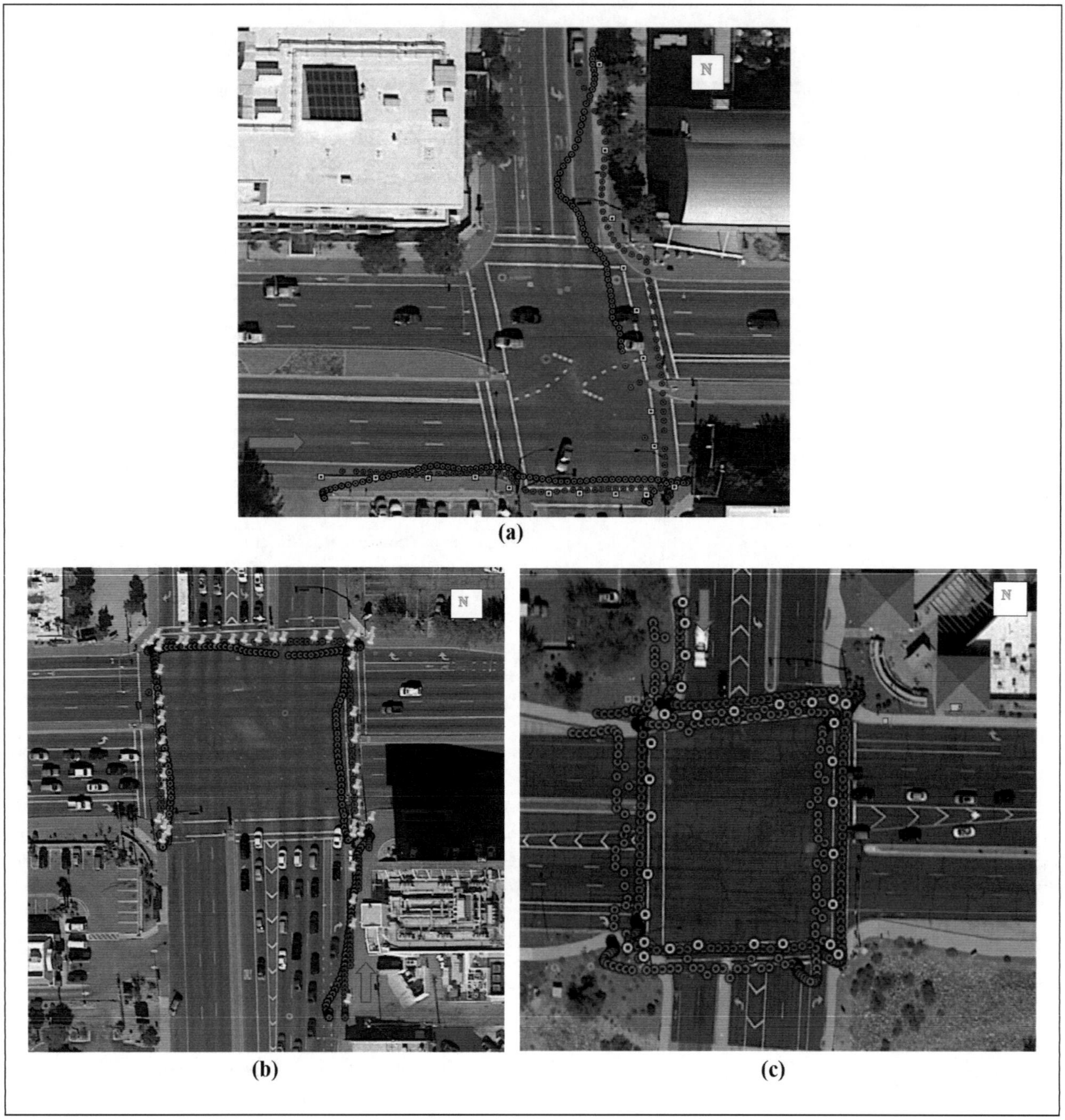

Figure 10. Phone trajectory vs true path vs estimated path for different test scenarios.

crosswalks and a path continuing to the north. The pedestrian uses the SWA application to send the pedestrian request wirelessly to the traffic controller through the RSU. Figure 10*a* shows the pedestrian location data collected by the smartphone (green points), and the true path that the pedestrian walked is shown in yellow points. The red points show the location estimation after applying the map-matching algorithm and the Kalman Filter. As the pedestrian reached the northeast corner of the intersection, where there are two large buildings very close to the sidewalk, multipath errors affected the GPS accuracy and the raw data located the pedestrian in the street instead of on the sidewalk. As Figure 10*a* shows, the estimation results are close to the true path.

The second experiment was at the intersection of Speedway and Campbell in Tucson, AZ. This intersection is one of the intersections in Tucson that Southern Arizona Association for the Visually Impaired use to teach blind, or visually impaired, pedestrians the crossing processes. A path starting from southeast of the intersection and passing through three crosswalks in a counterclockwise direction was selected. Figure 10*b* shows the pedestrian location collected by smartphone (green points), and the true path that pedestrian walked is shown in yellow points. The location estimation after applying the map-matching algorithm and the Kalman Filter is shown in red points.

The third experiment was at the intersection on Daisy Mountain Drive and Gavilan Peak Parkway in Anthem, AZ. This intersection is one of the intersections in the Arizona Connected Vehicle Test Bed in Maricopa County. A path starting from northwest of the intersection and passing through all four crosswalks in a counterclockwise direction was selected. At each crosswalk, the pedestrian uses the SWA application to send a pedestrian request. Figure 10*c* shows the pedestrian location collected by smartphone (green points), and the true path that pedestrian walked is shown in yellow points. The red points show the location estimation after applying the map-matching algorithm and the Kalman Filter.

Conclusion and Future Research

Avoiding conflicts between pedestrians and vehicles is an important issue in intelligent traffic systems. Wireless communications and connected vehicle technologies can have a significant role in improving pedestrian safety and provide accessibility and mobility for disabled pedestrians. In this paper the design, implementation, and test results of a new wireless-based situational awareness system are presented. The system provides wireless communications between pedestrians and infrastructure, which helps pedestrians, especially disabled and visually impaired pedestrians, to find the crosswalk and send pedestrian requests to a traffic controller. The SWA application also provides visualization of real-time signal status on their phone, as well as auditory messages and haptic alerts. Additionally, the system utilizes V2I and P2I communications to create a V2P pathway to provide alerts to pedestrians and drivers to avoid possible conflicts. The SWA application consists of a location estimation algorithm to provide accurate pedestrian location estimation. The system has been designed, implemented, and tested in the Arizona Connected Vehicle Test Bed, operated by the University of Arizona and Maricopa County Department of Transportation. The P2I communication has been tested at a signalized intersection equipped with RSUs. In addition, test results show that the location estimation algorithm implemented in the SWA application meets the pedestrian location-based system accuracy requirements. Future research includes testing the SWA system with visually impaired stakeholders to collect data and improve the application based on users' feedback.

Author Contributions

The authors confirm contributing to the paper as follows: study conception and design: Sara Khosravi, Byungho Beak, Larry Head, Faisal Saleem; data collection: Sara Khosravi, Byungho Beak, Larry Head, Faisal Saleem; analysis and interpretation of results: Sara Khosravi, Byungho Beak, Larry Head, Faisal Saleem; draft manuscript preparation: Sara Khosravi, Byungho Beak, Larry Head, Faisal Saleem. All authors reviewed the results and approved the final version of the manuscript.

References

1. Peden, M., R. Scurfield, D. Sleet, D. Mohan, A. A. Hyder, E. Jarawan, and C. D. Mathers. *World Report on Road Traffic Injury Prevention*. World Health Organization, Geneva, 2004.
2. World Health Organization. *Global Status Report on Road Safety 2013: Supporting a Decade of Action*. World Health Organization, Geneva, 2013.
3. Anaya, J. J., P. Merdrignac, O. Shagdar, F. Nashashibi, and J. E. Naranjo. Vehicle to Pedestrian Communications for Protection of Vulnerable Road Users. *Proc., 2014 IEEE Intelligent Vehicles Symposium*, IEEE, 2014, pp. 1037–1042.
4. Gandhi, T., and M. M. Trivedi. Pedestrian Protection Systems: Issues, Survey, and Challenges. *IEEE Transactions on Intelligent Transportation Systems*, Vol. 8, No. 3, 2007, pp. 413–430.
5. Harding, J., G. Powell, R. Yoon, J. Fikentscher, C. Doyle, D. Sade, M. Lukuc, J. Simons, and J. Wang. *Vehicle-to-Vehicle Communications: Readiness of V2V Technology for Application*. National Highway Traffic Safety Administration, U.S. Department of Transportation, Washington, D.C., 2014.
6. Bentzen, B., J. Barlow, and T. Bond. Challenges of Unfamiliar Signalized Intersections for Pedestrians Who Are Blind: Research on Safety. *Transportation Research Record: Journal of the Transportation Research Board*, 2004. 1878: 51–57.
7. Ponchillia, P. E., E. C. Rak, A. L. Freeland, and S. J. LaGrow. Accessible GPS: Reorientation and Target Location among Users with Visual Impairments. *Journal of Visual Impairment & Blindness*, Vol. 101, No. 7, 2007, p. 389.
8. Barlow, J. M., B. L. Bentzen, and T. Bond. Blind Pedestrians and the Changing Technology and Geometry of Signalized Intersections: Safety, Orientation, and Independence. *Journal of Visual Impairment & Blindness*, Vol. 99, No. 10, 2005, p. EJ720652.
9. DeRuwe, G., and R. Wall. Pedestrian Navigation and Integration with Distributed Smart Signal Traffic Controls. *Proc., IECON 34th Annual Conference of IEEE*, IEEE, 2008.
10. Ho, P.-F., and J.-C. Chen. WiSafe: Wi-Fi Pedestrian Collision Avoidance System. *IEEE Transactions on Vehicular Technology*, Vol. 66, No. 6, 2017, pp. 4564–4578.

11. Llorca, D. F., M. A. Sotelo, I. Parra, J. E. Naranjo, M. Gavilán, and S. Álvarez. An Experimental Study on Pitch Compensation in Pedestrian-Protection Systems for Collision Avoidance and Mitigation. *IEEE Transactions on Intelligent Transportation Systems*, Vol. 10, No. 3, 2009, pp. 469–474.
12. Keller, C. G., T. Dang, H. Fritz, A. Joos, C. Rabe, and D. M. Gavrila. Active Pedestrian Safety by Automatic Braking and Evasive Steering. *IEEE Transactions on Intelligent Transportation Systems*, Vol. 12, No. 4, 2011, pp. 1292–1304.
13. Chen, Z., C. Wu, N. Lyu, G. Liu, and Y. He. Pedestrian-Vehicular Collision Avoidance Based on Vision System. *Proc., 17th International Conference on Intelligent Transportation Systems (ITSC)*, IEEE, 2014, pp. 11–15.
14. Wang, T., G. Cardone, A. Corradi, L. Torresani, and A. T. Campbell. WalkSafe: A Pedestrian Safety App for Mobile Phone Users Who Walk and Talk While Crossing Roads. *Proc., Twelfth Workshop on Mobile Computing Systems & Applications*, 2012, p. 5.
15. Wu, X., R. Miucic, S. Yang, S. Al-Stouhi, J. Misener, S. Bai, and W. Chan. Cars Talk to Phones: A DSRC Based Vehicle-Pedestrian Safety System. *Proc., 80th Vehicular Technology Conference (VTC Fall)*, IEEE, 2014, pp. 1–7.
16. Yanagisawa, M., E. Swanson, and W. G. Najm. *Target Crashes and Safety Benefits Estimation Methodology for Pedestrian Crash Avoidance/Mitigation Systems*. National Highway Traffic Safety Administration, 2014.
17. Dhondge, K., S. Song, B.-Y. Choi, and H. Park. WiFi-Honk: Smartphone-Based Beacon Stuffed WiFi Car2X-Communication System for Vulnerable Road User Safety. *Proc., 79th Vehicular Technology Conference (VTC Spring)*, IEEE, 2014, pp. 1–5.
18. Hussein, A., F. García, J. M. Armingol, and C. Olaverri-Monreal. P2V and V2P Communication for Pedestrian Warning on the Basis of Autonomous Vehicles. *Proc., 19th International Conference on Intelligent Transportation Systems (ITSC)*, IEEE, 2016, pp. 2034–2039.
19. Jiménez, F., J. E. Naranjo, and F. García. An Improved Method to Calculate the Time-to-Collision of Two Vehicles. *International Journal of Intelligent Transportation Systems Research*, Vol. 11, No. 1, 2013, pp. 34–42.
20. Kong, J., M. Pfeiffer, G. Schildbach, and F. Borrelli. Kinematic and Dynamic Vehicle Models for Autonomous Driving Control Design. *Proc., 2015 IEEE Intelligent Vehicles Symposium (IV)*, IEEE, 2015, pp. 1094–1099.
21. Rajamani, R. *Vehicle Dynamics and Control*. Springer Science & Business Media, New York, 2011.
22. Khosravi, S., K. L. Head, and F. Saleem. Location Estimation of Smart Phone Equipped Pedestrians by Integrating Connected Vehicle Technologies and Map Matching. Presented at 96th Annual Meeting of the Transportation Research Board, Washington, D.C., 2017.

The Standing Committee on Paratransit (AHB15) peer-reviewed this paper (18-04707).

Sage Publications USA
sagepub.com

ISBN 978-1-5108-7824-2

CURRAN ASSOCIATES INC.
proceedings.com

Environment and Energy 2018

Transportation Research Record: Journal of the Transportation Research Board

Volume 2672, Issue 24

SAGE Publishing